Amarjit S. Basra, PhD

Seed Quality: Basic Mechanisms and Agricultural Implications

Pre-publication REVIEWS, COMMENTARIES, EVALUATIONS . . .

"Seed quality is basic to good and healthy crops. Rewards from recent advances in breeding techniques as well as from the opportunities now available for producing novel genetic combinations will be proportional to advances in seed technology. It is in this context that the present book edited by Dr. Amarjit Singh Basra is a timely and valuable contribution. All aspects of seed quality are dealt with in a thorough manner by experts in the different areas. The chapter on variety identification is particularly important in the context of the recent GATT agreement on varietal protection. *Seed Quality* is a must for all those involved in crop improvement and agricultural advancement."

Professor M. S. Swaminathan
Director, Centre for Research on Sustainable Agricultural and Rural Development, Madras, India

More pre-publication
REVIEWS, COMMENTARIES, EVALUATIONS . . .

"This book is a collection of reviews by experienced researchers on seed quality in relation to retention of viability and agronomic value. The physiological and biochemical mechanisms associated with differences in viability and vigor are broadly covered and restricted to the major food crops with the exception of the chapters on recalcitrant seeds, seed treatment, and seed quality in relation to crop performance.

The authors' interest in seeds in relation to improving the technology of seed production, storage, and crop establishment is clearly evident. The approach adopted is possibly unique among recent texts on seeds. *Seed Quality* is a useful source of reference and ideas for practicing seed physiologists and technologists. This is a text for the specialist."

Dr. David Gray
Horticultural Research International,
Wellesbourne, Warwick,
United Kingdom

Food Products Press
An Imprint of The Haworth Press, Inc.

NOTES FOR PROFESSIONAL LIBRARIANS AND LIBRARY USERS

This is an original book title published by Food Products Press, an imprint of The Haworth Press, Inc. Unless otherwise noted in specific chapters with attribution, materials in this book have not been previously published elsewhere in any format or language.

CONSERVATION AND PRESERVATION NOTES

All books published by The Haworth Press, Inc. and its imprints are printed on certified ph neutral, acid free book grade paper. This paper meets the minimum requirements of American National Standard for Information Sciences–Permanence of Paper for Printed Material, ANSI Z39.48-1984.

Seed Quality
Basic Mechanisms and Agricultural Implications

FOOD PRODUCTS PRESS

An Imprint of The Haworth Press, Inc.
Robert E. Gough, PhD, Senior Editor

New, Recent, and Forthcoming Titles:

The Highbush Blueberry and Its Management by Robert E. Gough

Glossary of Vital Terms for the Home Gardener edited by Robert E. Gough

Seed Quality: Basic Mechanisms and Agricultural Implications edited by Amarjit S. Basra

Statistical Methods for Food and Agriculture edited by Filmore E. Bender, Larry W. Douglass, and Amihud Kramer

World Food and You by Nan Unklesbay

Introduction to the General Principles of Aquaculture by Hans Ackefors, Jay V. Huner, and Mark Konikoff

Managing the Potato Production System by Bill B. Dean

Marketing Livestock and Meat by William Lesser

The World Apple Market by A. Desmond O'Rourke

Understanding the Japanese Food and Agrimarket: A Multifaceted Opportunity edited by A. Desmond O'Rourke

Marketing Beef in Japan by William A. Kerr et al.

Seed Quality

Basic Mechanisms and Agricultural Implications

Amarjit S. Basra, PhD
Editor

CRC Press is an imprint of the
Taylor & Francis Group, an **informa** business

CRC Press
Taylor & Francis Group
6000 Broken Sound Parkway NW, Suite 300
Boca Raton, FL 33487-2742

First issued in paperback 2019

ISBN-13: 978-1-56022-850-9 (hbk)
ISBN-13: 978-0-367-40181-8 (pbk)

Library of Congress Cataloging-in-Publication Data

Seed quality : basic mechanisms and agricultural implications / (edited by) Amarjit S. Basra.
p. cm.
Includes bibliographical references and index.
ISBN 1-56022-850-4 (acid-free paper).
1. Seeds–Quality. I. Basra, Amarjit S.
SB118.25.S44 1994
631.5′21–dc20

93-6092
CIP

Visit the Taylor & Francis Web site at
http://www.taylorandfrancis.com

and the CRC Press Web site at
http://www.crcpress.com

Dedicated to my parents,
Sardarni Harbans Kaur Basra
and
Sardar Joginder Singh Basra

CONTENTS

ABOUT THE EDITOR

Amarjit Singh Basra, PhD, is a Botanist (Associate Professor) at Punjab Agricultural University in Ludhiana, India. Previously, he was a Senior Research Fellow in the Department of Plant Physiology at Wageningen Agricultural University in The Netherlands. The author of over 60 research and professional publications, Dr. Basra has received coveted scientific awards and honors in recognition of his original and outstanding contributions to plant science. He has made scientific trips to several countries and provides leadership in organizing and fostering cooperation in seed research at the international level.

Contributors

R. N. Basu, Vice Chancellor, Calcutta University, India.

H. F. Chin, Department of Agronomy and Horticulture, University Pertanian, Malaysia.

R. J. Cooke, National Institute of Agricultural Botany, United Kingdom.

Peter Coolbear, Seed Technology Centre, Massay University, New Zealand.

David L. Dornbos, Jr., Director, Product Development, Ciba Seeds, Illinois, USA.

W. E. Finch-Savage, Horticultural Research, International Wellesbourne, United Kingdom.

J. G. Hampton, Seed Technology Centre, Massay University, New Zealand.

Martin M. Kulik, United States Department of Agriculture/ARS, Maryland, USA.

Wallace G. Pill, Department of Plant and Soil Sciences, University of Delaware, USA.

Dale O. Wilson, Jr., University of Idaho, College of Agriculture, USA.

Preface

As high quality seed is the basis of higher agricultural productivity, this book focuses on various aspects of seed quality with the goal of integrating research at the basic and applied levels. Quality in seeds embraces all the physical, biological, pathological, and genetical attributes which contribute to the final yield of a crop. Such seeds are specifically bred and genetically pure; they are free from disease, vigorous, and high in germination percentage. The recent upsurge of worldwide interest in seed quality has accentuated a new awareness regarding its importance in crop production.

This book contains 11 chapters written by acknowledged experts, providing comprehensive information on all aspects of seed quality with a clear perspective and having a balance in presentations of basic and applied research. Researchers, students, and teachers in many agricultural and botanical disciplines–seed science and technology in particular–will find this book to be of immense use. It can also be used as a handbook for those involved in seed industry and seed testing services. The need for such a distinct source of information has long been felt by the scientific community and I hope that this book will serve to accelerate the pace of research and make the planners and practitioners aware of the value of seed quality for achieving increased food production.

I should like to thank the contributing authors and the staff at The Haworth Press, Inc. for their efforts and cooperation during the course of publication. I thank my wife, Ranjit, also a colleague in science, for her forbearance and support. To my daughter, Sukhmani, and son, Nishchayjit, I extend my appreciation of their tolerance and understanding quite disproportionate to their years.

Sri Guru Granth Sahib has been a continuous source of inspiration and peace!

Chapter 1

Seed Viability

R. N. Basu

As the most important basic input in agriculture, the seed should be of very good quality, and the most vital attribute of that quality is viability. Viability of seed was a crucial factor in the establishment and sustenance of the ancient agrarian civilizations of the world. The Vedic invocation, "May the seed be viable" (Yajurveda, approximately 2500 B.C.), focused the great concern of the early Aryan settlers in India on maintaining seed viability. Even today in many parts of the world, especially those with hot and humid climates, maintenance of seed viability poses serious problems.

Loss of viability of seeds in storage is also reported from countries blessed with conditions favorable for production and storage of seeds because of unforeseen pre- and postharvest conditions. An understanding of the basic tenets of seed viability is therefore essential to formulating sound seed production and storage strategies. Further, the growing concern for preservation of germplasm of cultivated plants as well as their wild progenitors has led to the development of long-term storage facilities in germplasm banks in may countries, to provide viable materials to plant breeders and biotechnologists for future crop improvement programs. From the philosophical standpoint, seed viability studies offer a very promising insight into the attributes of life, deteriorative senescence, and death.

The author thanks Dr. S. L. Basak, Dr. N. C. Chattopadhyay, Dr. K. Sur, Dr. P. Pal, and Mr. A. K. Bhattacharya for their help in various ways in the preparation of this chapter.

ORTHODOX AND RECALCITRANT SEEDS

Seeds can be broadly classified into two groups; the orthodox and the recalcitrant seeds, depending upon their storability vis-à-vis seed moisture content. The orthodox seeds are those whose longevity is extended by gradual reduction of moisture content, while the recalcitrant seeds cannot be dried below certain critical high moisture contents without reducing their viability (Roberts, 1973a; Chin, 1988). The common dry-stored agricultural and horticultural seeds, including the cereals, millets, pulses, oilseeds, and summer and winter vegetables, are orthodox in nature. A number of plantation crops, tropical fleshy fruit trees, and forest tree species produce desiccation-sensitive recalcitrant short-lived seeds for which commercial long-term storage methods are yet to be developed (Roberts, 1979).

OLD RECORDS ON SEED VIABILITY

Dry-stored orthodox seeds, if properly maintained, retain viability for a long period. However, substantial controversy exists over the reported life spans of seeds surviving under natural conditions. An excellent account of the literature on the subject has been given by Bewley and Black (1982).

Perhaps the most authentic reports on life spans of seeds are those from herbarium specimens and museum collections and carefully conducted seed burial experiments. Otherwise, life spans of seeds have most often been derived from circumstantial, indirect, and weak evidence. The authenticated life span of seeds from old collections of the Museum of Natural History in Paris ranges from 55 to 158 years (Becquerel, 1934), with the legume *Cassia multijuga* topping the list. Results of seed burial experiments revealed that seeds of *Verbascum blattaria* remained viable even after 90 years (Kivilaan and Bandurski, 1973). Seeds surviving in soil for long periods possess innate dormancy. When the dormancy is broken by physical or physiological means, the seed readily germinates but if it fails to emerge out of the soil, it will perish in due course.

The high values for life spans of *Nelumbium nucifera, Canna compacta* and several other materials are not unlikely if it is accepted that the seeds had very low moisture contents and the temperature over the years was uniformly low. The life span of 10,000 years for the frozen *Lupinus articus* seeds (Porsild, Harrington, and Mulligan, 1967) is also not unusual in view of the data provided by Ellis and Roberts (1980) that a mere 5 percent fall in germination of barley seed would require 3,000 years when stored at a seed moisture content of 5 percent and temperature of –20°C.

DEFINITION OF SEED VIABILITY

The property of the seed that enables it to germinate under conditions favorable for germination is termed "viability," provided that any dormancy in the seed is removed before the germination test.

In a standard germination test for a particular material (ISTA, 1976), the percentage of seeds germinating is a quantitative measure of viability of that lot. For a deeply dormant seedlot, the percentage of seeds showing proper positive response to the tetrazolium test would also stand as a quantitative estimate of seed viability.

Measurement of viability is an essential prerequisite to assessing the value of the seed for planting and for industrial purposes such as malting. The standard germination test for a seedlot, when properly conducted following the rules prescribed by the International Seed Testing Association (ISTA, 1976), gives a good measure of viability and satisfactorily foretells the suitability of a seedlot for sowing. Methods of viability and vigor testing are discussed in Chapter 3 of this book by Dr. J. G. Hampton.

The term "storability" has been used in the text synonymously with poststorage germinability. As such, both prestorage germinability and poststorage germinability (storability) are covered in the text under viability. The term "longevity" is reserved for seed life under natural conditions.

RELATION BETWEEN SEED VIGOR AND VIABILITY

Seed vigor is a qualitative attribute of the seed which is most intimately related to seed viability (Heydecker, 1972), and loss of viability is usually preceded by loss of vigor. Therefore, an understanding of vigor is basic to the elucidation of the mechanisms of loss of viability. Vigor is that quality of the seed which is responsible for rapid, uniform germination, increased storability, good field emergence, and ability to perform well over a wide range of field conditions (Perry, 1978; McDonald, 1980).

PREHARVEST, HARVEST, AND POSTHARVEST CONDITIONS AFFECTING SEED VIABILITY

The noticeable differences in seed quality of the same cultivar in different years in different agroclimatic zones, and even in the same year in adjacent plots, underscore the role of production environment and growth condition of the parent plant on the germinability of the seed. The conditions prevailing during harvest operations, the mode of harvesting, and the subsequent processing also determine the quality of seed in terms of pre- and post-storage germinability. Even when a good-quality seed has been produced and harvested, faulty storage techniques may prove detrimental to the maintenance of vigor and viability.

Preharvest Conditions and Seed Viability

The various factors which are known to affect the growth of the parent plant and which are likely to have some influence on the quality of seeds produced include (1) quality of the initial seed, (2) fertility status of the soil, (3) temperature and photoperiod, (4) moisture status of soil, and (5) herbicide or pesticide application.

Seed Quality

Theoretically, well-developed G1 seed (first generation seed) from which the parent plant is grown may be expected to produce G2 seed (second generation seed) of good quality. This would not involve genetic inheritance, as the improvements are perpetuated primarily through maternal tissues.

The size of the seed within a lot influences germinability in a number of materials. Size of seedlings from small seeds has been found to be less in a large number of crop seeds (Kittock and Patterson, 1962). Lighter perennial rye grass seeds produced 25 percent fewer seedling emergence than heavier grains sown at a depth of 25 mm (Arnott, 1975). In pearl millet, seedling emergence increased from 40 percent with small low-density seed to 62 percent with large high-density seeds (Lawan et al., 1986). Dharmalingam and Basu (1987) demonstrated that large and medium seeds of mungbean (*Vigna radiata*) were superior to small or ungraded seeds in terms of field emergence, crop growth, and final seed yields. Further, irrespective of seed size, seeds of normal color (green) were superior to off-color (dirty brown) seeds in respect of the aforementioned performances.

However, studies on the role of seed size on germinability, plant growth, and final crop performance have quite often given contradictory results and superiority of smaller seeds to bigger ones has been reported (Moore, 1943; Verma and Gupta, 1975; Kler and Dhillon, 1983). Greater mechanical injury to bolder seeds during harvesting and processing may be responsible for such observations. In groundnut (*Arachis hypogaea*), Punnuswamy and Ramakrishnan (1984) attributed the increased membrane permeability of large seeds to invisible cracks developed during seed processing.

Differences in planting methods may show differential effect of seed size, vigor, and maturity on crop performance. Thus in soybean the seed yield of row plots was not related to seed size or vigor, but in hill plots the seed yield per plot was related to both, with lower yields occurring for seedlots of smaller size and low vigor (Tekrony et al., 1987). Baalbaki (1989) noted that heavy and large seeds of two winter wheat cultivars consistently resulted in increased yields in comparison to light and small seeds. It was further noted that heavy, large, and high-protein seeds stored better than light, small, and low-protein seeds and showed greater post-aging germinability.

Recall in this connection the views of Black (1959) that in a limiting environment the larger plants from large seeds showed greater competition with each other and, as a result, the relative advantages of large seed size declined with time. Similarly, the adverse effect of immaturity of the seed tapered off, giving little difference in final yield. Nevertheless, the subject is definitely not closed and there is considerable scope for further work on the role of size, density, and maturity of G1 seeds on immediate viability, seedling emergence, crop growth, and yield and quality of G2 seeds.

Crop Nutrition

Soil condition is a key factor in crop husbandry. The structural and textural status of the soil, its fertility level, pH, microbial environment, etc., are all important determinants of seed performance in terms of establishment, growth of the plant, and final seed yield. Information on this subject is scarcely available and only a few aspects such as effects of mineral nutrition on viability have been studied to some extent by researchers.

Severe nitrogen deficiency in carrot, lettuce, and pepper resulted in very low seed yield, and a major proportion of the seed was abnormal (Harrington, 1960). However, the germinability of normal seeds from nitrogen-deficient plants was similar to the normal seeds of control plants. The relationship between high protein content and poststorage germinability (Fox and Albrecht, 1957; Baalbaki, 1989) in wheat suggests that cultural operations leading to high protein content in the seed should be of advantage in crop establishment from stored seed. Das (1990) recorded significantly better pre- and poststorage germination of okra (*Abelmoschus esculentus*) seeds from parent plants grown under high nitrogen level in the field than seeds from plants not supplied with additional nitrogen fertilizer. However, such results differ from those of Sneddon (1963) on sugarbeet, in which there was significant reduction in the germinability of fruits from plants receiving a high dose of nitrogen nutrition (260kg/ha) compared to the fruits from control plants which did not receive nitrogen. The situation in sugarbeet is perhaps a special one, however. According to Austin (1972), the reduced germinability of high-nitrogen fruits as observed by earlier workers (Tolman, 1943; Sneddon, 1963; Scott, 1969) could be due to large accumulation of germination inhibitors in the fruits of high-nitrogen plants.

Phosphorus reserves in the seed play very important roles in the metabolism of germinating seeds. The major phosphorus reserve in the seed, phytic acid, in addition to its nutritional role, is believed to act as a natural antioxidant (Graf, Empson, and Eaton, 1987). As such, deficiency of phosphorus would be expected to adversely affect pre- and poststorage germinability. Thus, Austin (1966a) recorded reduced germinability of watercress seeds from phosphorus-deficient plants. Deficiency of phosphorus in the parent plant, however, did not affect germination of seeds of carrot, lettuce, and pepper (Harrington, 1960). Application of additional phosphate fertilizer to soil of average fertility did not show any significant effect on germinability of okra seeds before or after accelerated aging or natural storage (Das, 1990). Austin (1966b) noted that phosphorus-deficient pea seeds from parent plants grown under extreme P deficiency when sown in the field produced smaller plants and gave lower yields than the corresponding control seeds, but such extreme deficiencies would be unlikely in commercial seed production.

The importance of potassium in plant nutrition is well known. Potassium phosphate (dibasic) and several other potassium salts, when used in the germination substrate or for seed soaking in hydration-dehydration treatments, were found to improve immediate and poststorage germinability (unpublished data of Goswami, 1990; Mukhopadhyay, 1990). Petruzzeli et al. (1982) demonstrated that K^+ could retain viability of wheat seed during aging and the effect of K^+ (and fusicoccin) on the proton extrusion capacity of aged embryos was related to the restorative effect on viability. Our own unpublished data also showed a close relation between germination improvement by potassium salt and a reduction in the production of volatile aldehydes (Mukhopadhyay, 1990; Ghosh, 1992). It appears, therefore, that besides nutritive effects, K may influence seed viability in a more specific manner which is as yet not very clear.

Severe K^+ deficiency in pepper resulted in a higher percentage of abnormal seed production and lower pre- and poststorage germinability than seed from corresponding nondeficient parent plants. But extreme K^+ deficiency in soil may not be normally encountered in commercial seed production. Vanangamudi and Karivaratharaju (1986) reviewed the research on potassium nutrition of seed crops of cereals, legumes, and vegetables in the Tamil Nadu Agricultural University in South India and showed that application of K^+ to the soil or as supplemental foliar application or seed treatment increased germination percentage and seedling vigor of the resulting seeds. Das (1990), applied moderate to heavy doses of muriate of potash to soil with moderate K^+ content without any significant effect on the yield or quality of okra seed produced. Germination and storability of seed obtained from parent plants receiving K fertilizer was similar to that from control plants.

Deficiencies of other essential major and minor elements may also affect the quality of seed and its viability but information on the subject is meager. Researchers also do not know whether such effects will be carried over to the next generation of seeds even when the parent plants are grown in soils nondeficient in such elements. Perhaps, as indicated by Austin (1972), seeds with moderate deficiency of mineral elements when grown in normal soil with average fertility level would show little difference in emergence and establishment. The situation could, however, be different in soils with acute elemental deficiencies or in otherwise stressed soils.

Temperature and Photoperiod

Temperature has a profound influence on development and ripening of seeds which may ultimately be reflected in seed viability. Extreme temperatures are undesirable for growth and development of seeds: very low temperature (0°C and below) damages ripening corn seed and the extent of freezing injury depends on the state of maturity (Rossman, 1949); very high temperature increases the rate of respiration, decreases grain weight, and may also enforce untimely drying with consequent adverse effects on immediate and poststorage germinability.

In lettuce, Koller (1962) noted that when the seeds matured at high temperatures, germination was less at 26°C in the dark than the corresponding low temperature matured seed. In *Anagallis arvensis*, seeds which ripened at 20°C day and 15°C night temperatures remained fully dormant for ten weeks after harvest but those ripening at 30°C day and 25°C night temperatures showed no dormancy (GrantLipp and Ballard, 1963). Temperature differences during ripening also altered the dormancy patterns of wheat (Van Dobben, 1947; Kramer, Pest, and Witten, 1952), *Bromus* (Laude, 1962), rice (Ghosh, 1962), and several other crops.

In mungbean (*Vigna radiata*), Dharmalingam (1982) showed that late summer sowing in Tamil Nadu (South India) resulted in the production of a very high percentage (35 percent) of hard seeds and therefore recommended early summer sowings to reduce hardseededness.

Germinability of seed has also been found to be influenced by the photoperiod during ripening of the seed on the parent plant. *Chenopodium amaranticolor* and *Ononis sicula* seeds maturing during long days showed coat-imposed dormancy due to thickening of the seed coat (Lona, 1947, as quoted by Austin, 1972).

Moisture Status of Soil

For good-quality seed, a relatively dry climate during the ripening phase is preferred. Except for crops like rice, too wet soil may create a microclimate nonconducive to seed health, especially in combination with a high relative humidity of the air. Even for a wet land crop like rice, a dry climate during grain ripening phase produces seeds of good vigor and viability.

The source of soil water is manifold, but for field crops rainfall and irrigation are the two major determinants of seed quality. While the latter can be controlled, the former, although vitally important for successful agriculture, is often responsible for poor seed quality.

Adequate soil moisture is essential for crop growth, and regulated water supply, depending on the need of the crop, provides seed of good quality. However, soil with high moisture due to high irrigation or high rainfall may produce grains of low nitrogen and protein content (Greaves and Carter, 1923; Shutt, 1935; Russell and Voelcker, 1936) which has been correlated, in the case of wheat, with poor germinability (Fox and Albrecht, 1957).

Drought during flowering might interfere with fertilization, and thus grain number is reduced (Salter and Goode, 1967). Weight and size of grain, which are usually correlated with viability, are reduced by drought during grain development and maturation (Asana and Basu, 1963). Extreme water deficit stimulates premature desiccation, and the quality of seed in terms of pre- and poststorage germinability is poor.

In groundnut (*Arachis hypogaea*), however, Ramamoorthy (1990) observed that mild water stress had no effect on vigor and viability of harvest-fresh seeds, but that the storability of the seed was significantly improved. This was associated with a lowering of free fatty acids and electrical conductivity of seed-steep water. Incidentally, the seed produced under the water stress showed enhanced protein but reduced oil contents.

Preharvest rains can cause very serious damage to seed quality. Besides pathological considerations, the possibility of *in situ* sprouting of nondormant seeds or even partial germination advancement is physiologically undesirable. The storability of seeds in which germination advancement has taken place to an extent which is not compatible with subsequent drying is considerably reduced. In laboratory experiments, it has been consistently shown that soaking-drying of very high-vigor seed is deleterious for the maintenance of vigor and viability in storage (Basu, 1976; Basu and Rudrapal, 1982). As such, seeds badly affected by preharvest rains should not be stored for planting purposes.

Herbicide and Pesticide Application to the Parent Plant

Herbicides and pesticides applied to the soil or to the growing crop may affect the development of seed and influence its quality, especially if the concerned herbicide or pesticide is not easily biodegradable. There exists little information on the effect of pesticide application, but one report shows an increase in protein content of wheat with subherbicidal doses of simazine (Ries, Schweizer, and Chmiel, 1968), which may be extrapolated to account for an improvement in viability.

Ramamoorthy (1990) systematically studied the effect of the herbicides fluchloralin, pendimethalin, and oxyfluorfen applied before emergence, with or without postemergence application of fluazifopbutyl, on viability of groundnut seeds. There was no effect of the herbicides on germinability before storage, but after storage the use of herbicides, except fluchloralin, resulted in better germinability and seed vigor. The physiological and biochemical bases for such improvements are yet to be elucidated.

Harvesting and Processing Conditions Affecting Viability

Whether performed manually or mechanically, the harvesting and processing operations may affect the viability of the seed. The former, though safer, is more expensive and even in developing countries slowly giving way to mechanized operations, particularly for large seed plots. Mechanical injury to the seed during harvesting, including threshing, cleaning, etc., could be a major reason of poor pre- and poststorage germinability.

Mere visual examination of the seed may not enable detection of mechanical injuries unless such injuries are very extensive. Examination of fractured seed coats and different structures of the seed, following growth tests, provided information on the nature of damages incurred. Moore (1972) listed the mechanical injuries detected in standard growth tests as detached seed structures, breaks within structures, abnormally shaped structures, scar tissues, infections, restricted growth, uneven placement of cotyledons, unnatural shrinkage of cotyledons, abnormally developed hypocotyls and primary roots, and dwarfed and twisted roots with blunt tips of dull appearance. Tetrazolium tests have proved very useful in detection and assessment of the extent of mechanical injuries (Bulat, 1969; Moore, 1969). Comparable information on internal injuries could also be obtained by the X-ray method (Kamra, 1966).

While the major injuries are immediately reflected on germinability, the adverse effects of even minor bruises are magnified upon storage, resulting in significant reduction in the vigor and viability of the stored seed. The field emergences of three categories of soybean seeds, namely nonbroken, lightly broken, and moderately broken, were 96, 72, and 52 percent, respectively (Moore, 1957), indicating the poor performance of even lightly injured seed subjected to the stress conditions in the field.

The extent of damage depends on the type of seed, its shape and size, thickness of seed coat, structure and position of the embryo, and outside factors such as seed moisture. The large-seeded legumes such as the peas and beans are more susceptible to mechanical injuries than small-seeded millets and cereals which escape serious injuries. Presence of firm coats, lemma, palea, etc., is of advantage in resisting mechanical injury. Although bold-seeded legumes are more susceptible, the hardseededness of many legumes provides protection against mechanical injury. In snap bean cultivars resistant to mechanical injuries, the seed coats are found to adhere tightly to the cotyledons (Atkin, 1958). The thin coat of flatseeded sesamum (*Sesamum indicum*) poses a problem even with manual harvesting and processing in India and a significant reduction of vigor and viability is encountered following storage. Although spherical seeds are more resistant to damage than irregular, flat, or elongated seeds, sorghum (*Sorghum vulgare*) seeds, which are small and roundish, suffer from mechanical injury because of their slightly protruded embryos (Moore, 1972).

As regards the effect of seed moisture content, it has been a general experience that very dry seeds suffer fractures or mechanical impact while in the same situation high-moisture seeds are just bruised. Different seeds have different optimum moisture levels for mechanized or rough manual handling operations. For storage of seeds susceptible to mechanical injury, one has to make a compromise between a relatively high seed-moisture content for avoiding the injury and a low seed moisture which is conducive to the maintenance of vigor and viability in storage.

Moore (1972) distinguished water damage caused by moistening and drying of mature bold-seeded legume seeds from the usual mechanical injury. Water damage, which is synonymous with imbibitional injury or soaking injury, may be attributed to the unequal expansion and folding of seed coats and cotyledons during rapid water uptake and can be detected by tetrazolium tests. Perhaps in rain-soaked and subsequently dried seeds the type of water damage discussed by Moore (1972) would greatly magnify in storage, leading to substantial loss of vigor and viability. The reduced germinability of soaked-dried soybean seed (Saha and Basu, 1984; Saha, Mandal, and Basu, 1990) would also exemplify such a situation.

Postharvest Storage Conditions Affecting Seed Viability

The role of different factors influencing seed viability must be clearly understood to design proper storage systems. The longevity of dry-stored orthodox seeds depends on seed moisture content, temperature, and oxygen pressure. Lowering seed moisture and temperature greatly extends longevity, and replacing oxygen with inert gases in hermetically sealed containers gives better poststorage germinability than the seed sealed in air. The effects of seed moisture, storage temperature, and oxygen on the viability of stored seeds have been reviewed over the years (Owen, 1956; Barton,

1961; James, 1967; Roberts, 1972a; Harrington, 1972; Ellis and Roberts, 1980; Bewley and Black, 1982). Various aspects of storage of orthodox as well as recalcitrant seeds are discussed in Chapters 6 and 7 of this book.

Seed Moisture and Drying

Because of the great importance of seed moisture, seed drying has become a major field of study for seed technologists and agricultural engineers. Commercial seed drying is a highly specialized job and it is not merely the final seed moisture but the initial seed moisture, the temperature of the drier, the rate of drying, etc., that determine viability (Nellist, 1981). Although too rapid drying reduced viability (Iljin, 1957), in recalcitrant *Avicennia marina* seeds rapid drying retained viability at a lower moisture content than those dried slowly (Farrant, Berjak, and Pammenter, 1985). Even within orthodox seeds, the optimum rate of drying varied with the kind of seed. Excessive seed drying is discouraged not only because of rising fuel and operative costs but also on physiological grounds. According to Schultz, Day, and Sinnhauber (1962), at about 5 percent moisture level the monomolecular water layer surrounding the macromolecules in seeds ceases to be continuous, and that may enhance lipid peroxidative activity.

In storage it is most unlikely that the seed will attain a moisture content sufficient for germination. But at moisture contents well below that critical level, a host of physiological and biochemical reactions, the most important of which is respiration, will be set in motion. Such a situation is undesirable for maintaining the near-cryptobiotic nature of the orthodox seed in storage.

When oxygen is not readily available, high seed moisture is still more injurious, as the products of anaerobic respiration such as ethanol and acetaldehyde are toxic to the seed. The catalytic activity of the hydrolyzing enzymes also increases with progressive rise in seed moisture level.

As drying and moisture-barrier storage are becoming increasingly expensive, scientists should perhaps look for alternative methodologies. In this context, the commercial feasibility of coating seeds with polymers such as polyvinyl alcohol, which effectively resisted the adverse effect of accelerated aging in soybean and showed less fungal growth than seeds without synthetic coat (West et al., 1985), needs to be critically studied.

An apparently paradoxical situation is the high storability of dormant lettuce seeds maintained at the fully imbibed state. The physiological basis of this improved storability has been discussed by Villiers and Edgcumbe (1975) in terms of activity of an endogenous repair system in the hydrated state.

Storage Temperature

A rise in temperature increases the rate of chemical and biochemical reactions, and a Q_{10} of two or more than two are frequently recorded. In an orthodox dry-stored seed, a prolonged quiescent phase in which enzyme-catalyzed metabolic reactions should be at a very low ebb is ideal. The dry seed is metabolically active, albeit very feebly (Chen, 1972). As such, a rise in temperature will interfere with the rest period. At very high temperatures, beyond the usual physiological range, inactivation of enzymes and disruption of the fine structure of bioorganelles take place notwithstanding the high temperature tolerance of seeds with very low moisture contents. Further, nonenzymatic deteriorative processes such as lipid autoxidation would be more at high temperatures, which, coupled with thermal release of such autoxidative products, reduces viability.

Oxygen Pressure

Evidences derived from carefully controlled experiments clearly show that the storability of seed significantly increases with a reduction in the partial pressure of oxygen (Roberts, 1972a). The discrepancies recorded in some of the early studies may be attributed to lack of proper control of other variables. The role of oxygen on viability is studied in ampoules and hermetically sealed containers, and in seeds of high moisture contents a cumulative effect of anaerobiosis, resulting in the formation of ethanol, acetaldehyde, etc., may give erratic results. It would be better to use low-moisture seeds in sealed containers for reproducible results in such studies. Further, one should desist from extrapolating the observations and looking for similarities with semisealed and open storage systems.

The likely involvement of lipid peroxidation in seed aging implicates a profound role of oxygen. Autoxidative or lipoxygenase-catalyzed lipid oxidation should be less in a storage environment devoid of oxygen. In this connection, the observations of Woodstock et al. (1983) and Gorecki and Harman (1987) on the beneficial effects of a number of antioxidants on seed longevity are of considerable physiological relevance. Phytic acid, which blocks iron-driven hydroxyl generation and suppresses lipid peroxidation (Graf, Empson, and Eaton, 1987) is another natural antioxidant, whose role in seed aging requires further scrutiny.

In this context, the role of the inert gases on viability may be mentioned. The work of Harrison (1966) with ten lettuce cultivars, while confirming the deleterious effect of oxygen on storability, revealed substantial differences in the beneficial effects of different inert gases. Roberts (1972a) suggested that a reduction of partial pressure of oxygen explained the beneficial effect of an inert gas like nitrogen; such a view was also expressed by Van Toai, McDonald, and Staby (1986) in explaining the superiority of nitrogen gas storage of soybean to the usual warehouse storage. At present, there is no ready explanation of the differential behavior of different inert gases (Harrison, 1966) but, if that occurs consistently with other materials, one may be inclined to cast doubt on the basic property of inertness of gases like argon and nitrogen, so far as the metabolism of the seed is concerned.

Seed Dormancy and Viability

The inability of a viable seed to germinate under conditions normally considered favorable for the purpose, namely requisite water, temperature, and oxygen is termed "dormancy." The two major classes of dormancy are coat-imposed dormancy and embryo dormancy. The seed coat (true seed coat as well as other covering structures associated with the seed) may prevent germination by preventing water uptake or gas exchange, mechanically restraining the growth of the embryo, chemically inhibiting germination, or acting as a barrier to photoreception. But even when the restrictions caused by the covering structures are removed, germination may fail due to embryo dormancy attributable to deficiencies or blocks within the embryo or cotyledons. In view of the fact that, on imbibition of water, the nondormant viable seed will either germinate, under otherwise favorable conditions, or will perish in due course, as opposed to fully imbibed dormant seeds, a relation between dormancy and viability may be conjectured.

Roberts (1963) examined the association between viability and innate dormancy of six rice cultivars differing widely in genotypes and seed morphology, and demonstrated the absence of any functional relationship between dormancy and viability in dry storage. His findings were thus in agreement with those of Barton (1961) who opined that viability was not directly related to longevity. To avoid genotypic differences, Punjabi and Basu (1989) employed a different approach. They used the dormant lettuce cultivar Suttons A1 which requires prechilling or kinetin treatment for germination. The loss of viability of dormant seeds was found to be similar to that of nondormant (dormancy broken by low temperature or kinetin treatment) seeds subjected to accelerated aging.

It appears that coat-imposed dormancy, in which the seed coat acts as an effective barrier to entry of water and air, as in many hard-seeded dry-stored legumes, will extend the storage life of a well-dried seed. Such a seed will follow the viability pattern suggested by the viability equations for sealed storage. A corresponding seed with seed coat permeable to air and water may take up moisture from the humid atmosphere and lose viability rapidly.

Thus we have two situations, viz. dry storage and imbibed storage, in which the relation between dormancy and viability would be different. While in dry storage, orthodox dormant and nondormant seeds show similar storability, but the dormant seed exhibits greater survival potential in imbibed storage, as necessary repairs of biochemical lesions on vital bioorganelles are undertaken in the hydrated seed (Villiers, 1973; Villiers and Edgcumbe, 1975; Osborne, 1981). Such a mechanism would possibly account for the very long viability periods of seeds buried in soil, as exemplified by seeds of *Verbascum blattaria* surviving in moist soil for 90 years (Kivilaan and Bandurski, 1973). Dormancy of many imbibed weed seeds in soil could, therefore, be a very pertinent survival stratagem (Osborne, 1981). For dry storage, the situation would be different.

PHYSIOLOGICAL, BIOCHEMICAL, AND CYTOGENETICAL CHANGES ASSOCIATED WITH LOSS OF VIABILITY

There is virtually no difference in the physical appearance of viable and nonviable dry seeds except that in a few cases loss of viability is associated with development of off-color.

Physiological and Biochemical Changes

Germination of seed brings about a large number of differences between seeds of different grades of viability. These include, for instance, delay of radicle emergence, slower seedling growth, greater number of abnormal seedlings, and decreased tolerance to stress conditions. A brief account of the major metabolic changes observed in the dry seed and in the imbibed seed placed for germination follows.

Older nonviable seeds are very often lighter than harvest-fresh seeds of similar moisture level. As major portions of most seeds are food reserves, it was natural for earlier researchers to postulate a metabolic depletion of the reserves as a cause of nonviability (Went and Muntz, 1949). However, even in nonviable seeds reserve food still remains to counter such postulates (Ching and Schoolcraft, 1968). Of course, it is not the total content but the availability to the embryo of specific and vital nutritive factors that would be more decisive. Quite often, the loss of food reserves could very well be due to depletion by storage microflora and insect pests.

The metabolic activity of the dry-stored orthodox seed may not be altogether absent (Chen, 1972; Edwards, 1976), and with rise in seed moisture in the air-dry range, seed metabolism, especially respiration, will progressively increase. In a viable low-vigor seedlot, the rate of respiration upon imbibition would be lower, as indicated by reduced oxygen uptake and lower dehydrogenase activity. The respiratory quotient is higher because of low oxygen uptake coupled with high carbon dioxide release, indicating partial anaerobiosis attributable partly to disruption in the membrane-bound electron transport system (Woodstock and Taylorson, 1981). Obviously, in terms of ATP production the seedlot with low germinability would be less efficient. ATP production and dehydrogenase activity have, therefore, been used as indices of seed vigor (Ching and

Schoolcraft, 1968; Kittock and Law, 1968). Kharlukhi and Agrawal (1984), working with seeds of green gram (*Vigna radiata*), chick pea (*Cicer arietinum*), and wheat (*Triticum aestivum*), provided strong evidence for the participation of pentose phosphate pathway preceding the loss of viability.

Besides the accumulation of toxic products of anaerobic respiration such as ethanol and acetaldehyde (Woodstock and Taylorson, 1981) and lipid peroxidation giving lipid hydroperoxides and volatile aldehydes (Wilson and McDonald, 1986a; Sur and Basu, 1990a, 1990b), Sirkar and coworkers implicated accumulation of a number of phenolics, supraoptimal concentration of indoleacetic acid and abscisic acid in the loss of rice seed viability (Dey, Sirkar, and Sirkar, 1967; Dey and Sirkar, 1968). Mukhopadhyay et al. (1983) recorded a large accumulation of the polyamine spermine in nonviable rice embryos. Roberts (1973b) indicated that accumulation of mutagens, though not clearly proved, could be a possible mechanism by which nuclear damage occurs in storage.

An increase in the concentrations of reducing sugars, amino acids, and free fatty acids, as reported for some dry-stored seeds, obviously depends on the activity of the hydrolytic enzymes for the concerned reserves. It is, however, unlikely that the activities of the enzymes would increase during dry aging due to *de novo* synthesis, as the progressive failure of the synthetic machinery of the cell during aging is a more or less established fact. An increase in activity could, therefore, be due to breakdown of compartments separating the enzymes and substrates and quite often due to microbial hydrolases. In many of the earlier studies, especially those employing accelerated aging techniques, proper precaution to eliminate fungal and bacterial contamination had not been taken. Presently, we do not have any evidence to show that in a stored seed, which is sufficiently dry to eliminate the possibility of microbial contamination, activity of any enzyme has increased through *de novo* synthesis in the dry state.

When placed for germination, the aged seed shows significantly lower activity, *in vivo* and *in vitro* of the hydrolytic enzymes responsible for mobilization of reserve food for the growing embryo than the corresponding nonaged seed. They include not only those known to be synthesized *do novo* but also the preexistent ones because of the damage to enzyme synthesis machinery as well as to the enzyme molecules themselves. Therefore, as in the case of dry seeds, greater activity of certain enzymes in aged seeds during germination than in nonaged seeds could be attributed to breakdown of intracellular compartmentation separating the enzymes and their substrates, and breakdown or removal of an inhibitor of the enzyme, and not due to an increase in the total amount of the enzyme protein. A significant reduction in the synthesis of requisite enzymes for the mobilization of reserves required for new growth, coupled with the inadequate supply of energy for the surge of reactions leading to germination, may determine the extent of germinability of the aged seed.

Germination requires active participation of the complex synthetic machinery of the cell, consisting of a vast array of enzymes, their factors and cofactors, hormonal regulators, the nucleic acids, and perhaps other unidentified factors, along with the apparatus to provide necessary energy for the various synthetic activities. A major impairment of the functional activity of any one of the components of that complex but closely integrated system may put a brake on the train of metabolic events, culminating in the failure of the seed to germinate. During dry storage any damage caused to the vital bioorganelles and macromolecules of the seed cannot be repaired and will go on accumulating. Even during hydration prior to germination, the physiological and biochemical lesions may be too extensive to be effectively repaired. It is well documented that the protein synthesizing machinery of the cell becomes progressively less functional with the advancement of seed aging (Roberts and Osborne, 1973; Cheah and Osborne, 1977; Sen and Osborne, 1977; Bewley and Black, 1982; Osborne, 1982).

Membrane Permeability

One of the earliest symptoms of deteriorative senescence is the loss of membrane integrity. Any disruption of the membrane structure will not only affect permeability of the plasmamembrane but also intracellular compartmentation and separation of metabolic systems. Many of the enzymes are intimately associated with the internal membranes and membrane-bound bioorganelles and any change in membrane structure will have far-reaching physiological and biochemical consequences. Based on earlier research on the subject, Bewley and Black (1982) listed the consequences of membrane damage, which include (1) breaks in the structure of plasmalemma and its contraction from the cell wall, (2) fragmented endoplasmic reticulum devoid of polyribosomes, (3) monosomes randomly dispersed in the cytoplasm, (4) absence of dictyosomes (or indistinct dictyosomes), (5) disintegration of mitochondria and plastids, (6) condensation of chromatin and lobed nucleus, (7) coalescence of lipid droplets, and (8) lysis of membranes of lysosomic structures.

Our knowledge on the mechanism of membrane damage during aging is still not very clear. The most probable reason appears to be oxidative breakdown of the unsaturated lipid moieties of the lipoprotein membrane, although damage to the protein moiety also cannot be ruled out. Free radical mediated lipid peroxidation has been involved as a basic reason of membrane damage and seed aging (Wilson and McDonald, 1986b). Furthermore, free radicals may inactivate proteins by reacting with specific amino acids and residues (Gardner, 1979; Wolff, Gardner, and Dean, 1986), bringing about conformational changes or oxidative modifications in proteins and thereby enhancing their susceptibility to proteolysis (Gardner, 1979; Mondal and Choudhury, 1982; Wolff and Dean, 1986). Although free radicals can fragment proteins under certain conditions, crosslinking and polymerization of membrane proteins have been found to be associated with free radical mediated lipid peroxidation (Hochstein and Jain, 1981). Decreased protein solubility because of polymerization and complex formation (Roubal and Tappel, 1966; Sundholm, Visappa, and Bjorkston, 1978) and polypeptide chain scission (Zirlin and Karel, 1969) would be some of the reasons for the loss of enzyme activity noted by many researchers (Chio and Tappel, 1969; Kanner and Karel, 1976).

Increased hydrolytic susceptibility can be induced by the hydroxyl and peroxy radicals formed during lipid peroxidation (Wolff and Dean, 1986). Thus free fatty acids released by lipolytic activity act as substrates for membranous and cytosolic lipoxygenase, thereby facilitating membrane damage.

The onset of free radical chain propagation reactions brings about cross linkages in nucleic acids and causes other aberrations. *In vitro* studies have shown that a number of activated oxygen species could be responsible for increasing single-stranded scissions in DNA (Lesko, Lorentzen, and Ts'o, 1980).

Cytogenetical Changes

As quoted by Kostoff (1935), de Vries (1901) was the first to observe increased mutations in plants grown from aged *Oenothera* seeds. Navashin (1933) reported the occurrence of chromosomal aberrations in root tip cells of plants grown from aged seeds of *Crepis tectorum*. Various types of chromosomal abnormalities have since been reported by many authors (see reviews by Roberts, 1972b, Roos, 1982). According to Peto (1933), the frequency of chromosomal aberrations are largely eliminated during the growth of the plant through diplontic selection and therefore are not passed on to the next generation.

The incidence of chromosome breakages during the anaphase of the first mitotic divisions in root tip cells of seeds of A1 generation (plants developed from seeds receiving an aging treatment are A1 generation plants, subsequent generations are termed A2, A3, and so on) was highly correlated with the loss of viability. Such chromosomal aberrations were associated with pollen abortion in A1 and phenotypic mutations in A2 and A3 plants. Since the first report by Cartledge, Barton, and Blackeslce (1936) that increased temperature and moisture increased the rate of mutation, most subsequent evidence lent support to the view that chromosome damage in seeds is accelerated by temperature, moisture content, and duration of storage. Chromosome breakage is probably also induced by increased concentration of oxygen (Moutschen-Dahmen, Moutschen, and Ehrenberg, 1959; Jackson, 1959; Roberts, Abdalla, and Owen, 1967).

The increased frequency of pollen abortion in A1 plants of barley, broadbeans, and peas under age-accelerating moisture and temperature regimes (Abdalla and Roberts, 1968) reflects the significant damage to the meiotic apparatus occurring long after seedling establishment. Rao and Roberts (1989) recorded that meiotic abnormalities in lettuce plants grown from aged seeds increased with a decrease in viability of the seedlot, but the average pollen and seed fertilities are not significantly different from the control.

The question of heritability of the different types of age-induced cytogenetical irregularities which may involve mutation (point mutation or alteration of DNA), different kinds of chromosome breaks, and changes in the RNA and cytoplasm is very important from the point of view of preventing genetic drifts and maintaining pure lines. Roos (1982) concludes that although age-induced chromosomal aberrations are largely eliminated during plant growth (diplontic selection), sectoral chimeras may survive and transmit their altered chromosomes to future generations and become stabilized under certain situations. Roos (1982) further points out that nonchromosomal mutations may also be eliminated at the pollen formation stage because of lethality of the mutation in the haploid condition. But some mutations may still survive and be transmitted to the progeny especially in view of the fact that a loss of seed viability to a level of 50 percent could result in 2 to 3 percent mutations (Abdalla and Roberts, 1969a). Mutation within the polygenic system controlling quantitative characters like yield has been reported in peas (Purkar, Mehra, and Banerjee, 1985).

As with other bioorganelles, it is unlikely that the damage to the chromosomes in general and the nucleic acids in particular could be repaired in the air-dry stored seed. The extent to which hydration prior to germination corrects the biochemical lesions, in the manner suggested by Villiers (1973) and Villiers and Edgcumbe (1975), perhaps depends on the nature and extent of damage incurred during the previous dry storage, and there is very little direct evidence regarding the participation of the endogenous repair system. As such, further studies on the biochemistry of the enzymatic repair system for the vital macromolecules in the early imbibition and germination phase are clearly necessary.

Microflora and Seed Deterioration

Dr. Martin Kulik's chapter on seed quality and microorganisms (Chapter 5) in this volume gives an elaborate treatment of the subject, but here it is appropriate to mention two points: (1) very often the loss of seed viability due to physiological reasons is confounded by pathological deterioration, especially because a number of biochemical manifestations of the two processes are overlapping; and (2) the quality of seed sown in the field has a pronounced effect on the soil microflora, and the germinability of the seed and its establishment as a seedling depends on the interaction between seed quality (viability and vigor) and microflora.

That fungus-free culture filtrate of several storage fungi reduces immediate germinability of the seed has been reported for a number of crop seeds (Mathur and Sinha, 1978; Gupta and Rana, 1981). We noted that the storability of wheat seed treated with culture filtrate of *Aspergillus flavus* and dried back showed a progressive reduction of germination percentage and seedling growth with increasing concentration of the filtrate (Saha and Basu, 1985). But even at concentrations which showed no immediate adverse effect on germinability, storability (postaging germinability) of the seed was significantly reduced, showing a latent effect of the toxin on viability. The adverse effect of the soaking-drying treatment on wheat seed storability was associated with a significant increase in membrane permeability and substantial reduction in dehydrogenase activity. Such observations emphasize the importance of studies on physiological and biochemical bases of deterioration of stored seed by storage microflora.

Seed of low vigor and viability when sown in the field exudes a lot of metabolites because of poor membrane integrity which would serve as food for the soil microflora. The volatile aldehydes emanated by low-vigor seed in the soil microenvironment are also good substrates for the seed-infecting fungi (Harman and Stasz, 1986). Unless properly treated with chemicals to counteract by the invading fungi or other microorganisms, the chances of seedling establishment from poor-quality seed will be significantly reduced.

CONTROL OF LOSS OF SEED VIABILITY

Enhanced longevity of the stored orthodox seed can be ensured by drying the seed to a low moisture level and storing it in a relatively cool and dry place. The Chapter on storage of orthodox seeds by Dr. D. O. Wilson (Chapter 6) deals with the various aspects of the problem in sufficient detail. In the account that follows, emphasis is placed on nonconventional approaches toward maintenance of vigor and viability in storage.

One such approach is the extension of maize seed viability by exposure to a source of free electrons in a cathodic field (Pammenter, Adamson, and Berjak, 1974), which is of great fundamental importance, as the involvement of free radicals in seed aging was convincingly demonstrated. The other approaches include the use of antioxidants employing the dry permeation technique and the "wet" and "dry" seed treatments, with special emphasis on those developed at the College of Agriculture of the University of Calcutta.

Physiological Seed Treatments

For the maintenance of vigor and viability, the seed can be given "dry" treatments, which do not require use of water, and "wet" treatments in which seeds are hydrated and then dehydrated to safe limits of moisture for storage.

Dry Treatments

The dry permeation technique employing solvents like dichloromethane and acetone has been successfully used for introducing antiaging chemicals into the seed (Tao et al.,1974; Basu, Pan, and Punjabi, 1979; Woodstock et al., 1983; Dey and Basu, 1985; Gorecki and Harman, 1987; Dey and Mukherjee, 1988). The chemicals include conventional natural and synthetic antioxidants and free radical controlling agents.

Direct use of volatile chemicals such as iodine at very low concentrations was reported by Basu and coworkers for extending seed longevity of mustard (Basu and Rudrapal, 1980), mungbean (Rudrapal and Basu, 1982), and rice (Pal and Basu, 1988). For commercial use, iodine may be premixed with carriers such as talc, French chalk, and calcium carbonate, and the mixture then used for dry-dressing of seed in closed containers. Rudrapal and Basu (1982) in mungbean and Dey and Mukherjee (1984) in soybean and sunflower recorded significant extension of storability using carriers impregnated with low concentrations of iodine.

For commercial use, calcium oxychloride (the ordinary bleaching powder which slowly releases chlorine) can be used for treating high-vigor seeds of wheat (Mandal and Basu, 1986), rice (Pal and Basu, 1988), and pea (Bhattacharya and Basu, 1990) seeds for extending vigor and viability in storage. Rudrapal and Nakamura (1988a) confirmed the beneficial effects of chlorine, iodine, and bromine on the postaging germinability of harvest-fresh eggplant (*Solanum melongena*) and radish (*Raphanus sativus*) seeds.

Most recently, very low concentrations of the alcohols, methanol ethanol, and isopropanol have been found to improve storability of pea (Bhattacharya and Basu, 1990) and several other leguminous seeds by Mukhopadhyay (1991).

Wet Treatments

Seed hydration treatments such as soaking, dipping, spraying, moisture equilibration, moisture equilibration-soaking, and moist sand conditioning, all followed by drying-back, significantly improve post-aging germinability of orthodox seeds (Basu, 1976; Basu and Rudrapal, 1982; Basu and Mandal, 1985; Basu, 1990). The efficacy of the treatment depends on the kind of seed and vigor of the seed at the time of treatment. Harvest-fresh seed with a very high vigor status is not responsive to the hydration-dehydration treatment for the maintenance of vigor and viability. The efficacy of the hydration-dehydration treatment has also been reported and confirmed by research conducted in many other laboratories (Savino, Deleo, and Haigh, 1979; Roy, 1982; Goldsworthy, Fielding, and Dover, 1982; Dey and Mukherjee, 1984; Burgass and Powell, 1984; Kuo and Chu, 1986; Vanangamudi and Karivaratharaju, 1986; Rao, Roberts, and Ellis, 1987; Doijode and Raturi, 1987; Rudrapal and Nakamura, 1988b; Ramamoorthy, et al., 1990; Dharmalingam, 1990).

The major consideration in the hydration-dehydration treatments is to allow the seed slowly and progressively to take up moisture from the environment to about 25 to 30 percent on a fresh weight basis to allow for certain preventive or restorative reactions to occur prior to drying-back without causing any injury to the seed. Hydration alone is most effective, but presence of chemicals such as potassium and sodium salts and several other antioxidant or antioxidant-synergistic chemicals has shown additional beneficial effects on post-aging seed performance.

Important Manifestations of Physiological Treatments

Usually little difference in germinability between treated and untreated seeds is noted immediately after treatment. However, with relatively prolonged soaking durations, the rate of emergence and early seedling growth may be initially higher than in the control. Highly significant treatment effects are noted when the seed is subjected to accelerated aging or stored under natural ambient temperature and humidity regimes. The post-aging germination percentage, seedling growth, and crop performance of the treated material in the field have been found to be consistently superior to the control (Basu and Rudrapal, 1982; Basu and Mandal, 1985; Basu, 1990). The hydration-dehydration treatments not only showed resistance to ageing stress but also to stress conditions imposed by heat (Rudrapal and Basu, 1980), salinity (Mandal and Basu, 1982) and X- and Y- irradiations (Punjabi and Basu, 1982; Sur and Basu, 1986).

Dasgupta et al. (1977) recorded a significant reduction in the incidence of chromosomal aberrations in wheat following soaking-drying treatment, especially with sodium phosphate (dibasic, 10^{-4}M) and aging.

The beneficial effects of soaking-drying treatment on storability have been found to be associated with better membrane integrity (in all the seeds studied so far), greater dehydrogenase and amylase activities (Rudrapal and Basu, 1979), reduced lipid peroxide formation (Rudrapal and Basu, 1982; Basu, 1990), and emanation of less volatile aldehydes during the early germination phase (Pal and Basu, 1989; Sur and Basu, 1990a, 1990b).

Mode of Action of Physiological Seed Treatments

The mechanism of action of the vigor and viability maintaining seed treatments is yet to be fully elucidated. Several possibilities, especially to account for the beneficial effects of the hydration-dehydration treatments, which have been explored for the purpose include (1) removal of toxic metabolites, (2) reduced moisture uptake by treated seed, (3) germination advancement and embryo enlargement during hydration, (4) enzymatic repair of biochemical lesions by the cellular repair system, and (5) counteraction of free radical and lipid peroxidation reactions.

The question of leaching of toxic metabolic products is not an important factor in view of the very high efficacy of moisture equilibration-drying treatment in which no liquid water is used. We do not, however, rule out metabolic removal of toxic products during the hydration phase, although evidence of participation of any scavenging agent, such as superoxide dismutase, is as yet lacking. The similar pattern of moisture uptake by treated and untreated seeds subjected to accelerated aging under high humidity regimes suggests that differential moisture uptake is not involved. Further, even under very dry aging, where moisture uptake does not occur, the moisture desorption pattern is also similar. Incidentally, noninvolvement of storage microflora is also indicated by the results of dry aging at a very low relative humidity and high temperature, which are unsuitable for growth of the storage microflora (Christiansen, 1972). Germination advancement and embryo enlargement, as recorded by Austin, Longden, and Hutchinson (1969) in presowing seed treatments of carrot, cannot account for the increased storability as the hydration-dehydration treatments are not very effective in high-vigor seeds. Again, in our presoaking experiments, prolonged duration of hydration resulting in germination advancement always reduced poststorage germinability. The work of Falkenstein and Steiner (1985) in wheat also confirms the reduced poststorage germinability of seeds in which presoaking caused advancement of germination.

Enzymatic repair of biochemical lesions on vital cell organelles during the hydration phase, as proposed by Villiers and Edgcumbe (1975), may explain the extended longevity of the hydrated-dehydrated seed. The fact that the hydration-dehydration treatments are not effective in very high-vigor seeds, which obviously do not require any repair and restoration, lends further support to the idea of participation of a repair mechanism for correction of damages to organelles suffered by low- and medium-vigor seeds. However, a number of observations suggest alternative possibilities. I examine three of these concerning the beneficial effect of the moisture equilibration-drying (MED) treatment: (1) immediately after the MED treatment, there is no improvement of germinability, but large improvements are recorded after aging, suggesting a preventive rather than a curative action of the treatment; (2) there is a significant reduction in the leakage of metabolites after moisture equilibration (before drying), possibly due to proper orientation and arrangement of the membrane bilayer components, but after drying-back the leakage of metabolites would be similar to the untreated control seed, indicating the temporary nature of the beneficial hydration effect whereby the membrane reverts back to the original state upon drying. We studied leakage of metabolites in organic solvents to eliminate the possibility of membrane repair in aqueous environment during the leaching effect but could not detect any difference in leakage between MED and control; (3) the beneficial effect of MED on post-storage germinability was recorded even when medium-vigor lettuce seeds were impregnated with the potent protein synthesis inhibitor, cycloheximide, implying the absence of *de novo* protein synthesis during moisture equilibration treatment. As such, either repair of bioorganelles involving protein synthesis was not required or, even if a particular repair was undertaken, the enzymes involved would be preexistent rather than newly synthesized.

Studies by Basu and coworkers (Rudrapal and Basu, 1982; Dey and Basu, 1985; Choudhuri and Basu, 1988; Saha, Mandal, and Basu, 1990) lend strong support to the concept of a reduced lipid peroxidation during storage as a basic reason of the improved storability of treated seeds. The significantly lower volatile aldehyde production by germinating hydrated-dehydrated seeds following accelerated aging (Pal and Basu, 1989; Sur and Basu, 1990a, 1990b) also demonstrates a reduction of lipid peroxidation by the viability-maintaining seed treatments. Endogenous antioxidants like α-tocopherol as well as synthetic ones such as butylated hydroxytoluene have already been shown to extend seed viability (Woodstock et al., 1983; Gorecki and Harman, 1987) possibly through reaction with lipid radicals. Physical recombination of free radicals because of very close proximity of recombination in the presence of water, as in the case of irradiated lettuce seeds (Ehrenberg, 1961), also controls chain propagation reactions. The very large hydration effect, which is preventive in the sense that subsequent radical induced damage is effectively reduced, can be satisfactorily explained through such early recombination of free radicals.

The improved storability of seeds following dry treatments cannot be explained in terms of repair of bioorganelles in the dry state. The effect of the halogens such as iodine and chlorine in high-vigor seed may be explained in terms of stabilization of bonds of polyunsaturated fatty acids (Basu and Rudrapal, 1980), making them more resistant to peroxidative attack, although their direct free radical scavenging action (Pryor, 1973) at appropriate concentrations may also determine their efficacy. The promoting effect of the alcohols at very low concentrations is difficult to explain, except that, like the other beneficial treatments, volatile aldehyde emanation by germinating treated seeds is substantially reduced (Bhattacharya and Basu, unpublished). It must be pointed out, however, that unlike the large and safe hydration effect, the chemical effects are concentration dependent. Even an otherwise safe chemical such as sodium chloride may act either as an antioxidant or as a prooxidant depending on the concentrations used (Nickerson, 1967). Certainly, a more concerned approach by researchers in the field will elucidate the mode of action of the physiological seed treatments, resulting in the development of still simpler methodologies for the maintenance of vigor and viability of orthodox seeds in storage.

Seed Viability and Crop Yield

A critical appraisal of the relation between seed viability and yield was made by Roberts (1972c), covering a large number of agriculturally important crops. Loss of viability affects crop yield by reducing plant stand per unit area because of poor germinability, and by adversely affecting vigor of the surviving seedlings, which may be ultimately carried over to the biological and agricultural yield.

Within limits, the field stand of a relatively less viable seedlot can be adjusted to the level of a fully viable lot by proper increment of the seed rate, giving almost identical plant density per unit area of land. Of course, even with proper adjustment, there may be the problem of patchy and nonuniform field stand in case of a more deteriorated seedlot. Ignoring such minor hazards, we face the agriculturally very important question, will the history of the seed with respect to viability influence the final yield in plots with identical plant populations?

The early literature on the subject, as reviewed by Barton (1961), was mostly confined to yield potential of seeds of different chronological ages. These limited studies obviously would not give a clearcut insight because the degree of deterioration and extent of viability depend on the storage condition, and not merely on the age of the seed.

Barton and Garman (1946) demonstrated decreased yields with loss of viability from tomato and lettuce plants produced from surviving seeds. The work of Harrison (1966) on lettuce took into consideration the impact of slow-aging and rapid-aging treatments. In the slow aging treatments, in which the lettuce seeds were stored for five to ten years in open storage or sealed in air or carbon dioxide at 18°C, even a small loss of viability substantially reduced yield from the surviving plants. On the other hand, when the seeds were rapidly aged by storing at 10 percent moisture content at 35°C, significant decreases in seedling growth could be recorded only when the viability level dropped below 50 percent. In the case of five-year-old onion seeds, stored in sealed containers for another four years in air, oxygen, nitrogen, carbon dioxide, and argon, it was shown that storage in nitrogen gave the highest viability as well as the highest yield. The viability percentage, however, did not differ significantly from that with argon and carbon dioxide.

The experiments of Abdalla and Roberts (1969b) with barley, peas, and broadbeans showed that seed deterioration associated with a reduction of viability up to about 50 percent had no effect on final yield. Another very important conclusion was the apparent functional similarity between different aging conditions in yield determination. Thus it was immaterial whether deterioration was more due to temperature or moisture or whether the rate of aging was slow or rapid. These results with barley, peas, and broadbeans, therefore, do not conform to the differential response of slow and rapid aging treatments in lettuce as reported by Harrison (1966).

More recent studies by Roberts and coworkers (Roberts, 1986; Khah, Roberts, and Ellis, 1989) on spring wheat have indicated certain conditional departures from the earlier postulates (Abdalla and Roberts, 1969a, 1969b). Yield of spring wheat was lower in crops established late in the season from low-vigor seed than from high-vigor seed, at similar densities, implying the importance of vigor under relatively unfavorable conditions (Roberts, 1986). The results of Khah, Roberts, and Ellis (1989) showed that if the direct effects of poor-vigor seed on yield through reduced establishment were avoided by adjusting sowing rates to recommended levels, then there would be little or no effect on final yield, provided that laboratory germinability was not below 85 percent, at which point there could be significant yield reductions.

According to Benjamin (1982), uniformity of size and earliness of harvest resulting from use of seed of high vigor and viability would be of considerable financial advantage in crops such as lettuce, carrot, and brassicas. Finch-Savage and McQuistan (1988) studied the performance of carrot seeds of different germination rates and concluded that fast-germinating seeds within a seedlot gave better emergence than the slow-germinating seeds. Obviously vigor of the seed was involved, but whether the same was reflected on yield of roots was not indicated. Alizaga, Alizaga, and Herrera (1987) obtained five vigor grades by subjecting soybean seeds to aging at 16 percent moisture content and 30°C for different durations, with each grade having more than 70 percent germination, and recorded a decrease in most growth parameters with decreasing seed vigor; seed yield from plants grown from aged seed was lower than that from unaged seed.

In this connection, some of my and my colleagues' results on viability of hydrated-dehydrated seed and crop yield in a range of agricultural and horticultural crops including wheat (Dasgupta, Basu, and Basu, 1976; Mandal and Basu, 1984, 1987), rice (Basu and Pal, 1978; Pal and Basu, 1982), barley (Punjabi, Mandal, and Basu, 1982), jute (Basu et al., 1978), sunflower (Basu and Dey, 1983), sugarbeet (Basu and Dhar, 1979), mungbean (Dharmalingam and Basu, 1989), tomato (Mitra and Basu, 1979), carrot (Kundu and Basu, 1981; Pan and Basu, 1985a), lettuce (Pan and Basu, 1985b), okra (Kundu and Basu, 1983), and cauliflower (Mitra and Basu, 1986) and also unpublished results on groundnut, pea, and soybean, will be helpful. The viability-maintaining seed treatments, even when they showed moderate differences in germination percentage, significantly increased agricultural yield. The yield differences recorded in these experiments may not be solely attributed to viability differences, however, but also to as yet undefined effects of physiological seed invigoration. As such, the situations indicated by the authors dealing with yield differences due to loss of viability under natural and accelerated aging may not be exactly comparable to those in the essays mentioned above.

The reported poor establishment and yield performance of plants grown from low-viability seedlots are attributable to physiological, biochemical, and cytogenetical imbalances in the seed. While many of the lesions may be corrected or eliminated during early germination and subsequent growth phases, the possibility of heritable genetic mutation being transmitted to the subsequent generations cannot be discounted. Therefore, it would be safe always to avoid low-viability seed stocks for seed production of economically important plants. Dr. W. E. Finch-Savage (Chapter 11) has given a critical account of the subject of seed quality on the subsequent crop performance.

CONCLUSION

Seed viability has long been a matter of great concern, and efforts to maintain germinability of stored seed date back to the beginning of the agrarian civilization. While a very long-term preservation of viability of our plant genetic resources is contemplated in gene banks under strictly controlled low temperature and low moisture regimes, a relatively short-term preservation is necessary for commercial seed and crop production purposes. The relation between seed vigor and viability is a very intimate one, and the planting of seed stock of high vigor is always advocated. As such, easily practicable vigor tests to correctly assess the planting value of seeds under varying field conditions would be of considerable advantage to the farmer.

A range of preharvest, harvest, and postharvest conditions affect the viability of the seed, and many of these factors have been identified here and corrective measures recommended to ensure quality. Understanding the relationship between storage environment, especially temperature and moisture, and viability, and the genotypic effects thereof, has sufficiently advanced to predict the longevity of a seed in storage, as well as what should be the appropriate moisture and temperature for the safe storage of a seedlot for a specific period. The rising cost of seed drying and storage suggests the necessity of an effective compromise between drying and refrigeration on the one hand, and seed viability on the other, for commercial purposes.

In many developing countries of the world, due to prevailing climatic conditions and technological shortcomings, even short-term storage of the current year's harvest until the next sowing poses serious problems. The facilities of artificial drying may not be readily available, and then one has to depend on sun-drying to reduce seed moisture low enough for retaining viability, and then for only about eight to ten months. Some inexpensive and simple standardized methods would be of great benefit to seed producers and farmers. Attempts to do this have led to the development of a number of physiological seed treatments which include the "wet" (requiring hydration of the seed) and "dry" (hydration not required) treatments for a wide range of agriculturally and horticulturally important crops. These treatments, depending on the kind of seed and its initial vigor status, would significantly extend storability and in most cases improve the subsequent crop performance. The dry treatments, though they are much simpler and can be given to harvest-fresh seed before bagging, are less responsive than the mid-storage wet treatments for which hydration and the more difficult job of drying-back are essential prerequisites. Nevertheless, for low-volume high-value seeds of vegetables, and especially very expensive hybrid seeds, the hydration-dehydration treatments are most confidently recommended. These treatments are also very effective for carry-over seeds (left-over seeds after sowing or post-season surplus seeds lying with seed merchants) which need to be stored until the next sowing season.

An elucidation of the mode of action of the physiological seed treatments depends on a clearer understanding of the physiology of seed aging itself, and that is where, notwithstanding serious research efforts by scientists, much controversy still exists. Seed physiologists have not yet been able to pinpoint the basic reasons for deteriorative senescence, although a large array of physiological, biochemical, and cytogenetical changes accompanying the loss of vigor and viability have been identified. Obviously, many of the changes are secondary, but intimate metabolic integration of events involving a multitude of factors and cofactors makes an effective separation of the basic causes from the effects thereof a formidable task. Albeit a hotly debated subject, the destabilization of membranes, including plasmalemma, endoplasmic reticulum, and membranes of vital bioorganelles, through free radical mediated lipid peroxidation reactions could be the likely perpetuator of the subsequent disruptive events leading to loss of vigor and, ultimately, the viability of the stored seed. Use of specific controlling agents to check such reactions would be expected to extend storability, and a critical search for such agents should be undertaken.

REFERENCES

Abdalla, F. H., and E. H. Roberts. Effects of temperature, moisture and oxygen on the induction of chromosome damage in seeds of barley. *Ann. Bot.* 32 (1968): 119-136.

Abdalla, F. H., and E. H. Roberts. The effects of temperature, moisture and oxygen on the induction of genetic changes in seeds of barley, broadbeans and peas during storage. *Ann. Bot.* 33 (1969a): 153-167.

Abdalla, F. H., and E. H. Roberts. The effect of seed storage conditions on the growth and yield of barley, broadbeans and peas. *Ann. Bot.* 33 (1969b): 169-184.

Alizaga, G., R. Alizaga, and J. Herrera. Evaluation of soybean (*Glycine max* L. Merr.) seed vigour and its relationship with emergence and yield. *Agronomia Costarricense* 11 (1987): 195-203.

Arnott, R. A. A quantitative analysis of endosperm dependent seeding growth in grasses. *Ann. Bot.* 39 (1975): 757-765.

Asana, R. D., and R. N. Basu. Studies in physiological analysis of yield. VI. Analysis of the effect of water stress on grain development in wheat. *Ind. J. Plant Physiol.* 6 (1963): 1-13.

Atkin, J. D. Relative susceptibility of snap bean varieties to mechanical injury of seed. *Proc. Amer. Soc. Hort. Sci.* 72 (1958): 370-373.

Austin, R. B. The growth of watercress (*Rorippa nasturtium-aquaticum* L. Hayek) from seed as affected by phosphorus nutrition of the parent plant. *Plant & Soil* 24 (1966a): 113-120.

Austin, R. B. The influence of phosphorous and nitrogen nutrition of pea plants on the growth of their progeny. *Plant & Soil* 24 (1966b): 359-368.

Austin, R. B. Effect of environment before harvesting on viability. In *Viability of Seeds*, ed. E. H. Roberts (London: Chapman and Hall, 1972), pp. 114-149.

Austin, R. B., P. C. Longden, and J. Hutchinson. Some effects of "hardening" carrot seed. *Ann. Bot.* 33 (1969): 883-895.

Baalbaki, R. Z. The effect of seed size, density and protein content on field performance, vigour and storability of two winter wheat varieties. *Sci. & Eng.* 50 (1989): 815B.

Barton, L. V. *Seed Preservation and Longevity* (London: Leonard Hill, 1961).

Barton, L. V., and H. R. Garman. Effect of age and storage condition of seeds on the yield of certain plants. *Contrib. Boyce Thompson Inst.* 14 (1946): 243-255.

Basu, R. N. Physio-chemical control of seed deterioration. *Seed Res.* 4 (1976): 15-23.

Basu, R. N. Seed invigoration for extended storability. *Proceedings International Conference on Seed Science and Technology*, New Delhi, February 1990, in press.

Basu, R. N., and G. Dey. Soaking and drying of stored sunflower seeds for maintaining viability and vigour of seedlings and increasing the yield potential. *Ind. J. Agric. Sci.* 53 (1983): 563-569.

Basu, R. N., and N. Dhar. Seed treatment for maintaining vigour, viability and productivity of sugarbeet (*Beta vulgaris* L.). *Seed Sci. & Technol.* 7 (1979): 225-233.

Basu, R. N., and A. K. Mandal. In search of low cost technologies for improving seed performance. In *Proceedings of the Symposium on Transferable Technology for Rural Development*, ed. D. K. Dasgupta (New Delhi: Associated Publishing Company, 1985), pp. 33-52.

Basu, R. N., and P. Pal. Seed treatment to maintain viability, vigour and yield potential of stored rice seed. *Int. Rice Res. Newslett.* 3 (1978): 5.

Basu, R. N., and A. B. Rudrapal. Iodination of mustard seed for the maintenance of vigour and viability. *Ind. J. Exp. Biol.* 18 (1980): 492-494.

Basu, R. N., and A. B. Rudrapal. Post harvest seed physiology and seed invigoration treatments. In *Proceedings of the International Conference on Frontiers of Research in Agriculture, Golden Jubilee International Conference, Indian Statistical Institute*, ed. S. K. Roy (Calcutta: Eka Press, 1982), pp. 374-397.

Basu, R. N., D. Pan, and B. Punjabi. Control of lettuce seed deterioration. *Ind. J. Plant Physiol.* 22 (1979): 247-253.

Basu, R. N., K. Chattopadhyay, P. K. Bandopadhyay, and S. L. Basak. Maintenance of vigour and viability of stored jute seeds. *Seed Res.* 6 (1978): 1-13.

Becquerel, P. La Longevite des grainess macrobiotiques transmise par Louis Mangin, C. R. Academy of Science, Paris, 1934. In *Seed Preservation and Longevity*, ed. L. V. Barton (London: Leonard Hill (Books) Ltd. Interscience Publications, 1961), pp. 1-6.

Benjamin, L. Some effects of differing times on seedling emergence, population density and seed size on root-size variation in carrot populations. *J. Agric. Sci.* 98 (1982): 537-545.

Bewley, J. D., and M. Black. Viability, dormancy and environmental control. In *Physiology and Biochemistry of Seeds, Vol.* 2, eds. J. D. Bewley and M. Black (Berlin: Springer-Verlag, 1982), pp. 1-59.

Bhattacharya, A. K., and R. N. Basu. Retention of vigour and viability of stored pea seed. *Indian Agriculturist* 34 (1990): 187-193.

Bhattacharya, A. K., and R. N. Basu. Reduction in volatile aldehyde production by viability maintaining treatments in pea (*Pisum sativum* L.) seeds. Manuscript in preparation.

Black, J. N. Seed size in herbage legumes. *Herbage Abstr.* 29 (1959): 235-241.

Bulat, H. Keimlingsanomalien und ihre Festfellung an ruhenden Samenim topographis chen Tetrazolium verfahren. *Saatgut-Wit* 21 (1969): 575-579.

Burgass, R. W., and A. A. Powell. Evidence for repair processes in the invigoration of seeds by hydration. *Ann. Bot.* 53 (1984): 753-757.

Cartledge, J. L., L. V. Barton, and A. F. Blackeslce. Heat and moisture as factors in the increased mutation rate from *Datura* seeds. *Proc. Amer. Phil. Soc.* 76 (1936): 663-685.

Cheah, K. S. E., and D. J. Osborne. Analysis of nucleosomal deoxyribo-nucleic acid in a higher plant. *Biochem. J.* 163 (1977): 141-144.

Chen, S. S. C. Metabolic activities of dormant seeds during dry storage. *Naturwissenschaften* 59 (1972): 123-124.

Chin, H. F. *Recalcitrant Seeds–A Status Report* (Rome: International Board for Plant Genetic Resources, 1988), pp. 1-28.

Ching, T. M., and I. Schoolcraft. Physiological and chemical differences in aged seeds. *Crop Sci.* 8 (1968): 407-409.

Chio, K. S., and A. L. Tappel. Inactivation of ribonuclease and other enzymes by peroxidizing lipids and malonaldehyde. *Biochemistry* 8 (1969): 2827-2832.

Choudhuri, N., and R. N. Basu. Maintenance of seed vigour and viability of onion (*Allium cepa* L.). *Seed Sci. & Technol.* 16 (1988): 51-61.

Christiansen, C. M. Microflora and seed deterioration. In *Viability of Seeds*, ed. E. H. Roberts (London: Chapman and Hall, 1972), pp. 59-63.

Das, C. S. Quality seed production and seed invigoration of okra (*Abelmoschus esculentus* L.). PhD dissertation, University of Calcutta, 1990.

Dasgupta, M., P. Basu, and R. N. Basu. Seed treatment for vigour, viability and productivity of wheat (*Triticum aestivum* L.). *Indian Agriculturist* 20 (1976): 265-273.

Dasgupta, M., K. Chattopadhyay, S. L. Basak, and R. N. Basu. Radioprotective action of seed invigoration treatments. *Seed Res.* 5 (1977): 105-118.

de Vries, H. *Die Mutationstheorie Band I.* Veit and Co. Leipzig (1901).

Dey, B., and S. M. Sirkar. The presence of an abscisic acid like factor in nonviable rice seeds. *Physiol. Plant.* 21 (1968): 1054-1059.

Dey, B., P. K. Sirkar, and S. M. Sirkar. Phenolics in relation to nonviability of rice seeds. In *Proceedings of the International Symposium on Plant Growth Substances* (Calcutta, 1967): 57-64.

Dey, G., and R. N. Basu. Physicochemical control of radiation and ageing damage in mustard seed. *Ind. J. Exp. Biol.* 23 (1985): 167-171.

Dey G., and R. K. Mukherjee. Iodine treatment of soybean and sunflower seeds for controlling deterioration. *Field Crops Res.* 9 (1984): 205-215.

Dey, G., and R. K. Mukherjee. Deterioration of maize and mustard seeds: Changes in phospholipid and tocopherol content in relation to membrane leakiness and lipid peroxidation. *Agrochimia* 32 (1988): 430-440.

Dharmalingam, C. Studies on quality seed production and control of seed deterioration in mungbean (*Vigna radiata* L. Wil.) and sunflower (*Helianthus annuus* L.). PhD Thesis, University of Calcutta, 1982.

Dharmalingam, C. Midstorage correction to prolong viability of rice seeds. *Int. Rice Res. Newslett.* 15 (1990): 21-22.

Dharmalingam, C., and R. N. Basu. Influence of seed size and seed coat colour on the production potential of mungbean cv. Co_2. *Seeds & Farms* 13 (1987): 16-20.

Dharmalingam, C., and R. N. Basu. Invigoration treatment for increased production in carried over seeds of mungbean. *Seeds & Farms* 15 (1989): 34-36.

Doijode, S. D., and G. B. Raturi. Effect of hydration-dehydration on the germination and vigour of certain vegetable seeds. *Seed Res.* 15 (1987): 156-159.

Edwards, M. Metabolism as a function of water potential in air-dry seeds of charlock (*Sinapis arvensis* L.). *Plant Physiol.* 58 (1976): 237-239.

Ehrenberg, A. Research on free radicals in enzyme chemistry and in radiation biology. In *Free Radicals in Biological Systems*, eds. M. S. Blois, Jr., H. W. Brown, R. M. Lemmon, R. O. Linblum, and M. Weissbluth (New York: The Academic Press, 1961), pp. 337-350.

Ellis, R. H., and E. H. Roberts. Improved equations for the prediction of seed longevity. *Ann. Bot.* 45 (1980): 13-30.

Falkenstein, G., and A. M. Steiner. Promotion and reduction of vigour, storability and field germination and yield by sprouting (pre-germination) in wheat (*Triticum aestivum* L.). *Landwirtschaftliche Forschung* (1985): 539-549.

Farrant, J. M., P. Berjak, and M. W. Pammenter. The effect of drying rate on viability retention of recalcitrant propagules of *Avicennia marina*. *South African J. Bot.* 51 (1985): 432-438.

Finch-Savage, W. E., and C. I. McQuistan. Performance of carrot seeds possessing different germination rates within a seed lot. *J. Agric. Sci.* 110 (1988): 93-99.

Fox, R. L., and W. A. Albrecht. Soil fertility and the quality of seeds. *Research Bulletin of Missouri Agricultural Station Number 619* (1957): 23.

Gardner, H. W. Lipid hydroperoxide reactivity with proteins and amino acids: A review. *J. Agric. Food Chem.* 27 (1979): 220-229.

Ghosh, A. N. Analysis of the effect of K and Na ions on the germinability of wheat (*Triticum aestivum* L.) seeds. MSc (Ag.) Dissertation, University of Calcutta (1992): 1-18.

Ghosh, B. N. Agrometeorological studies on rice I. Influence of climatic factors on dormancy and viability of paddy seeds. *Ind. J. Agric. Sci.* 32 (1962): 235-241.

Goldsworthy, A., J. L. Fielding, and M. B. J. Dover. Flash imbibition: A method for the reinvigoration of aged wheat seed. *Seed Sci. & Technol.* 10 (1982): 55-65.

Gorecki, R. J., and G. E. Harman. Effect of antioxidants on viability and vigour of ageing pea seeds. *Seed Sci. & Technol.* 15 (1987): 109-117.

Goswami, R. Comparative efficacy of K and Na salts in the maintenance of vigour and viability of wheat seed by hydration-dehydration treatments. MSc dissertation, University of Calcutta (1990): 1-19.

Graf, E., K. L. Empson, and J. W. Eaton. Phytic acid. A natural antioxidant. *J. Biol. Chem.* 262 (1987): 11647-11650.

GrantLipp, A. E., and L. A. T. Ballard. Germination patterns shown by the light sensitive seeds of *Anagallis arvensis*. *Aust. J. Biol. Sci.* 16 (1963): 572-584.

Greaves, J. E., and E. G. Carter. The influence of irrigation water on the composition of grains and the relationship to nutrition. *J. Biol. Chem.* 58 (1923): 531-541.

Gupta, P. C., and V. P. Rana. Effect of culture filtrate of some seed mycoflora on seed germination of solanaceous crops. *Seed Res.* 9 (1981): 192-193.

Harman, G. E., and T. E. Stasz. Influence of seed quality on soil microbes and seed rots. In *Physiological-Pathological Interactions Affecting Seed Deterioration*, ed. S. H. West (CSSA, Special Publication Number 12, Chapter 2, 1986), pp. 11-37.

Harrington, J. F. Germination of seeds from carrot, lettuce and pepper plants grown under severe nutrient deficiencies. *Hilgardia* 30 (1960): 219-235.

Harrington, J. F. Seed storage and longevity. In *Seed Biology*, ed. T. T. Kozlowski (London, New York: Academic Press, 1972), pp. 251-263.

Harrison, B. J. Seed deterioration in relation to seed storage and conditions and its influence upon seed germination, chromosomal damage and plant performance. *J. Nat. Inst. Agric. Bot.* 10 (1966): 644-663.

Heydecker, W. Vigour. In *Viability of Seeds*, ed. E. H. Roberts (London: Chapman and Hall, 1972), pp. 209-252.

Hochstein, P., and S. K. Jain. Association of lipid peroxidation and polymerization of membrane proteins with erythrocyte ageing. *Fed. Proc.* 40 (1981): 183-188.

Iljin, W. S. Drought resistance in plants and physiological process. *Ann. Rev. Plant Physiol.* 8 (1957): 257-274.

ISTA. International Rules for Seed Testing. *Seed Sci. & Technol.* 4 (1976): 3-49.

Jackson, W. D. The life-span of mutagens produced in cells by irradiation. *Proceedings of the 2nd Australian Conference on Radiation Biology, Melbourne* (London: Butterworths, 1959), pp. 190-208.

James, E. Preservation of seed stocks. *Adv. Agron.* 19 (1967): 87-106.

Kamra, S. K. Determination of germinability of melon seed with x-ray contrast method. *Proc. ISTA* 31 (1966): 719-729.

Kanner, J., and M. Karel. Changes in lysozyme due to reactions with peroxidizing methyl linoleate in a dehydrated model system. *J. Agric. Food Chem.* 24 (1976): 468.

Khah, E. M., E. H. Roberts, and R. H. Ellis. Effects of seed ageing on growth and yield of spring wheat at different plant population densities. *Field Crops Res.* 20 (1989): 175-190.

Kharlukhi, L. K., and P. K. Agrawal. Evidence for participation of pentose phosphate pathway during seed deterioration in storage. *Ind. J. Exp. Biol.* 22 (1984): 612-614.

Kittock, D. L., and A. G. Law. Relationship of seedling vigour to respiration and tetrazolium chloride reduction by germinating wheat seeds. *Agron. J.* 60 (1968): 286-288.

Kittock, D. L., and J. K. Patterson. Seed size effects on performance of dryland grasses. *Agron. J.* 53 (1962): 74-77.

Kivilaan, A. K., and R. S. Bandurski. The ninety year period of Dr. Beal's seed viability experiment. *Amer. J. Bot.* 60 (1973): 140-145.

Kler, D. S., and G. S. Dhillon. Effect of seed size on the growth of yield of soybean (*Glycine max* L. Merr.). *Ind. Bot. Rep.* 2 (1983): 51-55.

Koller, D. Preconditioning of germination in lettuce at the time of fruit ripening. *Amer. J. Bot.* 41 (1962): 841-844.

Kostoff, D. Mutations and the ageing of seeds. *Nature* 135 (1935): 107.

Kramer, C., J. J. Pest, and W. Witten. *Drouwgersten Klimaat* (Utrecht: Kemink & Zoom, 1952).

Kundu, C., and R. N. Basu. Hydration-dehydration treatment of carrot seed for the maintenance of vigour, viability and productivity. *Scient. Hortic.* 15 (1981): 117-125.

Kundu, C., and R. N. Basu. Control of seed deterioration and improvement of field performance of okra by hydration-dehydration treatment of stored seed. *Veg. Sci.* 10 (1983): 84-93.

Kuo, W. H. J., and C. Chu. Prolonging storability of rice (*Oryza sativa* L.) seeds by imbibition-dehydration treatment: Conditions of treatment. *J. Agric. Assoc. China* 133 (1986): 16-24.

Laude, H. Fresh seed dormancy in annual grasses. *Calif. Agric.* 16 (1962): 3.

Lawan, M., F. L. Barnett, B. Khaleed, and R. L. Vanderlip. Seed density and seed size of pearl millet as related to field emergence and several seedling and seedling traits. *Agron. J.* 77 (1986): 567-571.

Lesko, S. A., R. J. Lorentzen, and P. O. P. Ts'o. Role of superoxide in deoxyribonucleic acid strand scission. *Biochemistry* 19 (1980): 3023-3028.

Mandal, A. K., and R. N. Basu. Alleviation of salt stress during germination and early seedling growth of wheat by hydration-dehydration pretreatments. *Indian Agriculturist* 26 (1982): 121-128.

Mandal, A. K., and R. N. Basu. Preservation of wheat seed by hydration-dehydration treatment with seed protectants. *Pestology* 8 (1984): 20-23.

Mandal, A. K., and R. N. Basu. Vigour and viability of wheat seed treated with bleaching powder. *Seeds & Farms* 10 (1986): 46-48.

Mandal, A. K., and R. N. Basu. Midterm and presowing hydration-dehydration treatments for improved field performance of wheat. *Field Crops Res.* 15 (1987): 259-265.

Mathur, S. K., and S. Sinha. Effect of fungal metabolites on seed viability and seedling vigour of bajra (*Pennisetum typhoides*). *Seed Res.* 6 (1978): 181-187.

McDonald, M. B. Jr. Vigour Test Subcommittee Report. *AOSA Newslett.* 54 (1980): 137-140.

Mitra, R., and R. N. Basu. Seed treatment for viability, vigour and productivity of tomato. *Scient. Hortic.* 11 (1979): 365-369.

Mitra, R., and R. N. Basu. Maintenance of vigour, viability and yield potential of cauliflower seed. *Veg. sci.* 13 (1986): 34-40.

Mondal, R., and M. A. Choudhury. Regulation of senescence of excised leaves of some C-3 and C-4 species by endogenous H_2O_2. *Biochem. Physiol. Pflanzen.* 177 (1982): 403-417.

Moore, R. P. Seedling emergence of small seeded legumes and grasses. *Amer. Soc. Agron.* 35 (1943): 370-381.

Moore, R. P. Rough harvesting methods kill soybean seeds. *Seedmen's Digest* 17 (1957): 14-16.

Moore, R. P. History supporting tetrazolium seed testing. *Proc. ISTA* 34 (1969): 233-242.

Moore, R. P. Effect of mechanical injuries on viability. In *Viability of Seeds*, ed. E. H. Roberts (London: Chapman and Hall, 1972), pp. 94-113.

Moutschen-Dahmen, M., J. Moutschen, and L. Ehrenberg. Chromosome disturbances and mutation produced in plant seeds by oxygen at high pressures. *Hereditas* 45 (1959): 230-244.

Mukhopadhyay, A. Maintenance of vigour and viability of some leguminous seeds. PhD Dissertation, University of Calcutta, 1991.

Mukhopadhyay, A., M. M. Choudhuri, K. Sen, and B. Ghosh. Changes in polyamines and related enzymes with loss of viability in rice seeds. *Phytochemistry* 22 (1983): 1547-1551.

Mukhopadhyay, S. Influence of different factors on seed vigour bioassay. PhD Dissertation, University of Calcutta, 1990.

Navashin, M. S. Origin of spontaneous mutations. *Nature* 131 (1933): 436.

Nellist, M. E. Predicting the viability of seeds dried with heated air. *Seed Sci. & Technol.* 9 (1981): 439-455.

Nickerson, J. T. R. Preservation and antioxidants. In *Fundamentals of Food Processing Operations*, eds. J. L. Heid and M. A. Joslyn (USA: AVI Publishing Co., 1967), p. 233.

Osborne, D. J. Dormancy as a survival strategem. *Ann. Appl. Biol.* 98 (1981): 525-531.

Osborne, D. J. Deoxyribonucleic acid and integrity and repair in seed germination: The importance in viability and survival. In *The Physiology and Biochemistry of Seed Development, Dormancy and Germination*, ed. A. A. Khan (Amsterdam, Elsevier Biomedical Press, 1982), pp. 435-463.

Owen, E. B. The storage of seeds for the maintenance of viability. Commonwealth Agricultural Bureaux, Farnham Royal England, 1956.

Pal, P., and R. N. Basu. Control of rice seed deterioration by physico-chemical treatments. In *Rice in West Bengal* (Government of West Bengal Publication), 3 (1982): 159-166.

Pal, P., and R. N. Basu. Treatment of rice seed with iodine and chlorine for the maintenance of vigour, viability and productivity. *Indian Agriculturist* 32 (1988): 71-75.

Pal, P., and R. N. Basu. Volatile aldehyde production in relation to seed vigour of rice (*Oryza sativa* L.). *Indian Agriculturist* 33 (1989): 223-225.

Pammenter, N. W., J. H. Adamson, and P. Berjak. Viability extension of stored seed by cathodic protection. *Science* 186 (1974): 1123-1124.

Pan, D., and R. N. Basu. Mid-storage and pre-sowing seed treatments for lettuce and carrot. *Scient. Hortic.* 25 (1985a): 11-19.

Pan, D., and R. N. Basu. Absence of *de novo* protein synthesis in moisture equilibration-drying treatment for maintenance of lettuce (*Lactuca sativa* L.) seed viability. *Ind. J. Exp. Biol.* 23 (1985b): 375-379.

Perry, D. A. Report of the vigour test committee 1974-1977. *Seed Sci. & Technol.* (1978).

Peto, F. H. The effect of ageing and heat on the chromosomal mutation rate in maize and barley. *Can. J. Res.* 9 (1933): 261-264.

Petruzzeli, L., L. Lioi, S. Morgutti, and S. Cocucci. Effect of fusicoccin and K on the viability of aged wheat seeds. *Attidella Acafdemia Nazionale dei Lincei Rendiconti* 71 (1982): 37-43.

Porsild, A. E., C. R. Harrington, and G. A. Mulligan. *Lupinus arcticus* Wats. grown from seeds of pleistocene age. *Science* 158 (1967): 113-114.

Pryor, W. A. Free radical reactions and their importance in biochemical systems. *Fed. Proc.* 32 (1973): 1862-1869.

Punjabi, B., and R. N. Basu. Control of age and irradiation-induced seed deterioration in lettuce (*Lactuca sativa* L.) by hydration-dehydration treatments. *Proc. Indian Natl. Sci. Acad.* B 48 (1982): 242-250.

Punjabi, B., and R. N. Basu. Relation between dormancy and viability of dry stored lettuce (*Lactuca sativa* L. cv. Sutton's A1) seed. *Ind. J. Exp. Biol.* 27 (1989): 739-741.

Punjabi, B., A. K. Mandal, and R. N. Basu. Maintenance of vigour, viability and productivity of stored barley seed. *Seed Res.* 10 (1982): 70-73.

Punnuswamy, A. S., and V. Ramakrishnan. Effect of seed size on physico-chemical characteristics of seeds in groundnut (*Arachis hypogaea*) cv. Pol. 1 and TMV 2. *Madras Agric. J.* 71 (1984): 669-672.

Purkar, J. K., R. B. Mehra, and S. K. Banerjee. Quantitative genetical changes in wheat induced through seed ageing. *Seed Res.* 13 (1985): 172-187.

Ramamoorthy, K. Seed viability studies in groundnut (*Arachis hypogaea* L.) PhD Thesis, University of Calcutta, 1990.

Ramamoorthy, K., V. Palanisamy, D. Kalavathi, K. Vanangamudi, and T. V. Karivaratharaju. Maintenance of vigour and viability of stored gingelly (*Sesamum indicum* L.) seeds. *Trop. Agric.* 67 (1990): 209-212.

Rao, N. K., and E. H. Roberts. Seed ageing and meiotic chromosomal abnormalities in lettuce. *Cytologia* 54 (1989): 373-379.

Rao, N. K., E. H. Roberts, and R. H. Ellis. The influence of pre-storage and post-storage hydration treatments on chromosomal aberrations, seedling abnormalities and viability of lettuce seeds. *Ann. Bot.* 60 (1987): 97-108.

Ries, S. K., C. J. Schweizer, and H. Chmiel. The increase in protein content and yield of simazine treated crops in Michigan and Costa Rica. *BioScience* 18 (1968): 205-208.

Roberts, B. E., and D. J. Osborne. Protein synthesis and viability in rye grains. In *Seed Ecology*, ed. W. Heydecker (London: Butterworths, 1973): 99-114.

Roberts, E. H. An investigation of intervarietal differences in dormancy and viability of rice seeds. *Ann. Bot.* 27 (1963): 365-369.

Roberts, E. H. Storage environment and the control of viability. In *Viability of Seeds*, ed. E. H. Roberts (London: Chapman and Hall, 1972a), pp. 14-58.

Roberts, E. H. Cytological, genetical and metabolic changes associated with loss

of viability. In *Viability of Seeds*, ed. E. H. Roberts (London: Chapman and Hall, 1972b), pp. 253-306.
Roberts, E. H. Loss of viability and crop yield. In *Viability of Seeds*, ed. E. H. Roberts (London: Chapman and Hall, 1972c), pp. 307-320.
Roberts, E. H. Predicting the storage life of seeds. *Seed Sci. & Technol.* 1 (1973a): 409-512.
Roberts, E. H. Loss of seed viability: Chromosomal and genetical aspects. *Seed Sci. & Technol.* 1 (1973b): 515-527.
Roberts, E. H. Seed deterioration and loss of viability. *Adv. Seed Res. & Technol.* 4 (1979): 25-42.
Roberts, E. H. Quantifying seed deterioration. In *Physiology of Seed Deterioration*, eds. M. B. McDonald, Jr., and C. J. Nelson (Madison: Crop Science Society of America, 1986), pp. 101-123.
Roberts, E. H., F. H. Abdalla, and R. J. Owen. Nuclear damage and ageing of seeds with a model for seed survival curves. *Symp. Soc. Exp. Biol.* 21 (1967): 65-100.
Roos, E. E. Induced genetic changes in seed germination during storage. In *The Physiology and Biochemistry of Seed Development, Dormancy and Germination*, ed. A. A. Khan (Amsterdam, Elsevier Biomedical Press, 1982), pp. 409-434.
Rossman, E. C. Freezing injury of maize seed. *Plant Physiol.* 24 (1949): 629-656.
Roubal, W. T., and A. L. Tappel. Damage to proteins, enzymes and amino acids by peroxidizing lipids. *Arch. Biochem. Biophys.* 113 (1966): 5.
Roy, S. K. Maintenance of vigour, viability and yield, potential of stored wheat seed. *Seed Res.* 10 (1982): 139-142.
Rudrapal, A. B., and R. N. Basu. Physiology of hydration-dehydration treatments in the maintenance of seed viability in wheat. *Ind. J. Exp. Biol.* 17 (1979): 768-771.
Rudrapal, A. B., and R. N. Basu. Iodine treatment of mungbean seeds for the maintenance for vigour and viability. *Curr. Sci.* 49 (1980): 319-320.
Rudrapal, A. B., and R. N. Basu. Lipid peroxidation and membrane damage in deteriorating wheat and mustard seeds. *Ind. J. Exp. Biol.* 20 (1982): 465-470.
Rudrapal, D., and S. Nakamura. Use of halogens in controlling egg plant and radish seed deterioration. *Seed Sci. & Technol.* 16 (1988a): 115-121.
Rudrapal, D., and S. Nakamura. The effect of hydration-dehydration pretreatments on egg plant and radish seed viability. *Seed Sci. & Technol.* 16 (1988b): 123-130.
Russell, E. J., and J. A. Voelcker. Fifty years of field experiments at the Woburn Experiment Station, London, 1936.
Saha, R., and R. N. Basu. Invigoration of soybean seed for alleviating soaking injury and ageing damage on germinability. *Seed Sci. & Technol.* 12 (1984): 613-622.
Saha, R., and R. N. Basu. Effect of culture filtrate of *Aspergillus flavus* on the vigour and viability of stored wheat seed. *Ind. J. Mycol. Res.* 23 (1985): 19-26.

Saha, R., A. K. Mandal, and R. N. Basu. Physiology of seed invigoration treatment in soybean (*Glycine max* L.). *Seed Sci. & Technol.* 18 (1990): 269-276.

Salter, P. J., and J. F. Goode. *Crop responses to water at different stages of growth.* (Commonwealth Agricultural Bureaux, Farnham Royal, 1967).

Savino, G., P. Deleo, and P. M. Haigh. Effects of presoaking upon seed vigour and viability during storage. *Seed Sci. & Technol.* 7 (1979): 57-65.

Schultz, H. W., E. A. Day, and R. O. Sinnhauber. *Symposium on Foods: Lipids and Their Oxidation*, ed. H. W. Schultz (Westport, Connecticut, USA: The AVI Publishing Co., 1962).

Scott, R. K. The sowing and harvesting dates, plant population and fertilizers on seed yield and quality of direct-drilled sugarbeet seed crops. *J. Agric. Sci.* 73 (1969): 373-385.

Sen, S., and D. J. Osborne. Decline in ribonucleic acid and protein synthesis with loss of viability during early hours of imbibition of rye (*Secale cereale* L.) embryos. *Biochem. J.* 166 (1977): 33-38.

Shutt, F. T. The nitrogen content of wheat as affected by seasonal condition. *Trans. Roy. Soc. Canada, Section 3,* 29 (1935): 37-39.

Sneddon, J. L. Sugarbeet production experiments. *J. Natl. Inst. Agric. Bot.* 9 (1963): 333-345.

Sundholm, F., A. Visappa, and J. Bjorkston. Cross-linking of collagen in the presence of oxidizing lipid. *Lipids* 13 (1978): 755-757.

Sur, K., and R. N. Basu. Control of age and irradiation induced seed deterioration in rice (*Oryza sativa* L.) by hydration-dehydration treatment. *Seed Res.* 14 (1986): 197-205.

Sur, K., and R. N. Basu. Vigour rating of wheat seeds. *Seed Sci. & Technol.* 18 (1990a): 661-671.

Sur, K., and R. N. Basu. Reduced volatile aldehyde production in wheat by seed invigoration treatments. *Curr. Sci.* 59 (1990b): 799-800.

Tao, K. L., A. A. Khan, G. E. Harman, and C. J. Eckenrode. Practical significance of the application of chemicals in organic solvents to dry seeds. *J. Amer. Soc. Hort. Sci.* 99 (1974): 217-220.

Tekrony, D. M., T. Bustaman, D. D. Egli, and T. W. Pfeiffer. Effects of soybean seed size, vigor and maturity on crop performance in row and hill plots. *Crop Sci.* 27 (1987): 1040-1045.

Tolman, B. Sugarbeet production in Southern Utah with special reference to factors affecting yield and reproductive developments. *Tech. Bull. USDA* 845 (1943).

Vanangamudi, K., and T. V. Karivaratharaju. Effect of prestorage chemical fortification of seeds on shelf life of redgram, blackgram and greengram seeds. *Seed Sci. & Technol.* 14 (1986): 477-482.

Van Dobben, W. De involved von klimaat opde gevolishered von tarwe von schot. *J. Verst. Cent. Inst. Landbouwk, Onderz* (1947): 40-43.

Van Toai, T. T., M. B. McDonald, Jr., and G. L. Staby. Cultivar, fungicide treatment and storage environment interactions on carry over soybean seed quality. *Seed Sci. & Technol.* 14 (1986): 301-312.

Verma, R. S., and P. C. Gupta. Storage behaviour of soybean varieties vastly differing in seed size. *Seed Res.* 3 (1975): 39-44.

Villiers, T. A. Ageing and the longevity of seeds in field conditions. In *Seed Ecology*, ed. W. Heydecker (London: Butterworths, 1973), pp. 265-288.

Villiers, T. A., and D. J. Edgcumbe. On the cause of seed deterioration in dry storage. *Seed Sci. & Technol.* 3 (1975): 761-774.

Went, F. W., and P. A. Muntz. A long term test of seed longevity. *El Aliso* 2 (1949): 63-75.

West, S. H., S. K. Loftkin, M. Wahl, and C. D. Batich. Polymers as moisture-barriers to maintain seed quality. *Crop Sci.* 25 (1985): 941-944.

Wilson, D. O., Jr., and M. B. McDonald, Jr. A convenient volatile aldehyde assay for measuring seed vigour. *Seed Sci. & Technol.* 14 (1986a): 259-268.

Wilson, D. O., Jr., and M. B. McDonald, Jr. The lipid peroxidation model of seed ageing. *Seed Sci. & Technol.* 14 (1986b): 269-300.

Wolff, S. P., and R. T. Dean. Fragmentation of proteins by free radicals and its effects on their susceptibility to enzymatic hydrolysis. *Biochem. J.* 234 (1986): 399-403.

Wolff, S. P., A. Gardner, and R. T. F. Dean. Free radicals, lipids and protein degradation. *Trends in Biochemical Sciences* 11 (1986): 27-31.

Woodstock, L. W., and R. B. Taylorson. Ethanol and acetaldehyde in imbibing soybean seeds in relation to deterioration. *Plant Physiol.* 67 (1981): 424-428.

Woodstock, L. W., S. Maxon, K. Faust, and L. N. Bass. Use of freeze drying and acetone impregnation with natural and synthetic antioxidants to improve storability of onion, pepper and parsley seeds. *J. Amer. Soc. Hort. Sci.* 108 (1983): 692-696.

Zirlin, A., and M. Karel. Oxidation effects in a freeze-dried gelatinmethyl linoleate system. *J. Food Sci.* 34 (1969): 160-164.

Chapter 2

Seed Vigor

David L. Dornbos, Jr.

Seed vigor refers to both the ability and strength of a seed to germinate successfully and establish a normal seedling. Vigor is positively related to the ability of a seed population to establish an optimum plant stand, in both optimum and suboptimum soil environments, and therefore to maximize yield. Because soil conditions during planting are often not optimal, growers require seed with good germination ability and vigor.

A common goal among seedsmen and growers is to produce and have access to high-quality seed populations for planting. In recent years, considerable effort has focused on the measurement of seed vigor levels, and the relationship of these vigor levels to both stand establishment in the field and yield. Numerous tests have been developed and combined, and more will likely be developed, to measure seed vigor and effectively correlate vigor with field performance. To be useful to both seed laboratories and growers, a seed vigor test must be reliable, quantifiable, rapid, simple, inexpensive, and relevant to field conditions.

Seed vigor is gradually acquired in the seed production environment as the seed develops on the maternal plant, achieves a maximum at physiological maturity, then steadily declines thereafter until being planted in the subsequent growing season. To ensure the highest possible vigor of a seedlot, then, efforts must focus on (1) creating a positive growth environment so that vigorous seeds develop, (2) harvesting as soon as possible after physiological maturity, and (3) handling and storing the seed in such a way as to minimize damage and slow the deterioration process. Vigor tests can be used to investigate the relationship between seed development, deterioration, and conditioning on seedlot quality. Consideration must be given to the physical, physiological, and pathological relationships that contribute to or detract from the attainment of maximum seed vigor.

The objectives here are threefold: to identify the need for a concept of seed vigor, to develop the underlying mechanisms of seed vigor attainment and deterioration, and to describe the methods by which seed vigor information is used.

DEVELOPMENT OF THE CONCEPT OF SEED VIGOR

The Problem

To maximize yield, growers require the rapid establishment of optimum plant stands under many different environmental conditions. Hence, growers require one or more tests that can be applied to seed populations to accurately predict emergence and, therefore, yield potential in a field production environment.

Germination refers to the ability of a seed to produce a normal seedling in "favorable conditions." Experience has shown that when field conditions are near optimum at planting, standard germination test results accurately predict field emergence rates (Perry, 1977; Egli and TeKrony, 1979). Numerous experiences show that when conditions in the germination environment are less than optimum, the standard germination test is a poor predictor of field emergence, underestimating plant stand establishment (Table 1; DeLouche, 1973; TeKrony and Egli, 1977; Johnson and Wax, 1978; Yaklich and Kulik, 1979).

Suboptimal germination environments are increasingly the rule, rather than the exception, in production systems. As growers plant earlier in the spring to maximize available moisture, season length, and yield potential, and as interest in no-till and minimum tillage methods grow, seeds are increasingly planted in cool and wet soils. Lower emergence and seedling rates of corn (*Zea mays*) are related to the lower temperatures in soils of no-till production systems (Burrows and Larson, 1962; Griffeth et al., 1973; TeKrony, Egli,

TABLE 1. Comparison of laboratory germination and field emergence percentages of samples from 94 soybean seedlots. Emergence tests made at Mississippi State, MS in 1969 (DeLouche, 1973).

Field Emergence	Laboratory Germination			Total Number
	90-94	85-89	80-84	
--- % ---	No. of Samples			
90+	5	0	0	5
80-89	9	12	0	21
70-79	10	14	6	30
60-69	2	8	4	14
50-59	3	6	6	15
40-49	0	3	2	5
40–	0	4	0	4
Total Number	29	47	18	94

and Wickham, 1989). Soybeans (*Glycine max* L. Merr.) produced in conventional tillage systems outyielded those produced in no-till systems primarily because of 14 percent better emergence rates (Oplinger and Philbrook, 1992). Therefore, growers require access to high-vigor seed to plant and reliable vigor information to facilitate their predictive power as they strive to attain optimum plant populations in the production environment. Improved seed vigor tests and the accurate analysis of vigor test results are necessary to provide the high-quality seed and information necessary to fully exploit today's agricultural systems.

Viability and Germination versus Vigor

Seed viability, germination, and vigor each describe different aspects of the quality of a seed population. When understood collectively, these three characteristics accurately describe the quality of a seed population. Vigor data without germination data, or germination data without vigor data, are far less meaningful than when taken together.

Viability and germination have dual meanings, the specific meaning being dependent upon whether the audience is physiologically or technologically oriented. To the seed technologist, seed viability refers to the capability of a seed to germinate and produce a "normal" seedling. From a physiological perspective, however, viability refers only to whether or not a seed contains any tissues that are metabolically active, possessing energy reserves and enzymes capable of sustaining live plant cells. A seed may be viable to the physiologist, then, as it contains live tissues, but be nonviable to the technologist, because it is incapable of producing a "normal" seedling. Meaningful use of the term "viability," therefore, requires an understanding of the character of the audience.

Germination ability is also defined differently, depending upon whether it is from a technological or physiological perspective. First, as defined in the Rules for Testing Seeds of the Association of Official Seed Analysts (AOSA), to the seed technologist germination is "the emergence and development from the seed embryo of those essential structures which, for the kind of seed in question, are indicative of the ability to produce a normal plant under favorable conditions." The technologist is inclined to use the standard germination test described in the Rules for Testing Seeds (AOSA, 1978) to evaluate the viability status of a seedlot. Second, from the physiological perspective, germination is defined simply as protrusion of the radicle through the seed coat, indicating whether or not seeds within a population are alive. Because the technologist requires that a seed produce both a normal root and shoot under the germination conditions defined by the AOSA for each seed type, attainment of a high germination percentage by a seed population is more restrictive than from a physiological perspective. Seed germination according to the seed technologist is more relevant to the grower and field environment, whereas germination to the physiologist pertains more to investigative study.

The definition of seed vigor adopted by the AOSA (1983) follows: "Seed vigor comprises those seed properties which determine the potential for rapid, uniform emergence, and development of normal seedlings under a wide range of field conditions." Vigor is differentiated from germination ability in several ways. In contrast to the technologist's definition of germination, the definition of vigor identifies the desire to measure uniform and rapid seedling emergence, thereby focusing attention on seeds with stronger germination potential. Vigor also focuses on emergence potential in a variety of field conditions, both optimum and less than optimum, thereby gaining greater relevance to typical agronomic conditions.

Background of the Seed Vigor Concept

Seed vigor is a relatively new concept. Reference was made to the "driving force" or "shooting strength" of seeds late in the nineteenth century (Nobbe, 1976). Early in the twentieth century, the brick grit test was developed to evaluate cereal seed emergence in association with the term "triebkraft," or vigor (Hiltner and Ihssen, 1911).

Little was done to develop a concept of seed vigor until the cold test for corn was developed. The basis for a seed corn vitality test was developed to investigate the relationship between pericarp injury, moisture absorption, temperature, and fungal attack (Alberts, 1927; Dickson and Holbert, 1926; Reddy, 1933, 1935). The cold test was refined, applied to different seed types, and implemented on a broad and routine scale by 1950. Objectives of the cold test were to evaluate corn seed quality after being carried from one year to the next, to determine the effectiveness and dose relationships of seed protectants, to monitor and control seed emergence rates following conditioning, and to identify genetic sources favoring emergence in cool and wet soils (Isely, 1950). The cold test contributed extensively to an understanding of the need for, and practical relevance of, seed vigor.

Testing and terminology differences between European and American seed testing laboratories confused the interpretation of seed quality measurements for a time. As a consequence, in 1950, the International Seed Testing Association (ISTA) Congress was held in part to debate discrepancies in seed testing results between laboratories. One committee of the Congress, the Biochemical and Seedling Vigour Committee, later developed a definition of vigor and standardized the methods for measurement of vigor (Perry, 1978,

1981). A critical aspect of this process was definition of the limitations and distinctness of seed viability, germination, and vigor.

The importance of developing and conducting both seed germination and vigor tests has continued to increase since that time. To date, between 40 and 50 tests of seed vigor have been studied, of which a mere handful, still less accurate than desired, are actively employed in seed testing. A 1982 survey of North American seed testing laboratories indicated that 74 percent evaluated seed for vigor (TeKrony, 1983). Impetus to further define vigor tests and to utilize these tests for a variety of applications will increase in the foreseeable future.

Character of the Seed Vigor Tests

Seedling Growth and Evaluation Tests

Seed vigor tests are classified as seedling growth and evaluation, stress, or biochemical tests. Detailed procedures for the most readily accepted tests are presented in the *Vigor Testing Handbook* (AOSA, 1983).

Seedling growth and evaluation tests include the seedling vigor classification, seedling growth rate, and speed of germination tests. Of these, the seedling growth rate has received somewhat stronger acceptance because seedling dry weight represents a logical and relevant estimate of the rapid and uniform emergence desired by growers and avoids more subjective ratings. Seedling growth rate was an accurate predictor of field performance with pepper (*Capsicum annuum* L. var. annuum) (Trawatha, Steiner, and Bradford, 1990). One criticism of these tests is the potential for bias in favor of large seed size.

Stress Tests

Stress tests include accelerated aging, cold, cool germination, brick grit, and osmotic tests. Of these, the cold test has received extremely wide usage and acceptance with many crop types by seed companies and testing associations alike. The cold test has received a high level of acceptance because of its relevance to typical production environment soils. It emulates fairly well the placement of seeds in cool and wet soils that contain native microflora populations and predicts emergence rates in similar environments reasonably well for a single vigor test.

Biochemical Tests

Finally, the biochemical tests include the tetrazolium, conductivity, respiration, glutamic acid decarboxylase activity (GADA), and adenosine triphosphate (ATP) content tests. Of these, the tetrazolium test has received wide usage, but is now used less because of the large amount of time needed to evaluate seed samples and the difficulty in relating zones of damage to germinability and vigor. The biochemical tests of vigor have received less attention by seed companies and testing associations for routine use because significant resources are required for equipment, training, and sample evaluation.

Vigor and Field Performance

Relationship Between Vigor, Field Emergence, and Yield

The germination ability and vigor of a seedlot is related indirectly to performance in the field. It is unlikely that low seed vigor directly limits the ability of a plant to express its genetic potential for maximum seed yield. Numerous reports are recorded in the literature describing experiments in which low seed vigor contributed to reduced forage or grain yield, or had no effect (TeKrony and Egli, 1991). When yield loss occurs, seed vigor is indirectly related to yield by reducing final emergence, plant emergence rate, or by causing uneven emergence, and is dependent on the plant type or the form of crop being harvested.

The effect of seed vigor on yield depends on the form of crop harvested. Crops harvested during vegetative or early reproductive growth demonstrate a consistently positive relationship between seed vigor and yield (TeKrony and Egli, 1991). With meadow bromegrass (*Bromus biebersteinii*), total emergence was correlated with standard germination, whereas both total emergence and forage yield were correlated significantly with accelerated aging percentage (Table 2; Hall and Wiesner, 1990). The slower rate of emergence frequently associated with low-vigor seed contributes to the development of smaller plants and reduced vegetative yield relative to the use of high-vigor seed (Ellis, 1989).

Annual crops harvested for grain at maturity seldom demonstrate a relationship between seed vigor and yield under normal conditions (TeKrony and Egli, 1991). The photosynthetic capacity of a single leaf is likely to be affected not by seed vigor, but rather by genetic composition. If the optimum leaf canopy (a leaf area index [LAI] of six) of a population is reached, regardless of seedlot vigor, the photosynthetic capability of the canopy should be sufficient to maximize yield. Conversely, because low seed vigor contributes to

TABLE 2. Seed germination and vigor of corn seedlots planted in 1982, 1983, and 1984 (TeKrony, Egli, and Wickham, 1989).

Year	Standard Germination	Cold Test	Accelerated Aging	Total Weight	Vigor Index*
	---------- %	----------	----------	mg/seedling	
1982	88	68	-	47	5.3
	94	64	-	61	6.3
	94	94	-	83	9.7
	96	94	-	81	10.0
1983	89	48	73	62	5.2
	88	86	90	53	7.5
	98	90	92	74	9.2
	97	98	96	83	10.0
1984	93	45	82	39	4.5
	88	57	81	48	5.5
	94	97	96	71	9.2
	98	97	99	78	9.7

* Vigor Index is the mean of vigor ratings. The results of four laboratory tests, Standard Germination, Cold and Accelerated Aging tests, and total seedling growth weight, were converted to a vigor rating on a scale of zero through ten.

the development of smaller plants, it is less likely that an LAI of six will be attained by the time reproductive development is initiated, resulting in the possibility for yield loss. High seed vigor frequently increased grain yield when plant densities were less than the optimum for maximum yield, most often following late planting or emergence in extremely stressful germination environments (TeKrony and Egli, 1991).

Low-vigor seedlots can also contribute to the development of uneven seedlings in a plant stand and therefore to reduced final grain yields. Corn plant stands that failed to emerge uniformly yielded 11 percent less grain than those that emerged uniformly (Ford and Hicks, 1992) because initial differences in plant size caused by cold stress early in the season continued into later stages of plant development (Landi and Crosbie, 1982). Many factors can encourage uneven emergence, including limited soil moisture, irregular seed placement depth, soil compaction, excessive plant residue on the soil surface, and low seed vigor.

Predictive Capability of Tests for Vigor and Field Emergence

Reductions in yield are indirectly related to low seed vigor if seeds fail to emerge rapidly and uniformly to produce a plant population above a critical threshold. A survey of seed-testing laboratories in North America in 1982 indicated that 74 percent of the respondents evaluate the vigor of seedlots, primarily using the cold and accelerated-aging tests (TeKrony, 1983). In spite of the need to measure vigor, and in contrast to germination, a single universally applicable seed vigor test has not yet been identified for multiple crop types.

Considerable effort has been invested to identify a single seed vigor test capable of accurately predicting field emergence. When corn seed was planted in less than ideal conditions using conventional and no-till systems, the cold test was more closely related to final emergence than the standard germination or accelerated aging tests (Table 3, 4; TeKrony, Egli, and Wickham, 1989). Among single vigor tests of wheat (*Triticum aestivum* L.), glutamic acid decarboxylase activity, seedling root length, and tetrazolium vigor were the best predictors of field emergence, followed by seedling dry weight and the respiratory quotient (Steiner, Grabe, and Tulo, 1989). Other tests that were progressively poorer predictors of emergence included conductivity, standard germination, and seedling shoot length (Table 5). All measures of pepper seedling growth were good indicators of field performance with the cool-temperature germination percentage at 14-d representing the best single-test predictor (Trawatha, Steiner, and Bradford, 1990). While vigor tests are valuable predictors of field emergence, particularly relative to the standard germination test (Tables 2-4), a single best vigor test has eluded identification; no one vigor test has proven applicable to all seed types or to the large variety of potential conditions possible. Similarly, vigor tests cannot account for the variability encountered in field emergence.

Subsequently, an effort to improve the predictive capability of seed quality tests centered on combining two or more tests of germination and/or vigor, and then constructing vigor test indices. First, germination and seedling growth results were combined to gain predictive power. A vigor rating was calculated as an average of standard germination percentage, shoot length, and seedling fresh weight and used to estimate seedling emergence (Woodstock, 1969). Three indices of soybean seed vigor were calculated using the standard, accelerated-aging, and four-day germination tests and then compared to field emergence under adverse soil conditions (TeKrony and Egli, 1977). Prediction accuracy of field emergence was 8.3, 46.7, and 56.9 percent for standard, accelerated, and four-day germination tests, respectively, when each test was used alone, but ranged between 65.0 and 68.3 percent for the three indices used. Similarly, when standard germination, seedling vigor classification, seedling length, and tetrazolium staining tests were applied in various combinations with soybean seed field emergence, laboratory measurements were correlated with emergence in all situations (Yaklich and Kulik, 1979). No one best combination was identified for all agronomic situations, however. Predictive accuracy varied between 50 and 60 percent.

More recently, statistical methods were employed to select combinations of vigor tests to better estimate wheat seedling emergence. The best overall test was identified using multiple regression models to calculate the geometric mean of 14 vigor tests (Steiner, Grabe, and Tulo, 1989). Combined vigor tests that proved to be the best

TABLE 3. Percent emergence of four seedlots in four tillage systems in 1982, 1983, and 1984 (TeKrony, Egli, and Wickham, 1989).

Vigor Index**	Tillage Method			
	Conventional	No-till Cornstalks	No-till Wheat	No-till Tall Fescue
		1982		
10.0	89	87	67	81
9.7	93	91	73	82
6.3	85	78*	61	69*
5.3	49*	61*	37*	36*
		1983		
10.0	89	69	66	78
9.2	90	71	65	71
7.5	80	54*	55	49*
5.2	70*	42*	40*	51*
		1984		
9.7	95	88	77	90
9.2	91	88	77	87
5.5	75*	73*	73	69*
4.5	71*	75*	64*	70*

* Seedlots with significantly lower emergence than the seedlot with the highest vigor rating in the same tillage system (LSD (0.05) in 1982 = 7, 1983 = 16, and 1984 = 10).

** Vigor Index is the mean of vigor ratings. The results of four laboratory tests, Standard Germination, Cold and Accelerated Aging tests, and total seedling growth weight, were converted to a vigor rating on a scale of zero through ten.

TABLE 4. Seedlot means for standard germination, vigor tests, and field performance variables for ten seedlots of Regar meadow bromegrass-Bozeman, MT, 1986 (Hall and Wiesner, 1990, in part).

Seed Lot	Standard Germination	Accelerated Aging	Forage Yield	Total Seedling Emergence
	%	%	kg ha^{-1}	plants m^{-1}
863	94 cd*	76 e	5580 d	35 ab
867	86 c	43 bc	5240 cd	33 ab
865	90 cd	35 b	4150 ab	38 ab
862	94 cd	58 d	3930 ab	35 ab
869	60 b	15 a	3840 ab	28 a
8610	49 a	13 a	3350 a	34 ab

*Means followed by the same letter are not significantly different ($P<0.05$) according to Neuman-Keuls mean separation test.

predictors included seedling root length, glutamic acid decarboxylase activity, and seedling dry weight. Combinations of simple seedling growth measurements were as effective as the biochemical tests to estimate seedling emergence. Similarly, an index developed from the geometric mean of cool-temperature percent germination (14-d) and seedling dry weight predicted pepper seedling emergence most accurately when compared to the 14 single methods (Trawatha, Steiner, and Bradford, 1990). Still, these methods accounted for an average of just 36 percent of the total variation in pepper seedling emergence. Again, while improving the predictability of seedlot performance relative to single tests, no index has yet proven acceptable for broad-scale use by seed testing laboratories. A combination of seedling growth tests holds the greatest promise for accurately predicting field emergence.

TABLE 5. Simple correlation coefficients between 14 vigor tests and field emergence of Hyslop wheat (Steiner, Grabe, and Tulo, 1989).

Vigor test	r
GADA, umol CO2 g-1 h-1	0.78
Seedling root length, mm seedling-1	0.78
Tetrazolium vigor, %	0.77
Seedling dry weight, mm seedling-1	0.71
RQ, g-1	–0.70
Tetrazolium viability, %	0.68
Dehydrogenase, umol formazan g-1 h-1	0.59
Standard germination, %	0.53
ATP, umol seedling-1	0.53
Seedling shoot length, mm seedling-1	0.49
Conductivity, dS m-1 100 seeds-1	–0.41
QCO2, µl g-1	–0.21
QO2, µl g-1	0.08
Seed weight, mg	0.05

UNDERLYING MECHANISMS OF SEED VIGOR

Key Stages in Seed Development

Vegetative Growth

Seed vigor is acquired during the growing season as each individual seed of a population grows and develops. Maternal plant development is vegetative in nature initially, eventually shifting to a reproductive mode. Vegetative growth presumably has little or no effect on vigor as the cell structures of the seed are not yet being built nor are the storage compounds being deposited.

Reproductive Growth

Plant health during reproductive growth, particularly seed development, affects the maximum level of attainable vigor. It is unlikely that early reproductive growth, up to pollination, would impact subsequent quality, because the primary structures of the seed are not yet formed, nor is a substantial amount of photosynthate stored for later reproductive growth. Drought stress imposed on soybean plants during flowering (R2) exhibited no effect on germination percentage, accelerated aging percentage, or seedling axis dry weight (Table 6; Smiciklas et al., 1989, et al., 1992).

During early seed development, after pollination, the cells ultimately forming the seed actively undergo the processes of division and differentiation. It is possible that conditions conducive to abnormal development during this period might contribute to vigor loss. Seed viability, the presence of live tissues within the seed, is established prior to vigor. Biochemical structures such as mitochondria, golgi apparatus, DNA, RNA, and various metabolic proteins that are essential to vigorous cellular activity at the local level and, ultimately, germination activity at the macro level, gradually accumulate during early seed devclopment. The R4 stage of soybean reproductive development is characterized by rapid cellular differentiation of seed structures and rapid pod growth, but very little

TABLE 6. Influence of drought-stress treatment on soybean seed yield, germination, seedling dry weight, and leachate current at Ames, IA. Values are averaged over the 1985 and 1986 growing seasons (Smiciklas et al., 1992).

Drought-stress Treatment	Seed Yield	Germination	Dry Weight	Leachate Current
	plant-1	%	mg seedling-1	μA mg-1
Nonstressed	9.42	95.0	60.6	0.78
Full pod (R4)	8.39	91.8	56.3	0.88
Seed formation (R5)	5.73	80.8	55.2	0.90
Full seed (R6)	6.19	87.9	51.8	1.04
LSD (0.05)	1.22	5.2	2.1	0.17

seed dry weight accumulation (Dornbos and McDonald, 1986). Drought stress imposed on soybean plants at R4 had no effect on subsequent seed germination ability or vigor in one study (Smiciklas et al., 1989) but exhibited reduced seedling axis dry weight in a second, similar study (Table 6; Smiciklas et al., 1992).

Events inhibiting plant growth during seedfill can substantially reduce subsequent seed vigor. During seedfill, numerous storage compounds, proteins, lipids, and carbohydrates, are rapidly deposited (Dornbos and McDonald, 1986) as energy reserves and building blocks to eventually drive the germination process. Environmental stresses, such as drought and heat, imposed during seedfill reduced both the germination ability and vigor of peanut (*Arachis hypogaea* L.) (Ketring, 1984) and soybean (Dornbos, Mullen, and Shibles, 1989; Dornbos and Mullen, 1991).

Within the seedfill period, seed quality is affected more by stress at certain stages. Soybean plants exposed to drought stress at R5, beginning seed development, produced seed with germination percentages of 85 percent in comparison with 96 percent for nonstressed seed (Smiciklas et al., 1989). Stress at R6, rapid seedfill, did not reduce germinability. Neither seedling dry weight nor electrolytic conductivity were reduced by stress at R5 or R6. A subsequent study again indicated a larger loss of germinability at R5 than R6 (Smiciklas et al., 1992). In addition, drought stress at R6 reduced seedling dry weight more than at R4 or R5 (Table 6). Increased leachate conductivity resulted from stress only at R6, again indicating reduced vigor.

Activity of pathogens during reproductive growth can indirectly reduce seed vigor. Infection by *Phomopsis* spp. between R6 and R7 reduced visual ratings of seed quality more than the other periods of development (Table 7; Vaughan, Bernard, and Sinclair, 1989). Longer R6-R7 periods are associated with greater incidence and severity of *Phomopsis* spp. infection, simply because of longer exposure to the warm and wet weather events conducive to infection. Seedlots infected by *Phomopsis* spp. exhibit higher incidence of wrinkled seeds, defective seed coats, green and moldy seeds, and lower seed weight. Environmental and pathological events during seed development, and particularly rapid seedfill, contribute significantly to the acquisition of maximum seed vigor.

TABLE 7. Simple linear correlation coefficients between seed quality, seed weight and late reproductive growth period for 141 soybean cultivars (Vaughan, Bernard, and Sinclair, 1989)

Trait	Seed Quality (1-5)	Seed Weight
R6-R7	0.13**	0.46*
R6-R8	0.14**	0.44*
R7-R8	0.03	0.00
R7-Harvest	0.00	0.01

**,* indicates significant deviation of the correlation coefficient from 0 at alpha = 0.01 and 0.05, respectively.

Physiological Maturity

With the onset of maternal plant senescence, each seed approaches the stage of development called physiological maturity, defined by maximum seed dry weight. Soybean seed dry weight is maximum at R7 (Dornbos and McDonald, 1986), and is characterized by leaf, pod, and seed yellowing. Physiological maturity in corn is characterized by formation of the black layer. Corn and soybean seeds contain approximately 35 and 55 percent moisture, respectively, at physiological maturity, at which point the funiculus, the connection between maternal and seed tissues, deteriorates, leaving no direct pathway for movement of nutrients or moisture between maternal and seed tissues.

The Maturation Process and Harvest Maturity

Because maximum seed vigor is acquired by physiological maturity, it is desirable to harvest the seed as soon as possible thereafter. Practically, however, seed harvest must wait until harvest maturity, when moisture levels are sufficiently low to permit mechanical harvest without causing undue damage to the seed. During the maturation process, the period between physiological and harvest maturity, the seed is dependent on the maternal plant only for physical support and some level of protection from the elements by means of husk or pod tissues. Exposure to wide temperature and moisture fluctuations and high inoculum levels of pathogens in the field during maturation make the seed vulnerable to rapid vigor loss. The maturation process involves two types of changes, the visible seed desiccation process and the more subtle changes in biochemical composition and physiological activity.

The most obvious change that occurs during maturation is seed desiccation, the extent to which varies among seed type. Harvest of corn seed can begin very shortly after physiological maturity, at approximately 35 percent moisture. Because corn seed is harvested by picking the entire ear, mechanical harvest can be completed without damaging individual kernels before harvest maturity. Early ear-picking is preferable to minimize the period of time the corn seed is exposed to field weathering after physiological maturity, thereby maximizing vigor potential. Soybean seed is threshed with a conventional combine, hence drying from approximately 55 to 18 percent moisture is necessary to minimize mechanical damage. The rate of moisture loss is a purely physical process that is governed by relative humidity, air temperature, wind flow, and pod or husk impedance in the case of soybean and corn, respectively.

Although less apparent than moisture loss, numerous biochemical and physiological changes take place during the maturation process. Gross changes in biochemical composition include a decrease in sugar content from 2.5 to 1.7 mg/seed, an increase in oil content from 31.2 to 36.2 mg/seed by virtue of increases in both saturated and unsaturated fatty acids, and, although gross protein content remained constant, a change in the concentrations of proteins in specific electrophoretic bands (Dornbos and McDonald, 1986).

One of the physiological changes associated with desiccation is the shift in organization of cellular membrane lipids comprising the seed from a highly organized fluid laminar structure operative during seed development (Abdul-Baki and Baker, 1973) to a form associated with dry, metabolically inactive seeds in storage at 12 to 14 percent moisture (Edwards, 1976). Disruption of the normal sequence of biochemical and physiological events associated with the maturation process could compromise seed vigor.

Deterioration Events That Influence Seed Vigor

Physical

Following physiological maturity, the stage of seed development associated with maximum potential vigor, the progressive loss of seed vigor is inevitable. Rates of vigor loss vary according to genetic composition, environmental conditions, and pathogenic organisms to which the seed is exposed. Seed producers strive to develop methods of seed handling that slow the rate of seed vigor loss as much as possible. Between harvest and planting the following year, however, seedlots must be conditioned, stored, bagged, and shipped to be ready for planting. There is considerable opportunity for seed vigor loss during this period. Deterioration can occur through physical damage, physiological decline, and pathological infection.

Physical or mechanical damage to the seed can be incurred during harvest, conditioning, movement of seed from bin to bin, improper storage, or careless handling of seed bags. Physical damage may be obvious, in the case of cracked, split, or discolored seeds, or covert, as physiological bruises to the embryo or seed coat fractures. Large-seeded legumes such as soybean are particularly vulnerable to damage (McDonald, 1985).

Seed moisture levels are critical to minimizing physical damage during any operation. Seed moisture levels that are too low cause extensive cracking and splitting. Seed moisture levels that are too high cause less obvious bruises which become apparent upon use of the tetrazolium test. Depending upon location and extent of injury, seed viability or vigor may be reduced.

Harvest of seed with combines can damage seed, depending upon type of combine, cylinder speed, seed moisture content, variety, and seed orientation. The percentage of splits was significantly greater for a conventional cylinder than for either a single or double rotary threshing mechanism (Newbery, Paulsen, and Nave, 1980). Reduced cylinder speed lessens mechanical damage, regardless of cylinder type. Again, large-seeded legumes are more susceptible to mechanical damage than caryopsis seed types. If impact occurs at a point between the cotyledons, damage to the embryonic axis or an increase in the proportion of splits is likely. An optimum moisture content exists for each seed type at which mechanical damage is minimized. Soybeans, for example, are most resistant to damage when moisture content exceeds 10 percent but is less than 14 percent (Henderson, 1965).

The ultimate goal of seed conditioning is to remove all contaminants and damaged seed without imposing additional physical damage (McDonald, 1985). Even minimal drops of five feet can reduce soybean seed vigor. While soybean seed conditioning enhanced overall seed germination by 3 percent, certain devices were more effective than others in cleaning seed without initiating additional damage (Misra, 1982). The gravity separator and air-screen cleaner are particularly efficient conditioning devices, improving germination 2.1 and 1.5 percent, respectively. In contrast, spiral separators did not improve germination, possibly because the operation increased physical damage by 0.6 percent.

Vigor tests have proven extremely valuable to seed processors as tools to critically evaluate the effectiveness of each piece of conditioning equipment and of an overall system. Seed samples can be taken from a seedlot at each stage of harvest, conditioning, and transport and submitted to a battery of germination and vigor tests. This procedure permits identification of each handling step that causes undue damage to the seedlot or fails to remove damaged seeds. Destructive or nonproductive steps can then be modified to improve future operations.

Physically damaged seedlots are capable of substantially reducing field emergence, and therefore yield, if planted. Soybean seeds with undamaged seedcoats exhibited 76 percent germination, whereas fewer than 30 percent of seeds with damaged seedcoats were capable of germination (Stanway, 1974). Germination of "Forrest" soybean seeds with damaged seedcoats was reduced 40 percent relative to those with undamaged seedcoats, contributing to 15 percent lower field emergence (Stanway, 1978). Damage to the seedcoat, obvious or subtle, is a common form of physical damage. Seedcoat damage promotes the rapid leakage of cellular contents upon imbibition. Biochemical leakage not only represents a loss of energy and building-block resources necessary to drive vigorous germination, but these compounds may further sustain the growth of soil microorganisms capable of causing pathogenic infection.

Physiological

Just as normal physiological processes during seed development and maturation are required to maximize vigor, impairment of normal physiological processes because of seed deterioration contributes to nonvigorous germination. Vigorous germination represents the culmination of normal and active physiological functioning at the cellular level. Proper physiological functioning in seeds is correlated with vigorous germination, providing the basis for several seed vigor tests. Abnormal physiological functioning precedes, and therefore indicates, reduced vigor.

A progressive, highly ordered cascade of events is hypothesized to explain the physiological basis of seed deterioration (DeLouche and Baskin, 1973). The initial phase of seed deterioration is membrane degradation, which causes and is followed by impairment of ATP synthesis, reduced respiration and biosynthesis rates, poor seed storability, and, ultimately, reduced emergence, development of abnormal seedlings, and reduced germinability. If membrane degradation is an initial stage of vigor loss, methods that detect membrane deterioration would be preferable to serve as early indicators.

Substantial evidence supports the idea of membrane deterioration as an early, and possible causative, factor in seed vigor loss. Low-vigor seed is strongly associated with membrane deterioration. Numerous studies have demonstrated the presence of membrane degradation in seeds of low vigor (Koostra and Harrington, 1969; Villiers, 1973; McDonald, 1975; Stewart and Bewley, 1980). When exposed to high temperature and increased oxygen pressure, the polar lipids, including the phospholipids of membranes, are susceptible to nonenzymatic peroxidative reactions (Priestley, Werner, and Leopold, 1985). Membrane lipid peroxidation is also suspected to occur in dry seeds during storage, resulting in the formation of hydroperoxides, oxygenated fatty acids, and free radicals (Wilson and McDonald, 1986). Free radicals from the lipid peroxidation process denature DNA, hinder protein translation and transcription, and oxidize certain amino acids, the cumulative effect of which can reduce seed vigor (Priestley, 1986). Peroxidation or any other degradation process that causes changes in membrane lipid composition would likely impair membrane fluidity, and therefore functionality. Physiological processes involving membranes or membrane-bound proteins, such as ATP, RNA, protein synthesis, or respiration, are inhibited by nonfunctional membranes. Reduced membrane fluidity also contributes to the leakage of cellular constituents because affected cells would be slow to reposition the lipid bilayers during imbibition which serve as effective barriers to solute loss. Nutrient leaching from damaged or aged seeds formed the basis for conductivity vigor tests with a variety of seed types (McDonald, 1975).

Seed membrane lipid deterioration is correlated with reduced mitochondrial activity. Naturally aged soybean seeds exhibit reduced accelerated aging germination percentage, but maintained standard germination percentage (Ferguson, TeKrony, and Egli, 1990a). While triglyceride fatty acid composition from excised axes of aged seeds was not altered, the percentage of unsaturated fatty acids from membrane lipids decreased slightly, suggesting peroxidation-derived changes in lipid composition (Ferguson, TeKrony, and Egli, 1990b). Conductivity of the leachate from the excised axes increased early during storage, prior to detectable changes in whole-seed vigor. State 3 respiration of mitochondria from the seed axes were also found to decrease throughout storage (Ferguson, TeKrony, and Egli, 1990a). Because decreased respiration rates preceded the detection of whole-seed vigor loss, and were correlated with increased leachate conductivity and changes in membrane lipid composition, these studies support a membrane-based physiological model of progressive seed deterioration.

Numerous other physiological processes are associated with seed deterioration as well. Both natural and accelerated aging processes exhibit declines in ATP synthesis (Ching, 1973), RNA synthesis (Van Onkelen, Verbeek, and Khan, 1974), and respiration (Woodstock and Grabe, 1967; Ferguson, TeKrony, and Egli, 1990a). Decreases in both number and efficiency of mitochondria have been reported in germinating, low-vigor soybean seed axes (Woodstock, Furman, and Solomos, 1984). In addition to lipid peroxidation, active lipases and lipoxygenases are present in deteriorating seeds (St. Angelo and Ory, 1983). Genetic changes also occur as seeds deteriorate. Aged seeds demonstrated a higher incidence of chromosomal aberrations in *Crepis tectorum* L. root tips (Navashin,

1933), delayed mitosis, and a reduced mitotic index in *Pisum sativum* L. seeds (Murata, Roos, and Tsuchiya, 1980).

Pathological

Infection of plants and seeds by various plant pathogens, either fungal, bacterial, or viral in nature, in the field or in storage, can reduce vigor directly through such mechanisms as enzymatic degradation, toxin production, and growth regulation. Seed vigor loss because of infection by pathogens can also occur indirectly to the extent that pathogenic infection limits the ability of the seed on that plant to develop normally.

Numerous pathogens attack seeds directly. The results of pathogenic infection can range across a broad spectrum from a dead seed to a seed carrying the pathogen but demonstrating no ill effect on vigor. Field fungi are capable of infecting seeds during development, maturation, or after maturation but before harvest. Many fungi are very specific, being capable of initiating infection on few plant species, at one stage of plant development, and with pathogenicity highly dependent upon weather conditions. Glume blotch, for instance, is more capable of infecting wheat during late, rather than early, growth stages. *Aspergillus flavus* infects corn kernels to a greater degree when silks are yellow-brown than later when they are brown. Cold and rainy weather promotes infection of rye by *Claviceps purpurea* (ergot) by delaying fertilization, thus increasing the length of time that the plants are susceptible to inoculum. Bacterial blights and anthracnoses are promoted by high humidity and warm temperatures. Preharvest loss of seed quality is increased by fungal invasion subsequent to warm and humid conditions with soybean (Nicholson, Dhingra, and Sinclair, 1972) and cotton (Roncadori, Brooks, and Perry, 1972). Storage fungi are capable of initiating infection following seed maturation, primarily as a function of weather or storage conditions. Most of these pathogens are ubiquitous in soil as well-adapted soil saprophytes and are capable of colonizing a large variety of plant types and tissues. The presence of moisture on the seed surface, resulting either from rainy weather during the harvest season or storage at too high a moisture content, and warm temperatures can promote such pathogens as *Fusarium*, *Alternaria*, *Helminthosporium*, and *Cladosporium*, causing such diseases as wheat scab, cereal black head mold, and corn ear rots (Wiese, 1977). Most storage fungi are very weakly pathogenic such that attack would seldom occur without optimum conditions for infection. An example of one strongly pathogenic storage fungal organism is *Phomopsis longicolla*, the cause of soybean seed decay. *Phomopsis* spp. pathogenicity is promoted by warm, rainy weather following maturity, and can cause substantial reductions in seed vigor (McGee, 1986).

Seed deterioration contributes to a loss of vigor, and eventually germination. Deterioration of field bean (*Phaseolus vulgaris* L.) seeds, after being naturally and artificially aged, culminated in losses of germination ability and vigor (Table 8; Rodriguez and McDonald, 1989). Aged seed produced less top and root growth, and less seed yield (Table 9; Rodriguez and McDonald, 1989). Nitrogen fixation levels in root nodules of plants from aged seeds were lower than those from plants of high-vigor seed. High-vigor

TABLE 8. Percent germination of naturally and artificially aged field bean seeds determined by the standard germination (SG) and seedling vigor classification (SVC) tests for 1984 and 1985 (Rodriguez and McDonald, 1989).

Seed Quality	1984		1985	
	SG	SVC	SG	SVC
	% germination			
Naturally aged				
High quality	89	78	93	86
Medium quality	81	69	81	73
Low quality	72	61	73	67
Artificially aged				
1 d	81	72	84	77
2 d	58	46	65	54
3 d	44	33	52	38
LSD (P=0.05)	5	6	8	7

seeds were more likely to develop an optimum leaf area index for canopy photosynthesis by the time reproductive development began. The accelerated aging vigor test was developed to speed the deterioration process, aiding the identification of low-quality seedlots susceptible to aging in storage.

STATUS OF SEED VIGOR

Applications of Seed Vigor

Plant Breeding

Knowledge about the concept of seed vigor, how it is acquired during seed development and subsequently lost through deterioration, is relevant to plant breeding, seed production and quality control, and seed marketing efforts. Reliable vigor information can then be confidently applied by the consumer of seed, the grower, as planning and predictive tools for the upcoming growing season.

Potential exists for plant breeders to improve seed vigor characteristics of the varieties they develop. Seed vigor is governed by the general principle that the genetic composition of a variety interacts with the environmental conditions of the production environment to determine emergence rates and yield, the phenotype. Seed vigor is in part an inherited trait, quantitative in nature, that can ultimately impact yield and yield stability, and it is genetically controlled in many ways and by many genes.

The primary objectives of commercial breeding programs are yield, standability, maturity, and disease reaction. Seed vigor is seldom a priority objective in plant breeding programs, but is recognized as an important factor that can impact yield. One of the ratings routinely taken by plant breeders and agronomists is "seedling vigor," which refers to the ability of a variety to emerge and grow rapidly and vigorously. Varieties can exhibit considerable variation for seedling vigor. It is very difficult, however, to distinguish between the causes of low vigor; genetic or production environment.

Substantial genetically derived differences exist among experimental hybrids for seed vigor, differences that may or may not

TABLE 9. Seed quality effects on plant top and root dry weights of field bean seeds (Rodriguez and McDonald, 1989).

Seed	1984			1985		
Quality	V3	R6	R8	V3	R6	R8
	Top dry weight, g/plant					
Naturally aged						
High quality	10.3	19.2	36.4	14.2	31.4	47.0
Medium quality	8.5	12.8	24.2	13.0	29.8	41.8
Low quality	8.2	12.9	20.3	12.2	25.2	33.1
Artificially aged						
1 d	10.2	17.3	36.6	13.3	31.4	34.3
2 d	9.4	15.4	34.0	11.9	28.9	37.4
3 d	7.4	13.6	25.4	8.4	20.6	25.8
LSD (P=0.05)	0.6	2.9	5.3	2.8	6.2	9.6
	Root dry weight, g/plant					
Naturally aged						
High quality	1.7	2.7	2.6	2.6	3.3	3.8
Medium quality	1.6	2.1	2.4	2.3	2.5	3.7
Low quality	1.5	2.2	2.2	2.0	2.6	3.2
Artificially aged						
1 d	1.5	2.3	2.8	2.2	3.0	3.5
2 d	1.4	2.1	2.4	2.0	3.0	3.5
3 d	1.4	2.0	2.4	2.5	2.1	2.6
LSD (P=0.05)	NS	0.2	0.3	0.4	0.6	0.6

V3, R6, and R8 refer to sampling of field bean plants at three developmental stages. The first trifoliate leaf completely unfolded (about 30 days after planting for 1984 and 22 DAP for 1985) was V3. R6 was the plant showing the first opened flower (about 45 DAP for 1984 and 40 DAP for 1985), and R8 was the plant initiating pod fill with the oldest pods containing seeds (about 70 DAP for 1984 and 65 DAP for 1985).

affect final stand populations and yield (Table 10). Hybrid B, for example, is rated as having below average, but commercially acceptable, seed vigor. Stressful germination environments, particularly cool and wet soils, can result in Hybrid B having lower plant populations than other hybrids with greater vigor. Nonetheless, the yield record of Hybrid B is comparable to hybrids with better early vigor ratings, possibly because of greater ear flex capability. Conversely, hybrids with strong vigor ratings (i.e., Hybrid D) do not necessarily confer yield advantages, even in stressful germination environments, because many factors interact to determine final yield.

Occasionally, seed vigor problems can be overcome in the seed production process. For example, the seed of a particular experimental hybrid was determined to be genetically susceptible to formation of stress fractures during drying. When a longer drying procedure was used at a more moderate temperature, stress fracture formation was avoided, and good seed vigor was maintained.

Quantitative genetic traits that contribute to seed vigor include morphological characteristics such as split-coleoptile, husk coverage, and ear rot susceptibility of corn, and hypocotyl length of soybeans. Genetic variability for hardseededness in soybean, a trait of the legumes, is also associated with the ability to maintain seed vigor by withstanding field weathering following maturation (Potts et al., 1978). Genetic variability for seedcoat etching, a trait indica-

TABLE 10. Final plant populations and yield of five genetically diverse corn hybrids averaged over eight locations and planted at 25,500 seeds/acre in 1989. (Dornbos, unpublished data).

Hybrid	Early Vigor Rating	Plant Population	Yield
		(plants/a)	(bu/a)
A	Unacceptable	21,700	121.3
B	Below Average	24,900	140.9
C	Average	24,650	144.9
D	Average	24,538	137.6
E	Excellent	25,425	148.7

tive of reduced vigor in large-seeded legumes, is associated with a single, incompletely dominant gene for susceptibility with 25 to 50 percent penetrance (Dickson, 1969). Selection among 41 navy bean lines for resistance to mechanical injury demonstrated that genetic differences exist (Barriga, 1961), likely because of differences in seedcoat thickness and seed density. Extensive screening has been conducted for seed vigor in low-temperature germination environments. Two tomato lines were identified that express improved germination ability at 10°C (Smith and Millett, 1964) through polygenic control (three to five genes) with dominance for the ability to germinate at low temperatures (Ng and Tigchelaar, 1973). Genetic controls clearly exist for many seed traits that ultimately impact seed germination ability and vigor.

An indirect trait that can affect seed vigor is maturity. Soybean varieties that mature too early in the growing season are subject to hot and dry weather during maturation, culminating in reduced seed quality (Green et al., 1965). Selection of appropriate seed production location and planting date can minimize the potential liability of maturity on vigor.

Seed Production and Quality Control

Plant breeders routinely accumulate information relevant to the production of seed for each advanced experimental in their breeding program. This information is conveyed to the agronomists who manage the foundation seed production process and, if the variety is advanced, commercial seed production. Plant varieties vary tremendously in their producibility. Producibility refers to such characteristics as seed drying and conditioning strategies, inbred-inbred planting splits in hybrid seed production, average yield of seed females, and seed quality history. Various methods are employed during the seed production process to ensure that seed of the highest possible quality is produced, conditioned, and finally delivered to the grower. Harvest and conditioning procedures, drying temperatures, and seed drying rates are developed for each seed product to maximize standard germination ability and vigor.

Seed production fields are subject to more stringent agronomic tolerances than grain production fields to minimize the possibility that impairment of plant growth will reduce seed yield or vigor.

Lower threshold tolerances for disease incidence and insect levels are used to guide management practices. In addition, very conservative plant populations are used to encourage the development of large, dense seed.

To police quality control, the seedlots produced are identity preserved and tested for seed quality factors in accordance with appropriate seed laws and company policy. Seed laws require that standard germination and seedlot identification information be printed on the bag tag, but not vigor test results. Referee testing indicated that vigor test results among different laboratories currently are not sufficiently reproducible within acceptable limits to warrant the publication of vigor test results on the bag tag (AOSA, 1983). Again, no one vigor test has met with broad spectrum approval within the seed industry.

Nonetheless, many seed companies require some form of vigor testing in their quality control programs to ensure that the seedlots to be released for distribution meet quality standards. The cold test is most commonly used among seed testing laboratories for in-house quality control.

Quality testing methods and standards vary greatly among seed suppliers. Those companies that utilize vigor testing for in-house quality control of seed often combine the results of standard germination and vigor tests to guide the decision-making process. For example, for the germination percentage of 95 percent to be printed on the tag for corn, two criteria need to be met: the standard germination test result must equal or exceed 95 percent, and the cold test result must equal or exceed 80 percent. The minimum standard germination percentage to be printed for corn seed is 90 percent, by company policy, and then only when standard germination and cold test results equal or exceed 90 and 85 percent, respectively. Typical industry standards are similar, but slightly less, for soybean. Frequently, the legal minimum standard germination percentage for soybean is 85 percent.

The tetrazolium test is a second vigor test used regularly by members of the seed industry. This test is usually reserved as a diagnostic tool for a variety of seed types when freeze or mechanical damage is suspected. Vigor tests, in conjunction with the standard germination test, are clearly recognized by many members of the seed industry as essential to assuring that high-quality seed is made available to growers.

Marketing and Promotion

Seed is seldom marketed or promoted on the basis of specific vigor test results. One reason for this is that although vigor tests conducted within a laboratory accurately indicate that quantitative differences exist between seedlots, and most laboratories can rank different seedlots similarly, quantitative test results are not reproducible when compared between laboratories (AOSA, 1983). Also, while uniform vigor test methods were recently recommended (AOSA, 1983), specific procedures have not been broadly accepted. Different laboratories currently use a variety of methods for the same vigor test for logistic and preference reasons, contributing to the variability in test results. Finally, no one or two tests have been formally endorsed as acceptable measures of vigor among seed testing laboratories.

The American Seed Testing Association recommends that until vigor testing becomes more refined and the seed user is better able to interpret and understand vigor test results, vigor ratings should not be included in laboratory test reports or on seed labels (AOSA, 1983). Further, if any vigor rating is used for marketing purposes, it must be stated that the vigor test results are based on unofficial procedures which are not adequately standardized between laboratories.

Numerous seed companies do promote the sale of "high-quality seed" without specifically referring to vigor test results. Very few companies publish vigor test results as part of a marketing effort, although seed company representatives may describe the types of tests conducted and seed company standards for marketability, like those described in the previous section, to engender greater confidence in the quality of seed a company can supply. Many growers may not even be aware that vigor tests are conducted as part of a seed quality assurance program.

Consumers

The growers' perception is that the risk of poor emergence is distributed among grower, seed, and agricultural chemical suppliers. Assuming that only high-quality seed is offered for sale, growers seldom inquire about the tests conducted to confirm quality. Most are aware of the germination percentage on the bag tag when the seed is delivered and may make a quality judgment at the point of seed delivery; they do occasionally reject delivery of seed if the bag tag indicates germination percentages lower than a personal minimum standard. The growers' assumption of high-seed quality may be heavily weighted on this percentage, possibly contributing to a false sense of security if the seed is subsequently planted in a stressful germination environment. The standard germination percentage alone is a poor predictor of emergence in non-optimum environments.

Final judgment by virtually all growers relates to actual emergence rates in the production field. If adequate stands fail to be established according to the growers' expectation, the seed company may be held liable for the quality of their product, possibly resulting in litigation for the costs of seed and, if unresolved, lost yield. Most seed companies contract replant seed discounts of approximately 50 percent for seed that fails to adequately emerge for any reason.

Regulation

Standardization Issues

To the extent that no attempt is made to promote seed products based on specific vigor test results, regulation is not necessary. Trends indicate increased emphasis on the use of vigor test results in marketing programs, however, to persuade growers when making purchasing decisions. In anticipation of the use of specific vigor test results for marketing and promotion, a regulatory mechanism should be developed to protect both seed producers and users by standardizing test procedures and the interpretation and presentation of appropriate test results.

Should vigor information be presented on the legal tag? Eight of 35 regulatory survey respondents replied that they would not permit any reference to vigor on seed labels, even if it was quantitative and a test procedure was identified (AOSA, 1983). Yet, the need is growing for publication of appropriate seed vigor information, because seed vigor does strongly impact yield by indirectly affecting plant size and final emergence rates. If seed companies desire to promote reliable vigor information and growers desire vigor information to help them make informed choices about which seedlot to plant, it seems reasonable that a proper forum to present vigor information should be developed. If vigor information is to be included on seed labels, regulation is needed to ensure fairness to both seed producers and users.

A first essential step is that one or more tests be accepted as reliable measures of seedlot vigor. Currently, no seed vigor test has met with broad acceptance. It is possible that no one test will suffice for all seed types, but rather that various seed types will be ascribed to the most pertinent seed vigor test. A similar situation existed with the standard germination test. Numerous variations of a basic standard germination method were developed and appropriate methods assigned to each seed type.

Once tests are adopted, consistent methods must be used by all seed testing laboratories before the possibility of achieving consistent test results will exist. Seed testing associations in various states of the Midwest currently utilize variable cold test methods, one of the most broadly accepted seed vigor tests. Some laboratories utilize sand with a thin layer of soil, or a mixture of sand and soil, whereas others use 100 percent soil. Microbial composition and concentration certainly vary among soil sources. Upon testing referee samples, it is no surprise, then, that the vigor test results that were reported differed significantly from laboratory to laboratory. To be useful, vigor test results must provide accurate, reliable, and meaningful information to the grower and seed testing laboratory.

Presentation of Test Results

Care must be taken to ensure that the results from accepted vigor tests and methods are published in a consistent manner that is reliable, unbiased, and meaningful. Growers need information, not simply data. Vigor test results will be meaningful only when compared to an accepted standard level, such as with the standard germination test. The minimum germination percentage result for corn is 85 percent in most states. A germination percentage of 95,

90, or 88 percent then have greater relevance to the grower when compared to the accepted minimum standard of 85 percent.

A basis for proper interpretation of informative vigor results by both seed producers and seed users also must be developed. Various misconceptions currently exist concerning seed vigor among growers. One is that high-vigor seed will germinate in even the harshest of germination environments. Growers need to understand that highly vigorous seedlots cannot guarantee performance, but only provide a greater likelihood of acceptable emergence in unfavorable environments. Regulatory agencies must develop proper guidelines for vigor result interpretation, thereby preventing abuse of the same information.

CONCLUSION

Access to seed of high germinability and vigor is essential to ensure the rapid and uniform establishment of plant stands that maximize yield potential in a wide variety of field conditions. Knowledge of seed vigor nicely complements knowledge of standard germination. Whereas germination percentage is an effective predictive measure of stand establishment in optimum environments, growers typically encounter stressful germination environments for which seed vigor measurements are necessary as predictive tools.

Seed vigor is gradually acquired during seed development subsequent to the ability to germinate. Factors inhibiting normal plant and seed development can minimize the maximum attainable level of vigor. Following physiological maturity, gradual vigor decline by means of physical, physiological, or pathogenic deteriorative processes is unavoidable. A critical goal in seed production is to minimize the rate of the deterioration process.

Seed vigor tests are applied in numerous capacities with the ultimate aim of improving the quality of seed that can be provided to the growers. Seed vigor is governed by biochemical and physiological processes, several of which can be screened by plant breeders to genetically improve seed quality. Vigor tests are routinely used by seed testing agencies to govern which seedlots are acceptable for planting and to define the quality levels of seed available to the marketplace.

While vigor testing provides an invaluable service to seed suppliers and growers by providing information about the strength of a seedlot's germination ability, specific test results are not sufficiently accurate for direct marketing and promotional purposes. Vigor test results of samples from one seedlot vary considerably among seed testing laboratories. Effort must be devoted to identify a single vigor test for each seed type and a common test procedure to minimize the variability in results. Such methods should permit accurate and reliable vigor data collection to support both marketing and regulatory efforts. The ultimate aim is to provide the grower with high-quality seed and useful information to facilitate the crop production process.

REFERENCES

Abdul-Baki, A., and J. E. Baker. Are changes in cellular organelles or membranes related to vigor loss in seeds? *Seed Sci. & Technol.* 1(1973): 89-125.

Alberts, H. W. Effects of pericarp injury on moisture absorption, fungus attack, and vitality of corn. *J. Amer. Soc. Agron.* 19(1927): 1021-1030.

AOSA (Association of Official Seed Analysts). Rules for testing seeds. *J. Seed Technol.* 3(1978): 1-126.

AOSA (Association of Official Seed Analysts). Seed vigor testing handbook. *AOSA Handbook* 32,(1983).

Barriga, C. Effects of mechanical abuse of Navy bean seed at various moisture levels. *Agron. J.* 53(1961): 250-251.

Burrows, W. C., and W. E. Larson. Effect of amount of mulch on soil temperature and early growth of corn. *Agron. J.* 54(1962): 19-23.

Ching, T. M. Adenosine triphosphate content and seed vigor. *Plant Physiol.* 51(1973): 400-402.

DeLouche, J. C. Seed vigor in soybeans. *Proc. 3rd Soybean Seed Res. Conf.* (ASTA, Washington, DC) 3(1973): 56-72.

DeLouche, J. C., and C. C. Baskin. Accelerated aging techniques for predicting the relative storability of seed lots. *Seed Sci. & Technol.* 1(1973): 427-452.

Dickson, J. G., and J. R. Holbert. The influence of temperature upon the metabolism and expression of disease resistance in selfed lines of corn. *J. Amer. Soc. Agron.* 18(1926): 314-322.

Dickson, M. H. The inheritance of seed coat rupture in snap beans, *Phaseolus vulgaris* L. *Euphytica* 18(1969): 110-115.

Dornbos, D. L., Jr. Unpublished data.

Dornbos, D. L., Jr., and M. B. McDonald, Jr. Mass and composition of developing soybean seeds at five reproductive growth stages. *Crop Sci.* 26(1986): 624-630.

Dornbos, D. L., Jr., and R. E. Mullen. Influence of stress during soybean seed fill on seed weight, germination, and seedling growth rate. *Can. J. Plant Sci.* 71(1991): 373-383.

Dornbos, D. L., Jr., R. E. Mullen, and R. M. Shibles. Drought stress effects during seed fill on soybean seed germination and vigor. *Crop Sci.* 29(1989): 476-480.

Edwards, M. Metabolism as a function of water potential in air-dry seeds of charlock (*Sinapis arvensis* L.). *Plant Physiol.* 58(1976): 237-239.

Egli, D. B., and D. M. TeKrony. Relationship between soybean seed vigor and yield. *Agron. J.* 71(1979): 755-759.

Ellis, R. H. The effects of differences in seed quality resulting from priming or deterioration on the relative growth rate of onion seedlings. *Acta Hortic.* 253(1989): 203-211.

Ferguson, J. M., D. M. TeKrony, and D. B. Egli. Changes during early soybean seed and axes deterioration: I. Seed quality and mitochondrial respiration. *Crop Sci.* 30(1990a): 175-179.

Ferguson, J. M., D. M. TeKrony, and D. B. Egli. Changes during early soybean seed and axes deterioration: II. Lipids. *Crop Sci.* 30(1990b): 179-182.

Ford, J. H., and D. R. Hicks. Corn growth and yield in uneven emerging stands. *J. Prod. Agric.* 5(1992): 185-188.

Green, D. E., E. L. Pinnell, L. E. Cavanah, and L. F. Williams. Effect of planting date and maturity date on soybean seed quality. *Agron. J.* 57(1965): 165-168.

Griffeth, D. R., J. V. Mannering, H. M. Galloway, S. D. Parsons, and C. B. Richey. Effect of eight tillage planting systems on soil temperatures, percent stand, plant growth and yield of corn on five Indiana soils. *Agron. J.* 65(1973): 321-326.

Hall, R. D., and L. E. Wiesner. Relationship between seed vigor tests and field performance of 'Regar' meadow bromegrass. *Crop Sci.* 30(1990): 967-970.

Henderson, J. Handling and cleaning soybean seed. *Proc. Soybean Prod., Harvesting, and Processing Mtg.*(1965): 37-40.

Hiltner, L., and J. Ihssen. Uber das schlechte Auflaufen und die Auswinterung des Getreides infolge Befalls des Saatgutes durch Fusarium. *Landw. Jahrbuch fur Bayern.* 1(1911): 20-60, 231-278.

Isely, D. The cold test for corn. *Proc. ISTA* 16(1950): 299-311.

Johnson, R. R., and L. M. Wax. Relationship of soybean germination and vigor tests to field performance. *Agron. J.* 70(1978): 273-278.

Ketring, D. L. Temperature effects on vegetative and reproductive developement of peanut. *Crop Sci.* 24(1984): 877-882.

Koostra, P. T., and J. F. Harrington. Biochemical effects of age on membranal lipids of *Cucumis sativus* L. seed. *Proc. ISTA* 34(1969):329-340.

Landi, P., and T. M. Crosbie. Response of maize to cold stress during vegetative growth. *Agron. J.* 74(1982): 765-768.

McDonald, M. B., Jr. A review and evaluation of seed vigor tests. *Proc. AOSA* 65(1975): 109-139.

McDonald, M. B., Jr. Physical seed quality of soybean. *Seed Sci. & Technol.* 13(1985): 601-628.

McGee, D. C. Prediction of *Phomopsis* seed decay by measuring soybean pod infection. *Plant Dis.* 70(1986): 329-333.

Misra, M. K. Soybean seed quality during conditioning. *Proc. Mississippi St. Seedsmen's Short Course* (1982): 49-53.

Murata, M., E. E. Roos, and T. Tsuchiya. Mitotic delay in root tips of peas induced by artificial seed aging. *Bot. Gaz.* 141(1980): 19-23.

Navashin, M. Origin of spontaneous mutations. *Nature* 131(1933): 436.

Newbery, R. S., M. R. Paulsen, and W. R. Nave. Soybean quality with rotary and conventional threshing. *Trans. Amer. Soc. of Agric. Eng.* 23(1980): 303-308.

Ng, T. J., and E. G. Tigchelaar. Inheritance of low temperature seed sprouting in tomato. *J. Amer. Soc. Hort. Sci.* 98(1973): 314-316.

Nicholson, J. F., O. D. Dhingra, and J. B. Sinclair. Internal seed-borne nature of *Sclerotinia sclerotiorum* and *Phomopsis* spp. and their effects on soybean seed quality. *Phytopathology* 61(1972): 1261-1263.

Nobbe, F. *Handbuch der Samenkunde*. Wiegandt-Hempel-Parey, Berlin, 1976.

Oplinger, E. S., and B. D. Philbrook. Soybean planting date, row width, and seedling rate response in three tillage systems. *J. Prod. Agric.* 5(1992): 94-99.

Perry, D. A. A vigor test for seeds of barley (*Hordeum vulgare*) based on measurement of plumule growth. *Seed Sci. & Technol.* 5(1977): 709-719.

Perry, D. A. Report of the vigour test committee 1974-1977. *Seed Sci. & Technol.* 6(1978): 159-181.

Perry, D. A., ed. *Handbook of vigour test methods*. Int. Seed Test. Assoc., Zurich, (1981).

Potts, H. C., J. D. Duangpatra, W. G. Hairston, and J. C. DeLouche. Some influences of hardseededness on soybean seed quality. *Crop Sci.* 18(1978): 221-224.

Priestley, D. A. *Seed aging*. Cornell Univ. Press, Ithaca, NY, 1986.

Priestley, D. A., B. G. Werner, and A. C. Leopold. The susceptibility of soybean seed lipids to artificially-enhanced atmospheric oxidation. *J. Exp. Bot.* 171(1985): 1653-1659.

Reddy, C. S. Resistance of dent corn to *Basisporium gallarum* Moll. *IA Agr. Exp. Sta. Res. Bull.* 167. (1933): 1-40.

Reddy, C. S. Pathogenicity of *Basisporium gallarum* to corn. In *Report on agricultural research.* IA Agr. Exp. Sta. Rept. for year ending June 30, 1935, p. 79.

Rodriguez, A., and M. B. McDonald, Jr. Seed quality influence on plant growth and dinitrogen fixation of red field bean. *Crop Sci.* 29(1989): 1309-1314.

Roncadori, R. W., O. L. Brooks, and C. E. Perry. Effect of field exposure on fungal invasion and deterioration of cotton seed. *Phytopathology* 62(1972): 1137-1139.

St. Angelo, A. J., and R. L. Ory. Lipid degradation during seed deterioration. *Phytopathology* 73(1983): 315-317.

Smiciklas, K. D., R. E. Mullen, R. E. Carlson, and A. D. Knapp. Drought-induced stress effect on soybean seed calcium and quality. *Crop Sci.* 29(1989): 1519-1523.

Smiciklas, K. D., R. E. Mullen, R. E. Carlson, and A. D. Knapp. Soybean seed quality response to drought stress and pod position. *Agron. J.* 84(1992): 166-170.

Smith, P. G., and A. H. Millett. Germinating and sprouting response of the tomato at low temperatures. *J. Amer. Soc. Hort. Sci.* 84(1964): 480-484.

Stanway, V. M. Germination response of soybean seeds with damaged seed coats. *Proc. AOSA* 64(1974): 97-101.

Stanway, V. M. Evaluation of 'Forrest' soybeans with damaged seed coats and cotyledons. *J. Seed Technol.* 3(1978): 19-26.

Steiner, J. J., D. F. Grabe, and M. Tulo. Single and multiple vigor tests for predicting seedling emergence of wheat. *Crop Sci.* 29(1989): 782-786.

Stewart, R. R. C., and J. D. Bewley. Lipid peroxidation associated with accelerated aging of soybean axes. *Plant Physiol.* 65(1980): 245-248.

TeKrony, D. M. Seed vigor testing–1982. *J. Seed Technol.* 8(1983): 55-60.

TeKrony, D. M., and D. B. Egli. Relationship between laboratory indices of soybean seed vigor and field emergence. *Crop Sci.* 17(1977): 573-577.

TeKrony, D. M., and D. B. Egli. Relationship of seed vigor to crop yield: A review. *Crop Sci.* 31(1991): 816-822.

TeKrony, D. M., D. B. Egli, and D. A. Wickham. Corn seed vigor effect on no-tillage field performance. I. Field emergence. *Crop Sci.* 29(1989): 1523-1528.

Trawatha, S. E., J. J. Steiner, and K. J. Bradford. Laboratory vigor tests used to predict pepper seedling field emergence performance. *Crop. Sci.* 30(1990): 713-717.

Van Onkelen, H. A., R. Verbeek, and A. A. Khan. Relationship of ribonucleic acid metabolism in embryo and aleurone to alpha-amylase synthesis in barley. *Plant Physiol.* 53(1974): 562-568.

Vaughan, D. A., R. L. Bernard, and J. B. Sinclair. Soybean seed quality in relation to days between development stages. *Agron. J.* 81(1989): 215-219.

Villiers, T. A. Ageing and the longevity of seeds in field conditions. In *Seed Ecology*, ed. W. H. Heydecker. The Penn. St. Univ. Press, Univ. Park, PA. 1973, pp. 265-288.

Wiese, M. V. *Compendium of wheat diseases*. The American Phytopathological Society. St. Paul, MN. 1977.

Wilson, D. O., Jr., and M. B. McDonald, Jr. The lipid peroxidation model of seed aging. *Seed Sci. & Technol.* 14(1986): 269-300.

Woodstock, L. W. Seedling growth as a measure of seed vigor. *Proc. ISTA* 34(1969): 273-280.

Woodstock, L. W., and D. F. Grabe. Relationship between seed respiration during imbibition and subsequent seedling growth in *Zea mays* L. *Plant Physiol.* 42(1967): 1071-1076.

Woodstock, L. W., K. Furman, and T. Solomos. Changes in respiratory metabolism during aging in seeds and isolated axes of soybean. *Plant & Cell Physiol.* 25(1984): 15-26.

Yaklich, R. W., and M. M. Kulik. Evaluation of vigor tests in soybean seeds: Relationships of the standard germination test, seedling vigor classification, seedling length, and tetrazolium staining to field performance. *Crop Sci.* 19(1979): 247-252.

Chapter 3

Methods of Viability and Vigor Testing: A Critical Appraisal

J. G. Hampton

The components of seed quality essentially fall into three categories: accurate description, hygiene, and viability and potential performance (Coolbear and Hill, 1988). Any complete assessment of the third category should consider

- the capacity of the seed to produce normal seedlings
- expected field emergence and uniformity
- potential storability

To obtain this information seed is tested for viability, and increasingly, for vigor. The concepts and underlying mechanisms of seed viability and seed vigor have been presented in Chapters 1 and 2. This chapter will deal with the measurement of seed viability and vigor and examine the methods available for determining seedlot potential and actual performance in the field and during storage.

VIABILITY TESTING

A viable seed is one that is alive and has the capacity to produce enzymes capable of catalyzing the metabolic reactions necessary for germination and seedling growth. Thus, seed viability denotes

I thank my colleagues Dr. Peter Coolbear and Professor Murray Hill for their encouragement and constructive criticism of the manuscript.

the degree to which seeds possess these properties. However, a seed may be viable but not capable of germinating because the germination process has been blocked by physical and/or chemical inhibitors, such as dormancy. Germination can only proceed once these inhibitors have been removed. Roberts (1972) therefore defined a nonviable seed as one which will not germinate when given near-optimal conditions even when it is nondormant, and a viable seed as one which will germinate under favorable conditions, provided any dormancy that may be present is removed.

To the seed industry a seed is either viable or nonviable depending on its ability to germinate and produce a normal seedling (Copeland and McDonald, 1985). However, equating viability with germination can create confusion, depending upon the definition of germination. Physiologically a seed can be said to have germinated once the radicle has protruded, and in this sense viability is germination in the absence of dormancy. In seed testing a seed may germinate (and is therefore viable), but produce a seedling which is classified as abnormal (ISTA, 1985; AOSA, 1988), so in this sense viability does not equal the germination as presented on the analysis certificate. For example, a seedlot may contain 98 percent viable seeds, but have a germination of only 75 percent because 23 percent of the seedlings produced were abnormal. In practice seed viability is used synonymously with germination capacity (the ability to germinate and produce a normal seedling) and germination testing is the principle and accepted criterion for seed viability (AOSA, 1983).

Germination Testing

Determining Field Planting Value

The ultimate objective of testing for germination is to gain information with respect to the field planting value of the seed (ISTA, 1985). Thus germination is defined by the International Seed Testing Association (ISTA) as "the emergence and development of the seedling to a stage where the aspect of its essential structures indicates whether or not it is able to develop further into a satisfactory plant under favorable conditions in soil." The "essential structures" are the root and shoot axes, cotyledons, terminal buds, and, for Gramineae, the coleoptile.

Field emergence ability is the major aspect of seed quality of concern to growers (Pieta Filho and Ellis, 1991) and high germination (greater than 90 percent) is obviously a prerequisite for seed to be sown, whether it be to provide pasture for grazing animals, to produce a harvestable crop, or to further multiply seed. While it is sometimes possible to compensate for reduced germination by increasing sowing rates to achieve a desired population, a point is reached (e.g., less than 85 percent germination in spring wheat [*Triticum aestivum* L.]; Khah, Ellis, and Roberts, 1989) where there can be deleterious effects on yield and quality even when an appropriate adjustment is made to the sowing rate (see also Willey and Heath, 1969; Roberts, 1972; Khah, Roberts, and Ellis, 1989; Chapter 11).

Seed Trading

The second objective of the germination test is to provide results which can be used to compare the value of different seedlots (ISTA, 1985). The outcome of the germination test may decide

1. the price paid to the grower (penalties are often imposed by the contract for lots which are below standard);
2. whether a seedlot is eligible to remain within a certification scheme (if the scheme includes standards for germination);
3. whether the seedlot can be sold (i.e., does it meet statutory or voluntary germination standards?);
4. whether two or more lots require blending to produce a single lot of acceptable germination;
5. whether the seedlot is suitable for export and/or internal distribution and sale.

The germination test result, in conjunction with the analytical purity result, provides the principal data upon which the seed trade buys, markets, and sells seed nationally and internationally.

Seed Storage

Seed is stored because ultimately it will be sown. However, the storage time may be weeks, months, years, or (in controlled environments) decades, depending on the reason for storage and the end use of the seedlot. Storage and seed deterioration are discussed in Chapters 6, 7 and 8. Germination results indicate whether a seedlot should be considered for storage, as even the best conditions can only maintain the quality of the seed placed in storage.

Requirements of the Germination Test

The basic requirements of the germination test are (1) to be objective, rapid and inexpensive, and (2) to be reproducible and uniformly interpretable. Seed testing associations (International Seed Testing Association and Association of Official Seed Analysts) have developed sets of standardized conditions for germinating seeds which are published as official rules (ISTA, 1985; AOSA, 1988). The optimum germination conditions are provided for hundreds of agricultural, horticultural, and silvicultural species. The rules also provide pretreatments which may aid germination of species where dormancy occurs (Table 1). The official rules of ISTA and AOSA often differ as to methodology and seedling evaluation, but the two organizations have recently begun work to eliminate such differences (Ashton, 1989).

The germination test is not objective as it is partly based on a subjective assessment of germinated seeds; the seed analyst must decide whether a seedling is "normal" (shows potential for continued development into a satisfactory plant) or "abnormal" (does not show the potential to develop into a normal plant). The ISTA rules define three categories of normal seedlings and list 13 defects of the primary root, 12 defects of the hypocotyl, epicotyl, and mesocotyl, 14 defects of the cotyledons, seven defects of the primary leaves, four defects of the terminal bud, 14 defects of the coleoptile and first leaf (Gramineae), and nine defects of the seedling as a whole. Any of these defects, alone or in combination, can render the seedling abnormal.

Despite subjective assessment, international "referee" testing has produced generally uniform results (Lovato, 1984; Scott, 1989), particularly where skilled and experienced seed analysts test species with which they are familiar. For example, when 18 ISTA accredited seed testing stations tested *Pinus contorta* L., germination results from all stations were within accepted levels of variance (i.e., they were within tolerance) (Miles, 1963); when 71 stations tested *Avena sativa* L., results from only eight were out of tolerance (Scott, 1989). Problems arise from inexperience with either the

TABLE 1. Some methods for germination testing[1]

Species	Prescriptions for:				
	Substrate	Temperature °C	First count (days)	Final count (days)	Additional directions including recommendations for breaking dormancy
Festuca rubra	TP[2]	20-30; 15-25	7	21	Prechill; KNO_3
Glycine max	BP;S	20-30; 25	5	8	–
Helianthus annuus	BP;S	20-30; 25; 20	4	10	Preheat; Prechill
Hordeum vulgare	BP;S	20	4	7	Preheat (30-35°C); Prechill; GA_3
Lathyrus sativus	BP;S	20	5	14	–
Lolium perennel	TP	20-30; 15-25; 20	5	14	Prechill; KNO_3

1. Data from "International Rules for Seed Testing," ISTA (1985)
2. TP = top of paper; BP = between paper; S = sand; KNO_3= 0.2% solution of potassium nitrate; GA_3 = gibberellic acid

species concerned or with germination testing, as is evidenced by the sometimes poor results presented by new stations seeking authorization to issue international seed analysis certificates (Scott, pers. comm.).

In seed testing, results are always wanted urgently and the germination test unfortunately is not rapid. While results for some crop species (e.g., *Brassica, Hordeum, Phaseolus*) are available in seven days, other agricultural and horticultural species require ten, 14, 21, or 28 days. Some tree and shrub species require up to 30 days after pretreatment for three to four months (e.g., *Rosa* spp., *Pyrus* spp.). Test periods have been prescribed to allow sufficient time for even weak seedlings to germinate (Copeland and McDonald, 1985). Test procedures require (1) precise temperature control and facilities for alternating temperatures, humidity control, and light, which may be provided through Jacobsen apparatus, germination cabinets, or room germinators (MacKay, 1972; ISTA, 1985); (2) suitable substrates for germinating the seeds (paper, sand, or rarely, soil); (3) facilities for breaking dormancy (low-temperature cabinets or rooms, heated cabinets, chemicals); and (4) skilled people who often undertake up to three years of "on the job" training before becoming qualified as seed analysts. Germination testing to internationally approved standards is therefore expensive and exerts considerable pressures on time, labor, and money (Matthews, 1981).

Relevance of the Germination Test

Field planting value. Germination testing has been developed and refined so that for any seedlot the maximum germination percentage is obtained and can be consistently reproduced. This is achieved through the use of artificial, standardized, essentially sterile media, in humidified, temperature-controlled germinators for periods sufficiently long to permit all those seeds capable of germinating to do so (AOSA, 1983). These tests are successful in two respects (Matthews, 1981): they are repeatable, and they provide information about the potential of the lot to germinate under optimum conditions. The problem of course is that field conditions deviate widely, and a germination test result may therefore vary markedly from actual field emergence.

The germination test reports the percentage of normal seedlings, abnormal seedlings (which are probably produced from those seeds close to death; Roberts, 1986), and dead or ungerminable seeds in a seedlot. When maximum viability is reached, a seedlot should in theory have a germination of nearly 100 percent, providing dormancy is not a factor. Loss of viability from this point results from the deterioration processes involved with seed aging, both pre- and post-harvest (Powell, 1988; Chapter 8). Symptoms of this aging include reduced rates of germination and emergence, decreased tolerance to suboptimal conditions, and poorer seedling growth (Powell, Matthews, and Oliveira, 1984). The further a germination result below 100 percent, the greater the deterioration of the lot, which is therefore generally less able to perform. Failure to recognize this deterioration has caused difficulties with interpretation of test results. For example, when a critical minimum germination is used as a criterion of quality and acceptability (e.g., 80 percent), there is a tendency to assume that all seedlots exceeding this minimum are equally good in the eventual test of quality; emergence in the field (Matthews, 1981). This is not usually the case (Delouche, 1981; Matthews, 1981). For example, Delouche (1973) compared laboratory germination with field emergence of 94 soybean (*Glycine max* L. Merr.) seedlots, all of which had a minimum germination standard of 80 percent. The field conditions were moderately adverse, and yet field emergence increased as germination increased (Table 2), i.e., the lower the germination the poorer the

TABLE 2. Comparison of laboratory germination and field emergence of 94 soybean seedlots which had a minimum germination of 80 percent.

Laboratory germination %	No. of lots	% of lots with field emergence >80%	% of lots with field emergence >70%
90-94	29	48	83
85-89	47	25	55
80-84	18	0	33

1. Calculated from the data of Delouche (1973).

performance, even though all lots met a minimum germination recommendation. When germination of a seedlot is less than a high standard (e.g., 90 percent), the test result alone indicates that the quality of the lot is suspect, i.e., deterioration has occurred (Hampton and Coolbear, 1990), and there is often a good correlation between germination and field emergence (Naylor and Syversen, 1988; Thomson, 1979). However, performance differences in field emergence among seedlots which the germination test indicates are of similarly high quality (Table 3) have resulted in a growing lack of confidence in the germination test as a measure of the plant-producing ability of seed under field conditions (AOSA, 1983), and the realization that a more sensitive differentiation of potential seed performance is required.

Seed trading. The early development of seed testing in Europe followed what Wellington described as a "commercial" concept, because seed laws required seed to be tested in state seed testing stations and "test results were mainly used by merchants to compare the value and fix appropriate prices for different seed lots, for which reproducibility of the results was regarded as essential, and the highest results desirable" (Wellington, 1965). This concept still remains in force in the seed trade, and Ellis and Roberts

TABLE 3. Field emergence performance of herbage seedlots which germination data indicate are of similar quality.[1]

		Trifolium pratense			
Seedlot	1	2	3	4	
% germination	90	90	90	90	
% field emergence	76	56	78	80	
		Lolium multiflorum			
Seedlot	1	2	3	4	5
% germination	96	95	94	92	94
% field emergence	90	67	78	79	87

1. From Hampton (1991). Species autumn (*T. pratense*) or spring (*L. multiflorum*) sown in replicated adjacent rows.

(1980b) concluded that "if a measure of seed quality is required solely for trading purposes, then the standard germination test will, as it has for many years, still suffice." Whether this statement will remain correct for much longer will depend upon market demand for further seedlot quality information (Hampton and Coolbear, 1990).

Seed storage. The initial quality of seeds determines their potential longevity under any storage conditions (Roberts and Black, 1989), and germination testing and seed moisture content are traditionally used to provide the data upon which storage decisions are based. Thus a seed store manager would correctly conclude that a seedlot with a germination of 95 percent should be able to be stored for longer under the same conditions of temperature and humidity than a seedlot of the same species and cultivar with a germination of 75 percent. However, germination data prior to storage provide no estimate of seedlot longevity, as lots with similarly high germination can differ markedly in their storage performance (Table 4).

Limitations of the Germination Test

The germination test as prescribed by ISTA (1985) and AOSA (1988) has a number of limitations and criticisms:

1. It adequately meets only one out of the three objectives–germination results continue to be the basis of seed trading, but often fail to relate to subsequent seedlot performance in the field or during storage (Matthews, 1981; AOSA, 1983; Powell, 1988; Hampton, 1991).
2. It is not yet completely standardized–for example, the ISTA rules (ISTA, 1985) state that "the substrate must at all times contain sufficient moisture," yet do not specify the amount of water to be used. Phaneendranath (1980) has demonstrated that the germination of *Zea mays* L. and *Sorghum bicolor* L. Moench. decreased significantly as water per paper towel increased. Similarly, the methods give temperature options for some species (e.g., Table 1), but if the purpose is to "obtain the most regular, rapid and complete germination" (ISTA, 1985) by providing optimum conditions for germination, by definition only one temperature (or alternating temperatures) can be optimal.

3. It is largely a subjective assessment where germination is based on the definition of a normal seedling. No account is taken of the strength or weakness (Fiala, 1978) of seedlings. In practice, "only the dead, badly diseased and irrevocably lame (seedlings) are eliminated–the weak, semi-lame, and robust are given equal weight" (AOSA, 1983).
4. Statistically nonsignificant percentage viability differences at high levels of viability between seedlots may mask significant differences of another quality component (Ellis and Roberts, 1980b).

Despite these problems, the germination test remains the principal and accepted criterion for seed viability, and, worldwide, several million tests are conducted annually. Providing the limitations are recognized, the germination test is a useful viability index (Copeland and McDonald, 1985), but a better understanding of the meaning of the results is required. In particular, the expectation that a germination result equals field performance must be corrected. Germination data should be used for an initial separation of seedlots: germination results less than an accepted standard (e.g., below 90 percent for cereals) by themselves indicate that the quality of the lot is suspect and is likely therefore to perform poorly in the field (or in storage) unless conditions are approaching the optimum for the species concerned. It is only at high germination values that the standard germination test is not adequate to indicate quality attributes of the seed (Roberts, 1984), when, because of the nature of the

TABLE 4. Storage performance of herbage seedlots which germination data indicate are of similar quality.[1]

	Trifolium pratense[2]				*Bromus willdenowii*[3]			
Seedlot	1	2	3	4	1	2	3	4
Prestorage germination (%)	90	90	90	90	97	98	96	90
Germination (%) after storage for 6m	74	87	76	86	98	90	96	88
12m	71	90	66	91	97	85	95	74

1. From Hampton (1991); 2. Ambient storage; 3. Stored at 11.6% SMC and 20°C

normal distribution on which the seed survival curve is based, a small difference in percentage germination represents a large difference in the progress of deterioration (Ellis and Roberts, 1980b). It is in these circumstances that a more sensitive differentiation of potential seed performance is required, and so vigor testing is necessary (Hampton and Coolbear, 1990).

Tetrazolium Testing

Rapid methods for biochemically determining seed viability began to be developed in the first half of this century (Lakon, 1928; Hasegawa, 1935), and Lakon (1942) first reported on the use of tetrazolium salts to test for viability. The topographical tetrazolium (TZ) test is based on the principle that when colorless triphenyltetrazolium chloride or bromide enters living plant tissue it is reduced to the red, stable, and nondiffusible triphenyl formazan (Smith, 1952) which is precipitated within live cells. In dead cells no reaction takes place and they remain colorless. Embryos may be classified as viable or nonviable according to the proportion and location of unstained, flaccid, or necrotic tissue–hence the term "topographical" tetrazolium test (Coster, 1988). For reviews of the development of the TZ test refer to Lakon (1953) and Moore (1969, 1973).

Objectives of Tetrazolium Testing

Providing a rapid estimate of seedlot viability. The TZ test is widely recognized as an accurate method of determining seedlot viability and is often referred to as a quick test since it can be completed in hours rather than the days or weeks required by the germination test. TZ test results for maize (*Zea mays L.*), for example, can be available within 24 hours, versus the seven days required for germination testing.

Primary method of germination testing. For species where dormancy breaking is particularly difficult, or where a long period of prechilling is required and a germination test cannot be completed within two months (e.g., *Malus* spp., *Rosa* spp.), TZ testing is accepted as a primary method of germination assessment (Coster, 1988).

Determining viability of dormant seeds. When a seedlot at the end of a standard germination test reveals a high percentage of apparently dormant seeds, a TZ test can be used to determine the viability of individual dormant seeds, or the viability of a working sample (ISTA, 1985).

Diagnosis. The TZ test is often used to supplement a germination test as a confirmatory diagnostic test and to interpret causes for the performance of seeds in other quality evaluations (Moore, 1969).

Requirements of Tetrazolium Testing

The TZ test should be (1) accurate, reproducible and interpretable, and (2) rapid, objective, simple, and economically practical. Standardized test conditions for some species have been accepted by both seed testing organizations (ISTA, 1985; AOSA, 1988), and handbooks of instructions for performing TZ tests and interpreting the results have been published (Grabe, 1970; Moore, 1985). However, many of the genera and families for which testing methods have been proposed have yet to be formally accepted in the "Rules for Seed Testing" (ISTA, 1985), usually because of problems with standardizing interpretation of results.

The accuracy of the TZ test relies on the skills of the individual analyst in the preparation of the test and evaluation of the results. An analyst must acquire a practical working knowledge and understanding of the location of major embryo structures and their major cell division areas; seedling characteristics that distinguish between normal and abnormal seedlings; embryo characteristics accountable for nonviable seed and for the production of normal and abnormal seedlings; causes for different staining intensities, color tints, and color patterns; and recognition of technique injuries. Accurate results depend upon the correct use of appropriate equipment, the correct preparation for staining, staining procedures, preparation for evaluation, and finally the evaluation itself (Moore, 1985). The evaluation of each seed is based upon the color, turgidity, and general appearance of tissues (sound; weak, viable; weak, nonviable; dead, nonstained), fractures, missing embryo tissue, insect damage, and abnormalities. The recognition and evaluation of different levels of tissue soundness is a factor of constant concern (Moore, 1985), and tests must be conducted by well-trained, very experienced staff with adequate time available to devote to the specialized analytical techniques involved in the method (Coster, 1988). Thus the testing involves subjective assessments which can create problems with reproducibility (Fiala, 1987b), particularly when inexperienced analysts are involved, but sometimes also among seed testing stations which routinely use the TZ test (Don et al., 1990).

The experienced analyst learns to distinguish between seeds with the capacity to produce normal seedlings and those that stain abnormally, and the results can agree closely with those of laboratory germination tests (MacKay, 1972; Coster, 1988). However, the TZ test does not indicate the presence of seed-borne fungi which may reduce germination, and may fail to detect subtle damage (e.g., that caused by frost, heating, phytotoxic chemicals) which will result in the production of abnormal seedlings. If the seeds' dehydrogenase enzyme systems have not been affected, the TZ test will not detect such damage, and germination results may differ for the same seedlot depending upon the test method used. For example, Don et al. (1990) reported TZ test results of 90 percent or greater, but germination test results as low as 4 percent in spring barley (*Hordeum vulgare* L.) seedlots which were subsequently found to have been sprayed with the herbicide glyphosate prior to harvest. Steiner and Fuchs (1987) showed that glyphosate had no effect on the reduction of tetrazolium salts in affected seed, and therefore seeds rendered nongerminable by glyphosate can produce a TZ staining pattern which enables them to be classified as viable (Don et al., 1990).

Most seeds require moistening to soften seed coats before staining, and many species require additional preparation to assure timely and adequate penetration of the staining solution into and throughout the vital tissues and structures of each seed, to accelerate the rate of staining, and to facilitate the ease, speed, and accuracy of evaluation (Moore, 1985). Preparation varies with species (ISTA, 1985; Moore, 1985), that for wheat (*Triticum aestivum* L.), for example, involving six to 18 hours moistening, and bisecting each seed longitudinally through the embryo and three-quarters of the endosperm before staining for two to four hours. These processes have been described as laborious and tedious, and preparation and evaluation (excluding sample division and replicate counting) takes some three times longer than for the laboratory germination test (Table 5). Because the test is time consuming and requires greater

TABLE 5. Preparation and evaluation time for standard germination and tetrazolium testing of wheat (*Triticum aestivum* L.).

	Mean timing (minutes)		
	Preparation	Evaluation	Total
laboratory germination[1]	5.3	16.4	21.7
tetrazolium test[2]	24.5	35.7	60.2

1. Adapted from Coster (1988)
2. Both methods used 400 seeds

analyst skills than the germination test, it is more expensive, and charges for TZ testing are usually more than double those for germination testing (Young, 1985).

Relevance of the Tetrazolium Test

The "International Rules for Seed Testing" (ISTA, 1985) contain approved procedures for TZ testing of 30 agricultural/horticultural species and 26 tree and shrub species, and TZ testing is listed as either the primary or an alternative viability evaluation method for 36 different tree and shrub seeds. Schwarzenbach (1991) reported that over 80 percent of ISTA accredited seed testing laboratories performed official TZ tests, but the data did not provide any indication of the mean number of tests performed per station.

By far the greatest use of TZ testing is "unofficial," in seed company quality control programs or as advisory tests for farmers and the seed trade. Rapid access to information on viability of a seedlot is economically important, particularly where there is only a short period of time between harvest and sowing (e.g., autumn cereals), or container space suddenly becomes available for exporting seedlots. In such cases there is usually insufficient time to obtain results from a standard germination test, and a TZ test at least allows decisions to be made. However, there is always a risk that the TZ test result will differ from the subsequent germination test result, and that if germination standards are involved, seedlots which the TZ test indicated would meet the standard may actually be rejected (Coster, 1988). The accuracy of the TZ test relies upon following approved procedures and, for official testing, examining 400 seeds, which is time consuming. Coster (1988) recently reported a protocol for a more rapid prediction of seedlot viability using a tetrazolium-based method in conjunction with sequential sampling (Banyai, 1978; Ellis and Whitehead, 1987), when the size of the sample to be examined is not fixed in advance (but may have a set limit). For autumn wheat seedlots which were required to meet a minimum germination standard of 85 percent, it was possible to make a pass/fail decision by examining on average 50 out of the 200 prepared seeds, which substantially reduced the workload and the time involved.

As a diagnostic test, TZ testing of individual seeds remaining after the completion of a germination test can provide valuable information, whether the seeds are dormant or have failed to germinate for other reasons (Moore, 1973; Young, 1985). The use of tetrazolium has been proposed as a test for seed vigor (Perry, 1981; Fiala, 1981; Moore, 1985); this use of TZ will be discussed later in this chapter.

The TZ test has been criticized for its procedural and interpretation difficulties (Justice, 1972), failure to detect phytotoxicity, heating injury, fungal infection, and dormancy (MacKay, 1972; Young, 1985; Don et al., 1990), requirement for specialized training and expertise, and the fact that it is usually more laborious and tedious to perform than a germination test (Moore, 1985). There are also problems with subjective assessment (Fiala, 1987a, 1987b). However, Moore (1985) considered that many expressed limitations of the test reflected an inadequate understanding of the test or inexperience with the methodology and interpretation of results. Coster (1988) emphasized the need for TZ tests to be conducted by experienced, highly trained analysts, and showed that, providing this was the case, agreement between TZ and germination of wheat seedlots was good. However, there were some differences between the methods, and although most of these were resolved by retesting, one or two samples produced unresolved differences, attributable to heat damage and sprouting.

Like any seed quality test, the TZ test has its limitations. However, provided it is carried out correctly, the merits of the test are that it can (1) be a reliable alternative to the germination test (e.g., for some cereal species; Coster, 1988), (2) provide rapid seedlot viability information for both nondormant and dormant seeds, and (3) in conjunction with the germination test, indicate causes for inferior seed quality and performance (Moore, 1985).

Other Biochemical Tests

A number of other rapid biochemical viability tests were developed either prior to or at a similar time to the tetrazolium test. These methods had varying degrees of success (Gadd, 1950), and have been largely supplanted by the tetrazolium method (Copeland and McDonald, 1985).

One of the first tests developed was the selenite test (Hasegawa, 1935), which was based on the reduction of colorless selenium salts to red elementary selenium by the dehydrogenase activity of living cells. Certain disadvantages of the selenite method, notably the slow development of a poisonous gas, prompted a continued search for a better reagent (AOSA, 1983), and Lakon (1942) replaced selenium salts with tetrazolium salts. Differential staining of live versus dead tissues when exposed to certain dyes has also been used to assess seedlot viability (Moore, 1969). Indigo carmine, which stains dead tissue blue but is incapable of penetrating live tissues, is considered useful for forest tree seeds (Gadd, 1950). Brucher (1948) recorded the peroxidase content of individual seeds by soaking them in a mixture of guaiacol and benzidine in 10 percent dilutions of saturated alcohol solutions, then treating with a reagent, and considered embryos that were slightly or well stained to be capable of germinating. However, the color disappeared very rapidly. Other oxidases, including catalase, have been tested, but with unreliable results (Gadd, 1950). Following the success of the TZ test, which is a dehydrogenase activity test, other chemicals (e.g., methylene blue, dinitrobenzine) have been assessed for dehydrogenase activity (Copeland and McDonald, 1985), but none has proved to be practical (Gadd, 1950). Jackisch, Lindner, and Sieg (1981) examined the use of acid fuchsin for determining cereal viability, but reported that the method was only suitable for seedlots with 85 percent or greater viability, and could not be recommended for seed damaged by heating, sprouting, chemical, or mechanical factors.

Excised Embryo Test

This method, which involves excising embryos and incubating them under conditions suitable for growth, was primarily developed by Flemion (1938, 1948) for determining viability of seeds of woody species. The objective of the test is to promptly determine the approximate viability of certain kinds of seeds which normally germinate slowly or show dormancy to such an extent that a complete standard germination test (including pretreatment) requires more than 60 days (ISTA, 1985). ISTA (1985) provides specific directions for *Sorbus*, *Euonymus*, *Fraxinus*, *Malus*, *Pyrus* and *Prunus* spp., and some *Acer* and *Pinus* spp.

The principle of the test is that after the prescribed incubation period, viable embryos either show evidence of growth, chlorophyll development, or remain firm and fresh, while nonviable embryos show signs of decay. This form of assessment does not reveal damage to the embryo, other than in the root-shoot axis, that might prevent normal germination of the intact seed (Copeland and McDonald, 1985). Problems can occur with mold infection and mechanical damage resulting from the excision, although usually embryos damaged in such a way can be distinguished from nonviable embryos by the localized discoloration of the tissue after 24 hours incubation. There is therefore a high degree of skill required in embryo excision and the process is time consuming. Schubert (1965) compared the excised embryo test with the TZ method for determining the viability of dormant tree seeds and concluded that the TZ method should receive preference. However, laboratories which test considerable quantities of tree and shrub seeds routinely conduct excised embryo tests (Copeland and McDonald, 1985).

Electrical Conductivity Measurement

Electrical conductivity measurement of bulk seed exudates, the basis of the conductivity vigor test (Matthews and Bradnock, 1968), has also been used to predict individual seed performance, either using standard conductivity meters and electrode dip cells (Siddique and Goodwin, 1985), or by means of an electronic "automated seed analyser" (Steere, Levengood, and Bondie, 1981). This machine has the capability of monitoring individual seed conductivity values, and the results were used to predict viability (Steere, Levengood, and Bondie, 1981) through the selection of a partition value whereby readings above this value represent nongerminable seeds, and readings below this value are considered germinable seeds. McDonald and Wilson (1979) found that the machine had the capability to predict high-quality soybean seed as high in germination and low-quality seed as low in germination, but failed to accurately forecast the germination of intermediate-quality seed. Matthews (1981) pointed out that the routine use of this test would assume that a common partition value would be applicable to all seedlots of a species regardless of variables such as cultivar and seed size, and McDonald and Wilson (1979) and Hepburn, Powell, and Matthews (1984) showed that within and between cultivars of soybean and pea, respectively, seed size did affect the results, and that therefore a single partition value could not be applied uniformly throughout a species. Hepburn, Powell, and Matthews (1984) also reported that, in both pea and soybean, measurement of leakage did not differentiate clearly between dead and living seeds either when the results were expressed per seed or per gram. Therefore, any partition conductivity value chosen as a basis for predicting germination within a species could not be considered as an absolute boundary between viable and nonviable seeds.

This approach to viability testing has great appeal (Matthews, 1981), and if the biological meaning of the conductivity readings could be clarified to cover all situations and provide consistent and repeatable assessments of germination, electronic instrumentation could very readily be used to automate germination tests.

VIGOR TESTING

Seed vigor is now accepted as an important seed quality component. The subject is complex, as illustrated by the fact that it took the ISTA Vigour Test Committee 27 years to agree upon a definition. However, that definition (Perry, 1978) is accepted and recognizes that vigor is not a single measurable property, but a concept describing several seed performance associated characteristics (Perry, 1981). It can encompass potential seed performance both in the field and in storage.

Perry (1981) stated that "a vigour test should provide a reproducible result which is more closely correlated with seed performance in the field under some conditions than the germination test." Hampton and Coolbear (1990) suggested that the removal of "in the field" would better cover all possibilities for requirements for vigor testing, especially the problems of evaluating seedlots with high storage potential.

The "Handbook of Vigour Test Methods" (Perry, 1981) suggested standard methodology for eight vigor tests, two of which were further modified in the Second Edition (Fiala, 1987a). All of these tests have been found to produce results which often, but not always, correlate better with field emergence than the standard germination test (Perry, 1984a; Fiala, 1987b). However, in almost 40 years of research and debate, only one test for one species, the conductivity test for peas, has met the requirements to be accepted internationally. Others suffer from problems of reproducibility when either strict control of the environment or subjective assessments are necessary. Yet others are carried out without a clear understanding of the variables and assumptions involved (Hampton and Coolbear, 1990).

Requirements for Vigor Testing

Meeting Market Demands

For many agricultural and horticultural crops, specified plant populations are currently recommended for maximizing yield and/or quality. The maximum economic yield of vining peas, for example, is reached at around 90 plants m^{-2}, and the achievement of this target population is of economic importance because of the cost of seed and the yield effects associated with variation from the optimal population (Gane, 1985). Precision sowing of cabbage (*Brassica oleracea* L.) and calabrese (*B. oleracea* convar. *botrytis* [L.] Alef. var. *cymosa* Duch) avoids the labor of transplanting (Perry, 1980), but is just as important operationally when cell transplanting is used. Synchronous emergence is crucial to obtain a stand of uniform maturity in crops where harvesting is a once-over operation. In such cases, the grower requires a more precise estimate of actual field emergence than can be provided by a germination test, and some form of vigor testing may meet this requirement.

Several local successes have been reported. In New Zealand, garden pea vigor test results expressed as expected field emergence (EFE; Scott and Close, 1976) provide better predictions than the standard germination test and allow sowing rates to be adjusted to achieve the target plant population required (Table 6). Other examples include the use of the cold test for maize in Austria (Fiala, 1978) and accelerated aging for soybean in the United States (TeKrony, 1985).

Marketing and Promotion

The current limitations of seed vigor testing, i.e., the failure of many methods to be inexpensive, rapid, uncomplicated, uniform, and reproducible (McDonald, 1980), mean that seed companies are understandably cautious about using seed vigor information in marketing and promotional programs, particularly internationally. However, nationally and in response to a competitive marketplace, there are companies that use vigor information as a marketing strategy. For example, in the United Kingdom at least two major seed houses market cereal seed using stress test results. This marks a change in the use of vigor information, because although many companies have used vigor tests in their quality assurance programs for over 20 years (AOSA, 1983), the majority continue to refrain from using such information for advertising purposes, realizing the dangers of misuse of information such as making claims for potential seed performance which are not subsequently displayed in the

TABLE 6. Relationship between germination, vigor, and field performance in the garden pea (after Hampton and Scott, 1982).

Seedlot	Percentage germination	Percentage EFE	Sowing rate $g.m^{-2}$	Percentage emergence recorded	Plants m^2 obtained*
1	98	90	41	89	147
2	94	79	42	77	143
4	99	79	43	80	150
5	96	65	49	68	147

*Target was 150 plants m^{-2}

field. Indeed AOSA (1983) concluded that such information should not be used in this way until the problems of standardizing test procedures and interpreting, presenting, and utilizing vigor test results have been resolved.

There is also a negative view of seed vigor testing by a small section of the seed trade that is concerned with the problems of selling low-vigor seed. In the absence of any other information, a seedlot of high germination may be sold readily whatever its vigor rating. If that information were available the seedlot would have to be downgraded.

The use of more precise information on seed quality (including vigor) is likely to become more commonplace as farmer and grower awareness continues to improve and the market becomes even more competitive for the seed producer. The future will produce a requirement for vigor information which can be used with confidence, firstly to improve seed quality and secondly to market and promote that fact (Hampton and Coolbear, 1990).

Official Testing

Through their collaborative programs (Fiala, 1987b; TeKrony, 1988), the Vigor Committees of ISTA and AOSA have been working toward the goal of routinely providing as much information as possible about the planting value of different seedlots under a wide range of conditions, but, as already noted, progress has been slow. Only for a relatively few species are vigor tests conducted on a regular basis. The New Zealand Official Seed Testing Station, for example, offers the EFE test for garden pea and accelerated aging tests for onion (*Alium cepa* L.), prairie grass (*Bromus willdenowii* Kunth), and crested dogstail (*Cynosurus cristatus* L.) (Anonymous, 1987). The amount of vigor testing work, if done at all, is usually only a very small proportion of a station's workload, for example, at the Official Seed Testing Station, Cambridge, UK, 73 vigor tests were carried out in 1985/86 out of a total of 3,487 advisory seed tests (Anonymous, 1987). In the near future at least, vigor testing at official seed testing stations may be used primarily for the purpose of diagnosis in commercial disputes where the station acts as independent arbiter (Hampton and Coolbear, 1990).

Research

Vigor testing is an important adjunct to research on seed production and seed quality and new vigor test methods have indeed been developed as a result of such work. From the seed technologist's and the seed physiologist's point of view, the ultimate goal must be to understand enough about the variables which limit any aspect of seed performance to be able to avoid vigor losses through the production system (Hampton, 1991). The process of unraveling these variables requires that accurate and repeatable methods of seed vigor assessment be available so that ideas and theories can be tested.

When Vigor Tests Are Appropriate

The objectives for vigor testing will differ for different applications. However, some basic requirements of a vigor test have been established (McDonald, 1980; Perry, 1984b):

1. to provide a more sensitive index of seed quality than the germination test;
2. to provide a consistent ranking of seedlots in terms of potential performance;
3. to be objective, rapid, simple, and economically practical;
4. to be reproducible and interpretable.

A discussion of seed vigor tests in the context of the latter three objectives has been presented by Hampton and Coolbear (1990), and the next section draws heavily on that review. However, the first criterion of a vigor test is that it should offer a better prediction of actual seedlot performance than the germination test, particularly (as previously discussed) for seedlots which the germination test indicates are of similar quality (e.g., Tables 3 and 4). It is in these circumstances that vigor testing is necessary.

The Status of Current Vigor Tests

Categorizing vigor tests into different groups is inevitably arbitrary, but three types of approaches to the problem can be readily distinguished (Hampton and Coolbear, 1990):

1. single tests based on some aspect of germination behavior
2. attempts to develop physiological or biochemical indices of vigor
3. multiple testing

Interpretation of results may also vary between

1. relative scores designed to rank different seedlots
2. presentation of an absolute vigor value

How well do these types of tests meet the objectives and requirements of vigor testing?

Single Tests Based on Germination Behavior

This is the most familiar approach. Methods include measuring the rate of germination, seedling growth and seedling evaluation, cold tests, the Hiltner test, accelerated aging, and controlled deterioration tests. All have been comprehensively described by Perry (1981), Fiala (1987a), and/or AOSA (1983). This is the type of test that should be easy to use, reproducible, and might be regarded as an extension of the routine activities of a seed testing laboratory (Hampton and Coolbear, 1990). However, recent referee testing studies reveal some of the problems with these techniques.

Methods which involve subjective assessments are always going to cause difficulties in reproducibility between laboratories (Perry, 1984b; Fiala, 1987b), while varying correlations with field emergence data emphasize the maxim that seed vigor is a multidimensional property of the seed. Thus, Perry (1984a) reported that the maize cold test was sufficiently reproducible to be an acceptable vigor test which generally related well to field emergence. Three years later, Fiala (1987b) noted that the maize cold test was poorly related to field emergence (Table 7). Fiala attributed this difference to the possibility that the field environments in the later series of tests provided more favorable conditions for seedling emergence than those reported in 1984. Comparative information on seed performance in the cold test is of little relevance in unstressed conditions (van de Venter and Lock, 1991). Where stress does occur, the cold test data usually correlate better with field emergence than do

TABLE 7. ISTA Vigor Test Committee collaborative results for the conductivity test for peas and the cold test for maize.

Source		Correlation coefficient (r) between means of the test and field emergence at different stations	Range of correlations at individual stations
1. Conductivity test: peas			
Perry (1984a)		−0.903*	−0.62 → −0.93**
Fiala (1987b)	1st sowing	−0.71*	−0.29 → −0.64***
	2nd sowing	−0.79*	−0.36* → −0.79**
2. Cold test: maize			
Perry (1984a)		0.996***	0.65[a] → 0.98***
Fiala (1987b)		0.63	negative → 0.86**

a. One station reported a negative correlation, but this was the lowest of 14 other returns.
*Significant at $P< 0.05$; **Significant at $P< 0.01$; ***Significant at $P<0.001$

germination data (Bekendam, Kraak, and Vos, 1987), but the relationship is highly dependent on the test procedures adopted and test variables; for example, the effects of fungicide treatment, temperature, and duration of the cold period on cold test results were dependent on the type of substrate used (Bruggink, Kraak, and Bekendam, 1991). Lack of uniformity in field soil is the greatest difficulty with the cold test (Copeland and McDonald, 1985), but the question as to whether soil is actually necessary at all is once more being debated (Nijenstein, 1988; Bruggink et al., 1991). The use of a standardizable test medium may move the cold test from a valuable "in-house" vigor test which can consistently rank seedlots (TeKrony, 1983) to one which achieves international acceptability in relevant environments.

One approach to vigor testing which is showing a great deal of promise is the aging test: either accelerated aging (AA) or controlled deterioration (CD). The present methodology of both tests is presented in detail in Fiala (1987a). In Fiala's (1987b) ISTA collaborative study, AA tests for soybean were found to have good predictive value, and similar results have been reported from AOSA referee tests (e.g., TeKrony, 1985). Perry (1984a) reported problems of reproducibility using the AA technique, and the test method has been criticized for its failure to hold seeds under constant conditions (Ellis and Roberts, 1980b; Matthews, 1980), because relative humidity and air temperature can vary, and initial seed moisture content affects the degree of aging that occurs within a specified time period. This latter parameter will not necessarily be related to seed vigor and may be affected by factors such as testa integrity and permeability, and also by the type of inner chamber and position of the seed within the outer chamber used for the test (Tomes, TeKrony, and Egli, 1988). A recent study of factors influencing the AA test for soybean (Tomes, TeKrony, and Egli, 1988) has identified methods for controlling several testing variables which markedly improved the standardization of this test (Hampton, Johnstone, and Eua-umpon, 1992). The CD test proposed by Matthews (1980) avoids some of the AA test variables by bringing seeds to a constant moisture content before being exposed to high artificial aging temperatures. Some good correlations have been obtained with field emergence and commercial storage potential using this test on a range of species (e.g., Powell and Matthews, 1981, 1985; Wang and Hampton, 1991), and it has been shown to be reproducible for some, but not all species (Powell et al., 1984). Further data are required to determine whether the CD test has any reliable predictive value for severe stress conditions (Naylor and Syversen, 1988).

The strengths of these single testing procedures are in their ability to rank seedlots for potential planting or storage value, providing the appropriate test is selected for anticipated local conditions (Hampton and Coolbear, 1990).

Physiological and Biochemical Tests for Vigor

Given the known predictive limitations and the length of time required for most germinative vigor tests, there has been continuing interest over the past 25 years in the potential of physiological or biochemical properties of seeds to act as indices of seed vigor. The most familiar of these are the electrical conductivity test and the tetrazolium (TZ) tests (Perry, 1981; AOSA, 1983). Other approaches often discussed include the measurement of respiratory capacity, ATP content, and glutamic acid decarboxylase activity (GADA), reviewed in AOSA (1983). With the exception of the conductivity test which has already been mentioned, it is difficult to escape the conclusion that the other methods in this category have little to recommend them (Hampton and Coolbear, 1990).

The conductivity test has become a routine method for ranking pea seedlots and, in general, has given a good correlation with the field emergence of high-ranking seedlots (Table 7). This test has tremendous advantages of simplicity and rapidity and meets most of the requirements for a good vigor test.

The amount of electrolyte leached from an imbibing seed will be a function of its soluble mineral ion content, any damage to the seed coat, proportion of nonviable cells, and, finally, its ability to repair and reorganize cell membranes in living cells. While the fourth parameter is directly related to seed performance, the others may not be. The impact of seed coat damage on seed performance, for example, may depend as much on position of the damage as on its extent *per se* (Coolbear and Hill, 1988). This distinction may not be revealed by a simple conductivity test. Initial seed moisture may be another source of variability in this test (Perry, 1984b), although Loeffler, TeKrony, and Egli (1988) found no significant differences in conductivity of soybean when seed moisture ranged from 11 to 18 percent. Conductivity measurements do seem to hold promise for a range of other crops, particularly large-seeded legumes. However, Bruggink et al. (1991) have demonstrated that for maize, beet (*Beta vulgaris* L.), and cabbage, the source of electrolyte leakage can considerably influence conductivity test results (e.g., cultivar differences in leakage from maize pericarps may interfere with conductivity measurements for vigor). There are arguments in favor of assessing the conductivity of individual seeds by, for example, using a multicell conductivity meter (McDonald and Wilson, 1980), but as conductivity is significantly altered by seed size, the weight of individual seeds would be required to correct for this variable.

The tetrazolium test as a vigor, rather than a viability, test has been extensively discussed at various times. However, as Perl, Luria, and Gelmond (1978) pointed out, the color reaction on which the TZ test is based depends not only on the activity of the dehydrogenase enzymes, but also on the multitude of other factors within a seed which can affect the reducing capacity of the tissue. Changing permeability of deteriorating cells to triphenyltetrazolium chloride may also give misleading results (Priestley, 1986). While in practice these problems are probably unimportant for the assessment of viability using TZ, they cast serious doubts on the value of this method as a test for seed vigor.

Assessment of ATP levels in seeds is another approach which has often been put forward as a method for predicting vigor. For example, Lunn and Madsen (1981) demonstrated that, allowing for variation between cultivars, initial ATP production was correlated with seed vigor and was a more sensitive index of seed deterioration than germination measurements. Perl (1987) has discussed the fallacies behind the assumption that ATP levels *per se* are of necessity related to vigor, in that the measurement is a function of both ATP synthesis and utilization. Only a small percentage of ATP synthesized (estimated at less than 5 percent) is actually measurable, the rest being utilized immediately on formation. Styer, Cantliffe, and Hall (1980), for example, found no correlation of ATP with vigor in a range of vegetable crops. It is possible, therefore, that if a simple and reproducible assay to measure ATP synthesis rather than net accumulation could be developed, it may give a better indication of seed vigor. However, care must be taken in choosing the time to make the measurements. ATP levels tend to rise very rapidly during early imbibition, presumably to fulfill demand for initial repair processes (Anderson and Gupta, 1986), but subsequent synthesis and utilization may not be at a constant rate (Priestley, 1986) as the germinating seed becomes organized and begins to initiate a wide range of metabolic activity for germinative growth. It is probable, too, that the initial production of ATP is not dependent on oxidative phosphorylation, but rather on a PEP carboxylase driven pathway utilizing malate, NAD, and AMP (Perl, 1987).

While mitochondrial damage in deteriorating seed will be directly related to vigor, oxygen uptake, the most commonly measured aspect of respiratory capacity suggested as a vigor index, may not always be related to the efficiency of ATP production. Abu-Shakra and Ching (1967) found evidence of considerable uncoupling of oxygen uptake and ATP synthesis in mitochondria extracted from aged soybeans. Oxygen uptake was higher and ATP production lower than in isolates from high-vigor material. In many kinds of seed, oxygen uptake is limited by the permeability of the seed coat (e.g., *Pinus radiata* D. Don; Rimbawanto, Coolbear, and Firth, 1989), thus mechanical damage may result in increased oxygen uptake. Further, there is evidence that chemical treatments of aged tomato (*Lycopersicon lycopersicum* [L.] Karst, ex Farw.) seed may increase the activity of key respiratory enzymes, such as cytochrome c oxidase, without causing any significant change in germination performance (Puls and Lambeth, 1974).

The last vigor test in this category is the assay for glutamic acid decarboxylase activity (GADA) which measures the evolution of CO_2 as glutamate is enzymically converted to aminobutyric acid. However, van de Venter, Grabe, and Currans (1989) have demonstrated that it is not possible to measure true GADA by means of the manometric procedures commonly used because it is not possible to differentiate between CO_2 evolved by respiration and that released by glutamic acid. Further, some correlations may be indirect ones via the association of the activity of glutamic acid decarboxylase with seed size which may itself be related to vigor (AOSA, 1983). The results of van de Venter, Grabe, and Currans (1989) do not invalidate reported relationships between GADA and seed vigor of wheat (Ram and Weisner, 1988; Steiner, Grabe, and Tulo, 1989). They do, however, question the belief that such studies show a direct relationship between seed vigor and glutamic acid decarboxylase activity.

Multiple Testing and the Search for an Absolute Vigor Score

In an effort to develop more accurate indices of vigor, a number of workers have developed vigor tests which involve assessments based on more than one technique and/or use a more detailed interpretation of data provided by the testing procedures. The most successful multiple testing approach to date has been the previously mentioned extension of the conductivity test in New Zealand to calculate EFE (expected field emergence) for peas, by a formula which also incorporates the results of the standard germination test and a correction for hollow heart (Scott and Close, 1976). This method has proved useful in local conditions (Table 6), but requires further evaluation with different cultivars in a wider range of environments (Hampton and Coolbear, 1990).

An attempt to tackle the difficult problem of compensating for vigor differences within a seed population has recently been presented by Barla-Szabò and Dolinka (1988), who have combined the results of the standard germination test with a complex stressing vigor test (CSVT). The CSVT evaluates seed performance into high vigor, medium vigor, and abnormal categories after a four-day water stress (25°C for two days, then 5°C for two days). The test results can be presented as a predicted emergence value, which is the average of the germination capacity of the seedlot and that percentage of the population shown to be stress resistant. Once again, this method needs to be validated for different cultivars under different planting conditions, but appears to have promise for both maize and wheat (Barla-Szabò, Dolinka, and Berzy, 1989).

The final approach to vigor testing which will be discussed is that proposed by Ellis and Roberts (1980a, 1981) as an extension of their fine modeling work on seed survival in storage. They have demonstrated that seed survival in constant, hygienic storage can be described by a linear relationship between probit percent viability and time. The slope of the line is defined by storage conditions and is independent of seedlot or cultivar, while the intercept on the y axis, the initial theoretical probit percentage viability, is a seedlot constant, K_i, and is a function of genotype and prestorage factors (Ellis and Roberts, 1981; Roberts and Ellis, 1984). They have suggested that the value, K_i (which can be calculated from the results of a series of controlled deterioration experiments), represents an absolute measure of seed vigor.

While there is little doubt that the concept of K_i is a valuable one in assessing the potential storability of a seedlot (Ellis, 1988), there are questions which can be raised about its applicability as an overall measure of vigor (Hampton and Coolbear, 1990). Firstly, K_i has the disadvantage of not being easily translatable into an estimate of planting value and may thus become no more than a sophisticated means of ranking seedlots. More importantly, using this relatively time-consuming method as an assessment of seed vigor relies on two key assumptions. Foremost of these is the assumption that all aspects of seed deterioration processes are part of the same continuum in which the loss of viability is merely an end point. Enough is known about the physiology of seed deterioration to suggest that this is an oversimplification. While it is difficult to assess the relative impact of different processes on losses of seed vigor and viability, it is clear that the process of deterioration is a matrix of interrelated events (Priestley, 1986).

The second assumption is that proportions of surviving seed, measured under optimal conditions after storage, are going to give an adequate picture of the potential performance of naturally low-vigor seeds (or those deteriorated postharvest) under stress conditions. This is an interesting hypothesis which has yet to be directly tested on an experimental basis. Previously mentioned results (Perry, 1984a; Naylor and Syversen, 1988) suggest that this assumption is unlikely to be validated.

The Future

Heydecker (1972) succinctly summarized the general characteristics of a practical vigor test when he said, "what we need is a reasonably accurate forecast of field quality." He then added, "we must admit we have not achieved this yet," and for most plant species this statement still remains correct. This discussion of the methods available for vigor testing has deliberately focused on those most commonly used, or those considered to have the greatest potential for further development and refinement, i.e., conductivity testing, aging tests (AA and CD), and possibly the combined stress test (Hampton and Coolbear, 1990). Because of this a number of other tests (e.g., seedling growth and evaluation, Hiltner, aleurone tetrazolium, germination speed, osmotic stress; AOSA, 1983; Fiala, 1987a) have not been evaluated in this chapter. All of these have their limitations (AOSA, 1983; Copeland and McDonald, 1985), but this is not to deny that they may have a place in local conditions. For example, the aleurone tetrazolium test is used to supplement the cold test for maize in Austria (Fiala, 1987a); a form of growth test, the slant-board test, is regarded as a useful predictor of seed vigor in small-seeded field vegetables and is widely used within the UK seed trade to assess the vigor of lettuce (*Lactuca sativa* L.) and to a lesser extent carrot (*Daucus carota* L.) (McCormac, Keefe, and Draper, 1990); Lichatschov (1981, and personal communication) has stressed the importance placed on seedling evaluation in determining cereal seedlot vigor in Russia.

In the future there will be an increasing market demand for better assessments of the planting value and/or storage potential of seedlots, particularly on a national or local basis, and this suggests that the focus of vigor testing should be on the development of a range of methods shown to be useful under anticipated local conditions. Internationally there is also a continuing requirement for vigor testing, and a clear need for international discussion and cooperation on this kind of work, particularly discussion and research centered on the extraneous sources of variation and hidden assumptions behind each test (Hampton and Coolbear, 1990). The time for simple collection of correlation data between vigor tests and field results is now over. Such studies must be supplemented by identification of production variables, planting condition variables, and their interactions.

CONCLUSION

Quality differences occur among seedlots and also within the population of seeds that constitute the seedlot. Detecting these differences is not an exact science because any form of test is dealing with living material which can alter in its response to test variables, and because many of the methods used rely on subjective assessment of some growth parameter. Despite this, it is possible, by adhering to prescribed standardized test conditions, to produce reproducible results whose variance is not greater than that expected from random sampling error.

The assessment of seed viability continues, and will continue, to rely on the laboratory germination test which indicates the potential of a seedlot to produce normal seedlings under optimum conditions for growth. When dormancy is excluded, the further the germination percentage is from 100 percent, the stronger the indication that the seedlot is of reduced quality, i.e., for physical and/or physiological reasons the seedlot has aged and is unlikely to perform well as field or storage conditions move away from optimum or good. Much of the criticism of the germination test has arisen from the use of the term "field planting value" in the objectives of the test, resulting in a misunderstanding of the meaning of germination test. For high-quality seedlots (i.e., those with a germination of 90 percent or greater), the germination test does not necessarily indicate field planting value, and it is particularly for these seedlots that a more sensitive indicator of seed quality, seed vigor testing, is required.

It is unlikely that any one aspect of seed behavior, whether germinative, physiological, or biochemical, will be a universally reliable index of all aspects of seed vigor (Hampton and Coolbear, 1990). It is unrealistic to expect such predictive utility from any single parameter, given the complexities involved and the many possible interactions of the seed with different postharvest and planting environments. The future of vigor testing is bright, but enthusiasm should be tempered with caution; vigor testing may not in the end provide the definitive answer to field planting value, but, at its best, rank high germinating seedlots in order of their anticipated performance in non-optimum conditions.

REFERENCES

Abu-Shakra, S. S., and T. M. Ching. Mitochondrial activity in germinating new and old soybean seeds. *Crop Sci.* 7(1967): 115-118.

Anderson, J. D., and K. Gupta. Nucleotide alterations during seed deterioration. In *Physiology of Seed Deterioration*, ed. M. B. McDonald and C. J. Nelson (Madison, USA: Crop Science Society of America, 1986), pp. 47-63.

Anonymous. Seed testing in England and Wales. In *Annual Report*, 1986 (Cambridge, UK: National Institute of Agricultural Botany, 1987), pp. 37-40.

AOSA. *Seed Vigour Testing Handbook*, ed. B. E. Clark, M. B. McDonald, and P. K. Joo (USA: Association of Official Seed Analysts, 1983). 88 pp.

AOSA. Rules for testing seeds. *J. Seed Technol.* 12 (1988): 1-109.

Ashton, D. B. Report of the germination committee 1986-1989. Seed *Sci.& Technol.* 17 (Supplement 1989): 83-85.

Banyai, J. Sequenzanalyse in der Saatgut prufing [Sequential analysis in seed testing]. *Seed Sci. & Technol.* 6 (1978): 505-515.

Barla-Szabò, A. G., and B. Dolinka. Complex stressing vigour test: A new method for wheat and maize seeds. *Seed Sci. & Technol.* 16 (1988): 63-73.

Barla-Szabò, G., B. Dolinka, and T. Berzy. Application of seed vigor tests for corn production. *Georgicon Agriculture* 2 (Supplement, 1989): 159-165.

Bekendam, J., H. L. Kraak, and J. Vos. Studies on field emergence and vigour of onion, sugar beet, flax and maize. *Acta Hortic.* 215 (1987): 83-94.

Brucher, A. Eine Schnellme node zur Bestimmung der Keimfähigkeit von Samen. *Physiol. Plant.* 1 (1948): 343-358.

Bruggink, H., H. L. Kraak, and J. Bekendam. Some factors affecting maize (*Zea mays* L.) cold test results. *Seed Sci. & Technol.* 19 (1991): 15-23.

Bruggink, H., H. L. Kraak, M. J. G. E. Dijkema, and J. Bekendam. Some factors influencing electrolyte leakage from maize (*Zea mays* L.) kernels. *Seed Sci. Res.* 1 (1991): 15-20.

Coolbear, P., and M. J. Hill. Seed quality control. In *Rice Seed Health*, ed. S. J. Banta (Manila, Philippines: International Rice Research Institute, 1988), pp. 331-342.

Copeland, L. D., and M. B. McDonald. *Principles of Seed Science and Technology* (Minneapolis, USA: Burgess Publishing, 1985). 321 pp.

Coster, R. M. The use of tetrazolium and sequential sampling as an alternative to germination testing for cereal seed. *Plant Varieties & Seeds* 1 (1988): 75-84.

Delouche, J. C. Seed vigour in soybeans. *Proc. Third Soybean Seed Res. Conf.* (1973): 56-72.

Delouche, J. C. Harvest and post-harvest factors affecting the quality of cotton planting seed and seed quality evaluation. *Proc. Beltwide Cotton Production Res. Conf.* (1981): 289-305.

Don, R., J. Bartz, G. F. M. Bryant, A. van Geffen, G. Lunn, P. Overaa, and A. M. Steiner. Germination and tetrazolium testing of treated barley seed samples from glyphosate treated crops in seven ISTA stations. *Seed Sci. & Technol.* 18 (1990): 641-651.

Ellis, R. H. The viability equation, seed viability nomographs, and practical advice on seed storage. *Seed Sci. & Technol.* 16 (1988): 29-50.

Ellis, R. H., and E. H. Roberts. Improved equations for the prediction of longevity. *Ann. Bot.* 45 (1980a): 13-30.

Ellis, R. H., and E. H. Roberts. Towards a rational basis for testing seed quality. In *Seed Production*, ed. P. D. Hebblethwaite (London, UK: Butterworths, 1980b), pp. 605-635.

Ellis, R. H., and E. H. Roberts. The quantification of ageing and survival in orthodox seeds. *Seed Sci. & Technol.* 9 (1981): 373-409.

Ellis, R. H., and J. Whitehead. Open, truncated and triangular sequential seed testing procedures. *Seed Sci. & Technol.* 15 (1987): 1-17.

Fiala, F. Einfluss der Kalibrierung auf den Anbawwert des Saatgutis von Hybridmaissorten. Jahrbuck 1977 der Bundersanstalt für Pflanzenbau and Samenprüfung in Wein (1978): 112-166.

Fiala, F. Aleurone tetrazolium test. In *Handbook of Vigour Test Methods*, ed. D. A. Perry (Zurich, Switzerland: International Seed Testing Association, 1981), pp. 61-65.

Fiala, F. *Handbook of Vigour Test Methods*, Second Edition (Zurich, Switzerland: International Seed Testing Association, 1987a). 72 pp.

Fiala, F. Report of the vigour test committee 1983-1986. *Seed Sci. & Technol.* 15 (1987b): 507-522.

Flemion, F. A rapid method for determining the viability of dormant seeds. *Contributions from the Boyce Thompson Institute* 9 (1938): 339-351.

Flemion, F. Reliability of the excised embryo method as a rapid test for determining the germinative capacity of dormant seeds. *Contributions from the Boyce Thompson Institute* 15 (1948): 229-241.

Gadd, I. Biochemical tests for seed germination. *Proc. ISTA* 16 (1950): 235-253.

Gane, A. J. The pea crop–Agricultural progress, past, present and future. In *The Pea Crop*, eds. P. D. Hebblethwaite, M. C. Heath, and T. C. K. Dawkins (London, UK: Butterworths, 1985), pp. 3-15.

Grabe, D. F. *Tetrazolium Testing Handbook* (USA: Association of Official Seed Analysts, 1970). 62 pp.

Hampton, J. G. Herbage seed lot vigour–Do problems start with seed production? *J. Appl. Seed Prod.* 9 (1991): 87-93.

Hampton, J. G., and P. Coolbear. Potential versus actual seed performance–Can vigour testing provide an answer? *Seed Sci. & Technol.* 18 (1990): 215-228.

Hampton, J. G., and D. J. Scott. Effect of seed vigour on garden pea production. *New Zealand J. Agric. Res.* 25 (1982): 289-294.

Hampton, J. G., K. A. Johnstone, and V. Eua-umpon. Ageing vigour tests for mungbean and Frenchbean seed lots. *Seed Sci. & Technol.* 20 (1992): 643-653.

Hasegawa, K. On the determination of vitality in seed by reagents. *Proc. ISTA* 7 (1935): 148-153.

Hepburn, H. A., A. A. Powell, and S. Matthews. Problems associated with the routine application of electrical conductivity measurements of individual seeds in the germination testing of peas and soybeans. *Seed Sci. & Technol.* 12 (1984): 403-413.

Heydecker, W. Vigour. In *Viability of Seeds*, ed. E. H. Roberts (London, UK: Chapman and Hall, 1972), pp. 209-252.

ISTA. International Rules for Seed Testing. *Seed Sci. & Technol.* 13 (1985): 299-355.

Jackisch, W., H. Lindner, and M. Sieg. Die Prüfung der Keimfähigkeit von Getreidearten mit Säurefuchsin [The use of acid fuchsin for testing the germination capacity of cereals]. *Seed Sci. & Technol.* 9 (1981): 557-565.

Justice, O. L. Essentials of seed testing. In *Seed Biology*, ed. T. T. Kozlowski (New York, USA: Academic Press, 1972). Vol. 3, pp. 302-370.

Khah, E. M., R. H. Ellis, and E. H. Roberts. Effects of laboratory germination, soil temperature and moisture content in the emergence of spring wheat. *J. Agric. Sci.* 107 (1989): 431-438.

Khah, E. M., E. H. Roberts, and R. H. Ellis. Effects of seed ageing on growth and yield of spring wheat at different plant population densities. *Field Crops Res.* 20 (1989): 175-190.

Lakon, G. Ist die Bestimmug der Keimfähigkeit der Samen ohne Keimversuch möglich. *Angewandte Botanik* 10 (1928): 470.

Lakon, G. Topographischer Nachweis der Keimfähigkeit der Getreidefrüchte durch Tetrazolium salts. [Topographical determination of the viability of cereal seed by tetrazolium salts]. *Ber. dfsch. Bot. Ges.* 60 (1942): 299-305.

Lakon, G. Zur Geschichte der "biochemischen" Keimprüfungs methoden. [History of "biochemical" germination testing methods]. *Sautgutwirtsch* 5 (1953): 180-183, 205-207.

Lichatschov, B. S. Die Bedeutung der Triebkraftbestimmung bei der Bewertung der Samenqualitat. [The importance of determining vigour in the evaluation of seed quality]. *Seed Sci. & Technol.* 9 (1981): 623-631.

Lichatschov, B. S. Personal communication, Novosibirsk, Russia (1990).

Loeffler, T. M., D. M. TeKrony, and D. B. Egli. The bulk conductivity test as an indicator of soybean seed quality. *J. Seed Technol.* 12 (1988): 37-53.

Lovato, A. Report of the referee testing committee 1980-1983. *Seed Sci. & Technol.* 12 (1984): 333-345.

Lunn, G., and E. Madsen. ATP–Levels of germinating seeds in relation to vigour. *Physiol. Plant.* 53 (1981): 164-169.

MacKay, D. B. The measurement of viability. In *Viability of Seeds*, ed. E. H. Roberts (London, UK: Chapman and Hall, 1972), pp. 172-208.

Matthews, S. Controlled deterioration: A new vigour test for crop seeds. In *Seed Production*, ed. P. D. Hebblethwaite (London, UK: Butterworths, 1980), pp. 647-660.

Matthews, S. Evaluation of techniques for germination and vigour studies. *Seed Sci. & Technol.* 9 (1981): 543-551.

Matthews, S., and W. T. Bradnock. Relationship between seed exudation and field emergence in peas and French beans. *Hort. Res.* 8 (1968): 89-93.

McCormac, A. C., P. D. Keefe, and S. R. Draper. Automated vigour testing of field vegetables using image analysis. *Seed Sci. & Technol.* 18 (1990): 103-112.

McDonald, M. B. Assessment of seed quality. *HortScience* 15 (1980): 784-788.

McDonald, M. B., and D. O. Wilson. An assessment of the standardisation and ability of ASA-610 to rapidly predict potential soybean germination. *J. Seed Technol.* 4 (1979): 1-11.

McDonald, M. B., and D. O. Wilson. ASA-610 ability to detect changes in soybean seed quality. *J. Seed Technol.* 5 (1980): 56-66.

Miles, S. R. Handbook of tolerances and measure of precision for seed testing. *Proc. ISTA* 28 (1963): 523-686.

Moore, R. P. History supporting tetrazolium seed testing. *Proc. ISTA* 34 (1969): 233-242.

Moore, R. P. Tetrazolium testing for assessing seed quality. In *Seed Ecology*, ed. W. Heydecker (London, UK: Butterworths, 1973), pp. 347-366.

Moore, R. P. *Handbook on Tetrazolium Testing* (Zurich, Switzerland: International Seed Testing Association, 1985). 99 pp.

Naylor, R. E. L., and M. J. Syversen. Assessment of seed vigour in Italian ryegrass. *Seed Sci. & Technol.* 16 (1988): 419-426.

Nijenstein, J. H. Effect of soil moisture content and crop rotation on cold test germination of corn (*Zea mays* L.). *J. Seed Technol.* 12 (1988): 99-106.

Perl, M. Biochemical aspects of the maturation and germination of seeds. *Adv. Seed Res. & Technol.* 10 (1987): 1-27.

Perl, M., I. Luria, and H. Gelmond. Biochemical changes in sorghum seeds affected by accelerated ageing. *J. Exp. Bot.* 29 (1978): 497-509.

Perry, D. A. Report of the vigour test committee, 1974-77. *Seed Sci. & Technol.* 6 (1978): 159-181.

Perry, D. A. The concept of seed vigour and its relevance to seed production techniques. In *Seed Production*, ed. P. D. Hebblethwaite (London, UK: Butterworths, 1980), pp. 585-591.

Perry, D. A. *Handbook of Vigour Test Methods* (Zurich, Switzerland: International Seed Testing Association, 1981). 72 pp.

Perry, D. A. Report of the vigour test committee, 1980-1983. *Seed Sci. & Technol.* 12 (1984a): 287-299.

Perry, D. A. Commentary on ISTA vigour test committee collaborative trials. *Seed Sci. & Technol.* 12 (1984b): 301-308.

Phaneendranath, B. R. Influence of amount of water in the paper towel on standard germination tests. *J. Seed Technol.* 5 (1980): 82-87.

Pieta Filho, C., and R. H. Ellis. The development of seed quality in spring barley in four environments. a. Germination and longevity. *Seed Sci. Res.* 1 (1991): 163-177.

Powell, A. A. Seed vigour and field establishment. *Adv. Seed Res. & Technol.* 11 (1988): 25-61.

Powell, A. A., and S. Matthews. Evaluation of controlled deterioration, a new vigour test for crop seeds. *Seed Sci. & Technol.* 9 (1981): 633-640.

Powell, A. A., and S. Matthews. Detection of differences in the seed vigour of seed lots of kale and swede by the controlled deterioration test. *Crop Res.* 25 (1985): 55-61.

Powell, A. A., S. Matthews, and M. deA. Oliveira. Seed quality in grain legumes. *Adv. Appl. Biol.* 10 (1984): 217-285.

Powell, A. A., R. Don, P. Haigh, G. Phillips, J. H. B. Tonkin, and O. E. Wheaton. Assessment of repeatability of the controlled deterioration vigour test both within and between laboratories. *Seed Sci. & Technol.* 12 (1984): 421-427.

Priestley, D. A. *Seed Ageing. Implications for Seed Storage and Persistence in the Soil* (Ithaca, NY: Comstock 1986). 304 pp.

Puls, E. E., and V. N. Lambeth. Chemical stimulation of germination rate in aged tomato seeds. *J. Amer. Soc. Hort. Sci.* 99 (1974): 9-12.

Ram, C., and L. E. Weisner. Glutamic acid decarboxylase activity (GADA) as an indicator of field performance of wheat. *Seed Sci. & Technol.* 16 (1988): 11-18.

Rimbawanto, A., P. Coolbear, and A. Firth. Morphological and physiological changes associated with the onset of germinability in *Pinus radiata* D. Don seed. *Seed Sci. & Technol.* 17 (1989): 399-411.

Roberts, E. H. Introduction. In *Viability of Seeds*, ed. E. H. Roberts (London, UK: Chapman and Hall, 1972), pp. 1-13.

Roberts, E. H. The control of seed quality and its relationship to crop productivity. *Proc. Australian Seed Res. Conf.* (1984): 11-25.

Roberts, E. H. Quantifying seed deterioration. In *Physiology of Seed Deterioration*, ed. M. B. McDonald and C. J. Nelson, (Madison, USA: Crop Science Society of America, 1986), pp. 101-23.

Roberts, E. H., and M. Black. Seed quality. *Seed Sci. & Technol.* 17 (1989): 175-185.

Roberts, E. H., and R. H. Ellis. The implications of the deterioration of orthodox seeds during storage for genetic resources conservation. In *Crop Genetic Resources: Conservation and Evaluation*, ed. J. H. W. Holden and J. T. Williams (London, UK: George Allen and Unwin, 1984), pp. 18-37.

Schubert, J. Vergleichuntersuchungen zür Prüfung der Excised-embryo Methode an Hund des Keim-und Tetrazolium tests bei *Fraxinus excelsior, Prunus axium* and *Pinus monticola*. *Proc. ISTA* 30 (1965): 821-827.

Schwarzenbach, H. U. *Survey on Testing Activity of ISTA Seed Testing Stations* (Zurich, Switzerland: International Seed Testing Association, 1991). Third edition, part 1.

Scott, D. J. Report of the referee testing committee 1986-1989. *Seed Sci. & Technol.* 17 (1989). Supplement 1: 107-117.

Scott, D. J. Personal Communication, Palmerston North, New Zealand (1991).

Scott, D. J., and R. C. Close. An assessment of seed factors affecting field emergence of garden pea seed lots. *Seed Sci. & Technol.* 4 (1976): 287-300.

Siddique, M. A., and P. B. Goodwin. Conductivity measurements on single seeds to predict the germinability of French beans. *Seed Sci. & Technol.* 13 (1985): 643-652.

Smith, F. G. The mechanism of the tetrazolium reaction in corn embryos. *Plant Physiol.* 27 (1952): 445-456.

Steere, W. C., W. C. Levengood, and J. M. Bondie. An electronic analyst for evaluating seed germination and vigour. *Seed Sci. & Technol.* 9: (1981): 567-576.

Steiner, A. M., and H. Fuchs. Keimfähigkeitsbestimmung und Tetrazoliumtessuchung bei mit Heibiziden und Pestiziden behandelfem Saatgut. [Germination and tetrazolium testing in seeds sprayed with herbicides or treated with pesticides]. *Seed Sci. & Technol.* 15 (1987): 707-716.

Steiner, J. J., D. F. Grabe, and M. Tulo. Single and multiple vigour tests for predicting seedling emergence of wheat. *Crop Sci.* 29 (1989): 782-896.

Styer, R. C., D. S. Cantliffe, and C. B. Hall. The relationship of ATP concentration to germination and seedling vigour of vegetable seeds stored under various conditions. *J. Amer. Soc. Hort. Sci.* 105 (1980): 298-303.

TeKrony, D. M. Seed vigour testing. *J. Seed Technol.* 8 (1983): 55-60.

TeKrony, D. M. An evaluation of the accelerated ageing test for soybeans. *AOSA Newsletter* 59 (1985): 86-96.

TeKrony, D. M. Report of AOSA seed vigour sub-committee. *AOSA Newsletter* 62 (1988): 81-84.

Thomson, J. R. *An Introduction to Seed Technology* (Glasgow UK: Leonard Hill, 1979). 252 pp.

Tomes, L. J., D. M. TeKrony, and D. B. Egli. Factors influencing the tray accelerated ageing test for soybean seed. *J. Seed Technol.* 12 (1988): 24-36.

van de Venter, H. A., and H. W. Lock. A comparison of seed vigour tests for maize (*Zea mays* L.). *South African J. Plant & Soil* 8 (1991): 1-5.

van de Venter, H. A., D. F. Grabe, and S. P. Currans. Analysis of gas exchange in the manometric GADA seed vigour test for wheat. *S. Afr. Sci.* 85 (1989): 524-525.

Wang, Y. R., and J. G. Hampton. Seed vigour and storage in "Grasslands Pawera" red clover. *Plant Varieties & Seeds* 4 (1991): 61-66.

Wellington, P. S. Germinability and its assessment. *Proc. ISTA* 30 (1965): 73-88.

Willey, R. W., and S. B. Heath. The quantitative relationships between plant population and crop yield. *Adv. Agron.* 21 (1969): 281-321.

Young, K. A. Seed quality: Tetrazolium testing. In *Annual Report, Official Seed Testing Station, 1984*, ed. J. G. Hampton and D. J. Scott (Palmerston North, New Zealand: Ministry of Agriculture and Fisheries, 1985), pp. 27-29.

Chapter 4

Production Environment and Seed Quality

David L. Dornbos, Jr.

A chronic problem facing the seed industry is the production of seeds that possess low vigor (TeKrony, Egli, and Balles, 1980). Environmental conditions during crop growth and development can strongly impact subsequent seed yield, germination ability, and vigor. While environmental stresses, such as temperature and moisture extremes, are known to reduce seed viability and vigor, physiological causes of seed quality loss are unclear.

At least four strategies can be employed to maintain yields of vigorous seeds: (1) sound agronomic management in seed production fields will minimize the severity of stress when it occurs; (2) seed production areas are spread over wide geographic regions to minimize the risk of locally unfavorable growing conditions; (3) many seed production areas are capable of being irrigated; and finally, (4) new cultivars are continually being developed with improved stability, the ability to maintain seed yield and quality in variable production environments.

In spite of these advances, the accurate prediction of mid- and long-term weather conditions is not possible, and stressful conditions will continue to occur in seed production areas. Therefore, to facilitate the continued improvement of seed production practices and new cultivars that minimize the negative effects of environment on seed yield and quality, it is essential to improve understanding of how environmental variables, particularly stress, affect the physiological processes that determine seed viability and vigor.

The overall objective of this chapter is to review the relationship between the seed production environment and subsequent seed quality. There are four specific objectives. First, to review a model that describes the sequence of events of viability and vigor acquisition by seeds during development, the attainment of maximum seed quality by physiological maturity (PM), and the subsequent loss of seed quality due to deterioration after maturity. Second, to define the impact of stress, particularly drought and high air temperature, on germination ability and vigor, and the relationship between seed quality, yield, and the yield components. Third, to suggest physiological causes of seed quality loss associated with environmental stress during seed development. Finally, to describe agronomic practices that can be exploited to maximize seed yield and quality by minimizing the deleterious effects of environmental stress. Because of constraints in the literature and the author's experience, the legumes, and soybean in particular, will be emphasized.

A PHYSIOLOGICAL MODEL OF SEED QUALITY ACQUISITION AND LOSS

Seed quality can be limited by environmental conditions both before and after PM, the stage of development at which the seed possesses its maximum mass, viability, and vigor. The acquisition and loss of seed viability and vigor is modeled diagrammatically in Figure 1. Seeds gradually attain viability and vigor during the developmental process as seed dry weight is accumulated. Viability, the least discriminating measure of seed quality, is quickly gained during seed development and strongly maintained after maturity relative to germination ability. Germination ability, in turn, is a less discriminating measure of seed quality than seed vigor, which is more slowly gained during development and more quickly lost after the onset of seed deterioration.

When immature soybean seeds were harvested at 62, 100, 143, and 167 mg seed^{-1} (PM), 67, 97, 99, and 99 percent of the seeds were viable after seven days and 6, 34, 73, and 99 percent of the seeds produced normal seedlings, respectively (Miles, TeKrony, and Egli, 1988). Immature *Phaseolus vulgaris* seeds are capable of germinating as early as 26 days past anthesis but have high percentages of abnormal seedlings and low vigor (Mariga and Copeland, 1989). Percentages of viable seed increased to a maximum at PM

FIGURE 1. Pattern of soybean seed quality increase during development and decrease during storage for desiccated seed (Miles, 1985).

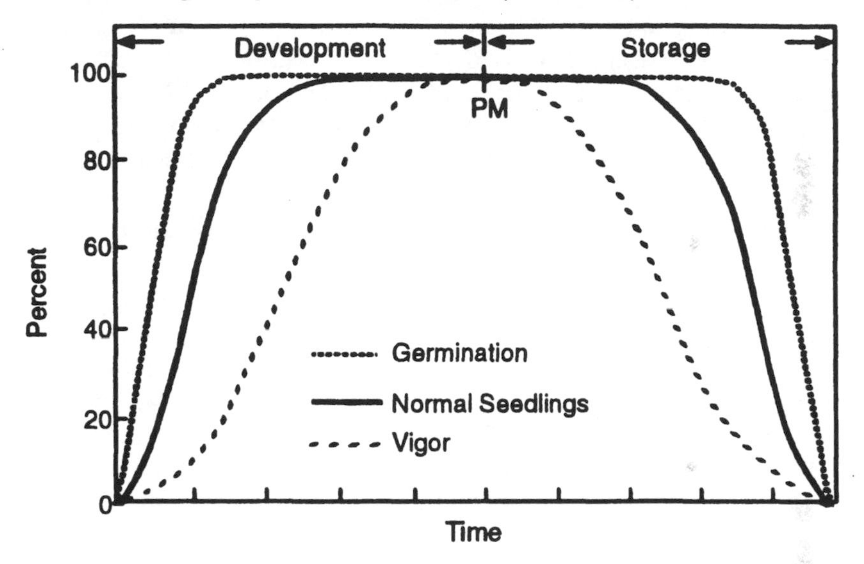

(TeKrony, Egli, and Phillips, 1980). Environmental stress occurring during seedfill limits the maximum level of seed mass, viability, and vigor attainable by PM.

Following PM, both seed viability and vigor are gradually lost as seed deterioration progresses. Seed vigor declined rapidly to levels significantly lower than those at PM within four to 39 days after harvest maturity (HM) (TeKrony, Egli, and Balles, 1980; TeKrony, Egli, and Phillips, 1980). Environmental conditions such as elevated temperatures, high humidity, and rainfall promote faster seed deterioration after HM.

ENVIRONMENTAL STRESS AND SEED QUALITY

Environmental stresses that occur during soybean development greatly reduce seed yield, but their effects on the ability of the seed to acquire maximum germination ability and vigor have received much less attention. It is important to develop a good understanding about the effect of stress during seed production on subsequent yield and quality. Only then can seed growers develop management and planning tools that will facilitate the production of adequate and commercially acceptable seed supplies in normal and stressful environments. The effect of stress on seed yield and quality depends on many factors, including its severity, duration, the stage of plant development during which it occurs, and the crop type affected.

Stress During Vegetative Development and Seed Quality

Drought and high temperature stress are normally less problematic during vegetative growth, as opposed to reproductive growth, in most production areas. Environmental stress that occasionally does develop during vegetative growth affects yield less than stresses occurring later in development. Yield losses depend upon crop type and the effect on plant population. Drought stress that occurred before R5 or beginning seedfill of soybean (Fehr and Caviness, 1977) did not affect yield (Meckel et al., 1984). In corn, however, stress during early vegetative growth can reduce yield because the developing reproductive primordia that determine kernel row number and the number of kernels per row are very sensitive to stress (Slayter, 1969). Presumably, stress during vegetative development would have very little effect on seed quality.

Stress During Reproductive Development and Seed Quality

From at least an anecdotal point of view, seed producers have often been frustrated by the fact that stressful production environments can confer seed of inferior quality. Typical seed quality results were obtained after evaluating soybean seed from 36 production environments between 1973 and 1977 (TeKrony, Egli, and Balles, 1980). At HM, seed quality declined to less than 80 percent germination in eight of the 36 environments and to inferior accelerated aging germination levels in 22 of 36 environments.

Very little information has been published regarding production environment effects on corn seed quality. Corn seedlots are submitted for extensive germination and vigor testing, however, and occasionally are rejected for sale, potentially because of quality-limiting production environments. Production of representative corn plants for seed in environmentally controlled chambers is difficult, limiting the study of environmental stress effects on seed quality.

Good plant health during reproductive development is extremely important to ensure maintenance of seed yield and quality. Plants progress through several distinct stages during reproductive development: flowering and pollination, seedfill, and maturation. Specific substages during reproductive development have been defined for crops such as corn, soybean, and sorghum. Stressful production environments can alter seed quality or yield as a function of severity, reproductive stage during which stress occurs, and genetic composition of the cultivar grown.

Effect of Stress Between Flowering and Beginning Seedfill

While stress during flowering and pollination can reduce yield and the number of seed produced, a direct effect on germination ability or vigor has not been identified. It is unclear if stress during flowering and pollination predisposes the ability of the ovule to develop into a viable and vigorous seed in a positive or negative way.

Similarly, little is understood about the interaction between the production environment and subsequent seed quality during the period between pollination and pod elongation in legumes or early caryopsis development in grasses. Although little change in seed size occurs during this period and sink strength is relatively small, biochemical activity and cell division is rapid.

Irrigation of soybean plants during flowering had no effect on the visual quality of soybean seed, whereas irrigation during pod elongation (between R3 and R4) improved the visual quality of the seed subsequently harvested (Korte et al., 1983). In contrast, other researchers found no effect of irrigation applied at R1.1, 2.5, or 3.7 on visual seed quality (Kadhem, Specht, and Williams, 1985). Seed quality was evaluated by making visual estimates, considering the proportion of seeds exhibiting wrinkling, greenishness, moldy seeds, and seeds with defective seed coats in the seedlot.

Stress During Seed Development and Seed Quality

Environmental stress during seedfill has often, but not always, been associated with reduced seed quality. Drought stress sufficiently severe to interrupt seed development caused light, shriveled soybean seed to be produced, seed likely to exhibit poor vigor (DeLouche, 1980). Irrigation during R4 improved visual seed quality while irrigation during seedfill tended to worsen visual quality (Korte et al., 1983), possibly by promoting a greater incidence of disease.

Numerous investigators have measured reductions in soybean seed germination ability and vigor because of drought stress during seedfill (Drummond, Robb, and Melville, 1983; Smiciklas et al., 1989; Dornbos, Mullen, and Shibles, 1989; Dornbos and Mullen, 1991). The most severely affected seed populations contained some small, flattened, and shrunken seeds, but many seed populations containing no abnormal physical characteristics also demonstrated reduced germination ability and vigor after scarification.

High air temperature during seed development can reduce visual seed quality, germination ability, and vigor. Seeds from soybean plants grown at 32°C were shriveled (Hesketh, Myhre, and Willey, 1973) and those from bean (*Phaseolus vulgaris*) exhibited reduced quality (Siddique and Goodwin, 1980). Soybean plants grown in pots buried in the field were removed at R5 and placed in growth chambers under a day/night temperature regime of 27/22 or 32/28°C until R8 (Keigley and Mullen, 1986). High temperature stress during seedfill increased the percentage of etched and discolored seeds in the population from 3.6 and 2.3 percent to 20.1 and 26.2 percent, respectively. These seedlots exhibited germination percentages of 73 and 56 percent, respectively, whereas control plants exhibited an 86 percent germination rate (Figure 2). Seedling axis dry weight (SADW) was reduced 38 percent from 63 mg seedling^{-1} to 39 mg seedling^{-1}. Both germination ability and SADW declined linearly as heat stress accumulated over time. In an independent study, high average daytime temperatures during the last 45 days of soybean seed growth were strongly associated with poor seedling vigor (Harris, Parker, and Johnson, 1965). Clearly, damage from excessive heat during seedfill can dramatically reduce seed quality. These reductions can be incurred early in the seedfill period.

High night temperatures are associated with low seed corn germination ability and vigor. Maximum daytime air temperature had

FIGURE 2. Germination and seedling vigor response to seed maturation under increased periods of high temperature (Keigley and Mullen, 1986).

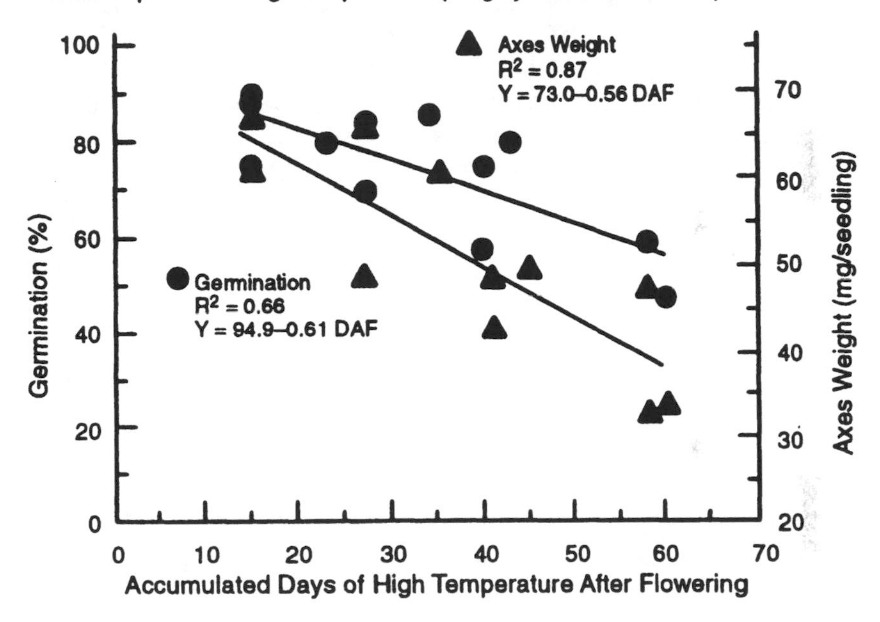

only a small effect on seed corn quality, whereas night temperature had a large effect (B. A. Martin, personal communication, Pioneer Hi-Bred International, Johnston, Iowa). A trend for consistently low germination ability and vigor of corn seed produced under irrigation in the southern seed production areas of the United States (North Carolina and Georgia) led to the closing of the involved seed production facilities. Seed produced at 30/21°C had similar appearance and weight as seed produced at 30/14 and 30/10°C, but both cold test germination and respiration rates were significantly lower (Reding, Martin, and Cerwick, 1990).

Environmental stress during seedfill has not always lead to reductions in seed visual quality, germination ability, and vigor. Germination ability and vigor tests of seed quality are progressively more discriminating estimates of seed quality than visual estimates (AOSA, 1983). Seed with good visual quality may be found to be of inferior quality when using germination or vigor tests. Irrigation during pod development and seedfill that increased yield had no effect on visual seed quality (Kadhem, Specht, and Williams, 1985). Visual seed quality estimates varied among years, however, presumably because of differences in seasonal rainfall and temperature, suggesting that weather did affect seed quality. Sandbench emergence rates of seeds from field-grown soybean plants were not reduced when water was withheld during three periods of seed development, but seed vigor was reduced when measured using the accelerated aging test (Yaklich, 1984). Soybean plants exposed to drought stress between R5 and R6 exhibited large reductions in yield, seed number, and mass, but not germination ability, accelerated aging germination, conductivity, or cold test germination in the greenhouse (Vieira, TeKrony, and Egli, 1991) or field (Vieira, TeKrony, and Egli, 1992). In contrast with earlier reports, only wrinkled and misshapen seeds were less vigorous. The authors suggested that drought stress during development had little effect on seed quality unless it was sufficiently severe to cause abnormally shaped seeds to develop.

Variability in Stress Effects on Seed Quality

These results indicate that tremendous variability exists among studies in the effects of stress during seed development on subsequent seed viability and vigor. The effects of stress during seedfill on seed quality range from no deleterious effect to 50 and 75 percent reductions in germinability and vigor, respectively. Reasons for the divergence of results are likely numerous, including the type of stress imposed and its severity, the method of stress imposition, genetic differences between the varieties tested, or the timing of stress during seed development. A major limitation has been the inability to nondestructively quantify stress intensity by measuring a physiological plant parameter.

Methods have been developed recently that allow investigators to quantify stress intensity in a nondestructive manner by measuring leaf temperature remotely with an infrared thermometer. Leaf temperatures that are elevated relative to air temperature indicate vegetation under stress (Blad and Rosenberg, 1976). The stress degree day (SDD) index was then introduced, defined as

$$SDD = A - B \text{ (summation of } SDD_i\text{)},$$

where A and B represent constants and SDD_i the midafternoon LT – AT difference summed over days i (Idso, Jackson, and Reginato, 1977). Stress degree days accurately related stress intensity to yield reductions with durum wheat (Idso, Jackson, and Reginato, 1977) and soybean (Harris, Schapaugh, and Kanemasu, 1984). Yield is negatively and linearly related to the number of SDD that accumulated during reproductive growth.

Germination and Vigor of Soybean Seed in Relation to SDD

Although it is qualitatively understood that environmental stress during seed development can reduce seed yield and quality, SDD can be utilized to investigate the quantitative relationship between stress intensity, the yield components, and seed quality. Drought stress was imposed throughout seedfill on soybean plants grown in pots in the greenhouse by differential watering, then stress intensity was quantified by accumulating SDD (the difference between leaf and air temperature) each day throughout seedfill (Dornbos, Mullen, and Shibles, 1989). The number of SDD accumulated during seedfill was larger for moderate and severe drought stress treatments, relative to control plants. Standard germination percentage decreased linearly by 6 percent in 1985 when regressed against SDD (Figure 3). A similar, but nonsignificant, trend occurred in 1986. Severe drought stress reduced SADW by 10 percent in 1985 and by 9 percent in 1986, and increased single-seed conductivity by 20 percent in 1986. Changes in conductivity measurements were not significant in 1985, but the trend was consistent with 1986 results. Changes in germination percentage, SADW, and conductivity were consistently linear when regressed against SDD, suggesting that as drought stress intensity increased, seed quality was progressively reduced.

While significant, the reductions in seed quality because of drought stress alone during seedfill were smaller than those reported earlier and in many seed production situations. Drought stress is commonly associated with high maximum temperatures in most production areas. Therefore, a similar study was conducted to identify the effect of simultaneous drought and high temperature

FIGURE 3. Standard germination, seedling axis dry weight, and conductivity of soybean seeds that were exposed to three levels of drought stress during seedfill in 1985 and 1986. Bars indicate LSD (0.05) (Dornbos, Mullen, and Shibles, 1989).

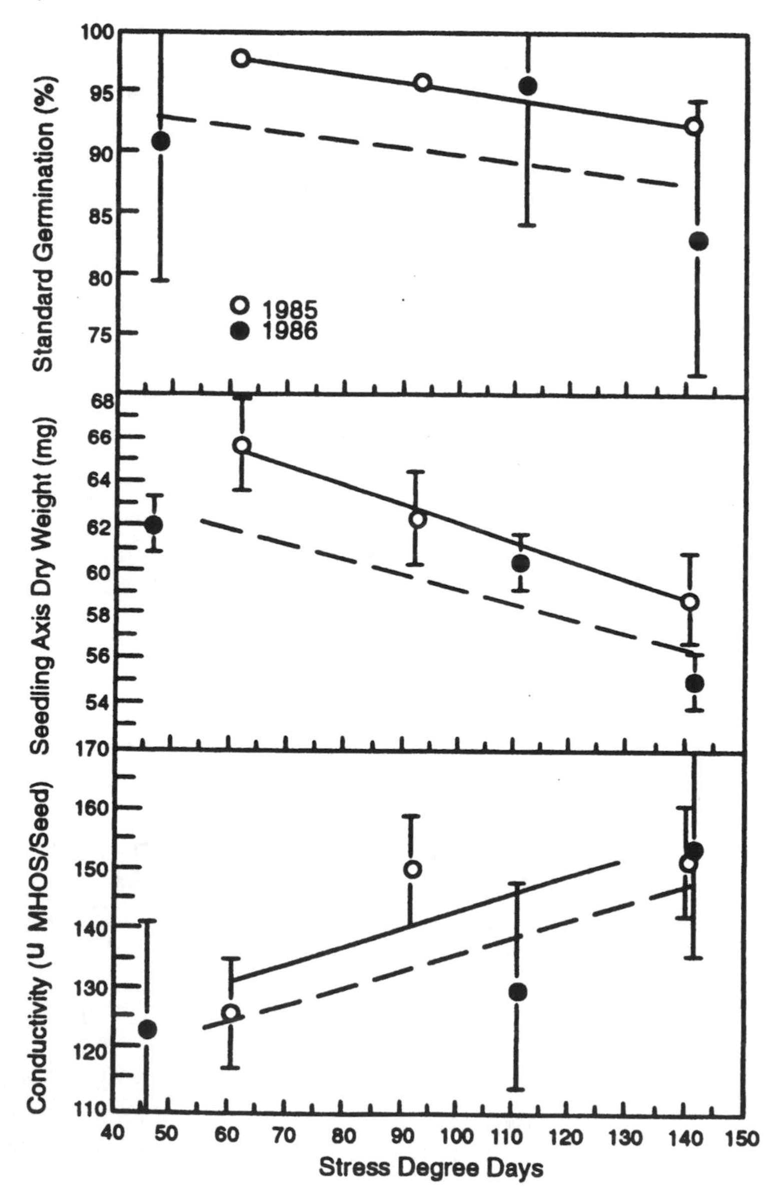

stress on seed quality. When drought occurred in conjunction with high daytime air temperature stress throughout seedfill, germination ability and vigor were reduced by a much greater extent than by drought alone (Figure 4). Again, the degree of seed quality loss varied between experiments. Germination percentage was unaffected by severe drought at 29°C in 1985, but was reduced by 9 percentage points at 27°C in 1986 (Dornbos and Mullen, 1991). Severe drought reduced the germination percentage by 26 and 4 percentage points at the high temperatures of 33 and 35°C in 1985 and 1986, respectively. Drought reduced the germination percentage linearly as a function of SDD at each air temperature.

High air temperatures during seedfill reduced the germination of control plants by 12 percentage points in 1985, but did not affect the germination percentage in 1986. While seed from well-watered plants at the control air temperature germinated at 91 and 90 percent in 1985 and 1986, respectively, the germination of seed from plants exposed to severe drought and high air temperature was reduced to 53 and 85 percent. Clearly, the combination of drought and high temperature during seedfill substantially reduced the germination percentage.

Seedling axis dry weight was measured to estimate seed vigor. Severe drought reduced SADW by 5 and 26 percent at 29 and 35°C in 1985, respectively, and by 15 and 11 percent at 27 and 33°C in 1986 (Figure 4). Drought stress again reduced SADW linearly at both optimum and high air temperatures when regressed against SDD. The high air temperature treatment reduced the SADW of well-watered plants 11 percent in 1985 and 21 percent in 1986. The SADW of seed from control plants decreased from 48 and 45 mg seedling^{-1} to 29 and 32 mg seedling^{-1} following severe drought at the high air temperature in 1985 and 1986, respectively. The combination of drought and high temperature during seedfill reduced SADW more than germination ability. The combined stresses also reduced seed quality by more than either drought or high temperature independently. Quantification of stress intensity and the subsequent effect of stress on germination ability and vigor will facilitate the comparison of results among studies and modeling efforts considering seed yield and quality in production environments.

FIGURE 4. Percent hardseed, standard germination percentage, and seedling axis dry weight of seed from Hodgson 78 soybeans exposed to control, moderate, or severe drought stress at an optimum and high air temperature of exps. I (1985) and II (1986).

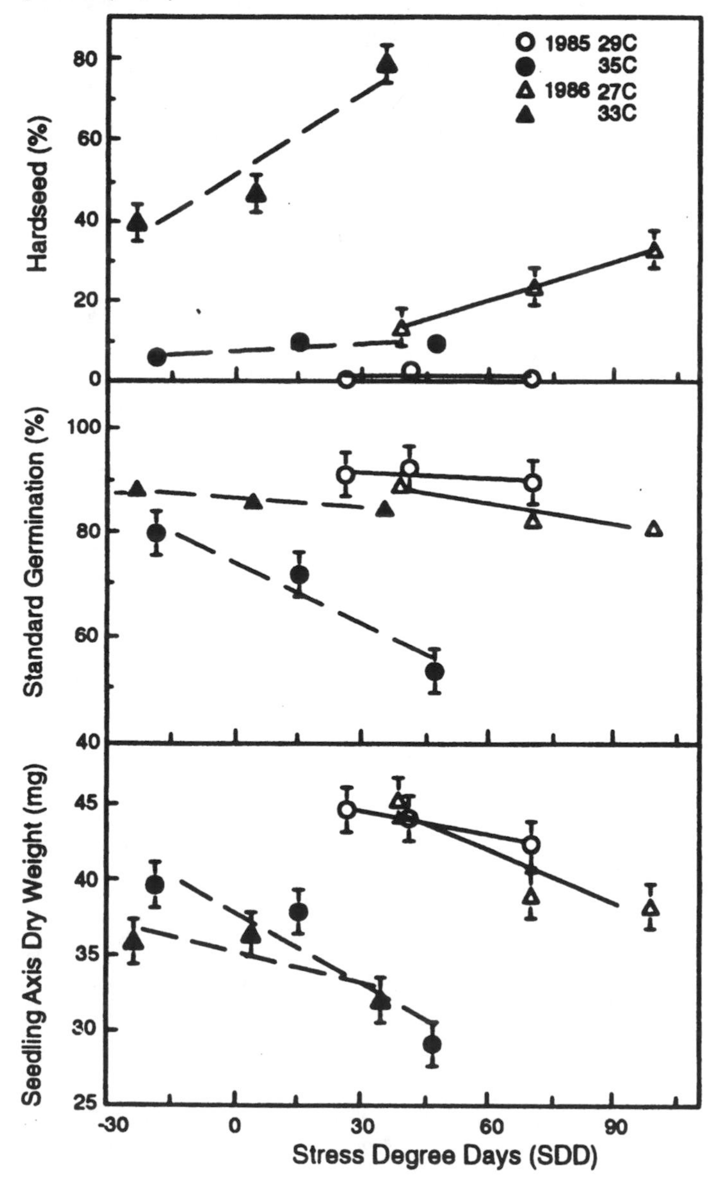

Stress and Seed Quality by Reproductive Stage

Environmental stress during seedfill has a more deleterious effect on seed quality than during vegetative or early reproductive development. Irrigation during pod development (R3-R4) improved visual seed quality, whereas seed of poorer visual quality were produced subsequent to irrigation during seed development (R5-R6) (Korte et al., 1983). In contrast, other investigators found seed quality was unaffected by irrigation at any stage of reproductive development (Kadhem, Specht, and Williams, 1985).

Drought stress imposed in the field under a rainout shelter at R2 and R4 had no effect on germination ability (Smiciklas et al., 1989). In contrast, drought stress during R5 reduced the germination percentage by 10 percentage points both years (Table 1). Stress during R6 reduced the germination percentage by 8 percentage points in 1985, but the trend was not significant in 1986. Typically abnormal seedlings were short with split hypocotyls. Seed vigor, evaluated using the seedling growth rate to measure SADW and conductivity tests, was unaffected by stress at any developmental stage. Stress during late pod or early seed development, between R4 and R5, has the largest negative effect on soybean seed quality.

Effect of Environmental Stress after Physiological Maturity

Environmental conditions between PM and HM can reduce soybean seed quality. Early planted soybeans produced seed with lower standard germination and field emergence rates because maturation, the drydown period between PM and HM, more typically occurred during hot and dry weather (Feaster, 1942; Green et al., 1965). When maturation occurred either before or after a hot and dry period, higher-quality seed was produced. Low-quality seedlots contained seed with greater proportions of green cotyledons and wrinkled seedcoats.

Seed of six cultivars produced for five years lost vigor rapidly after PM but maintained germination ability (TeKrony, Egli, and Balles, 1980). Seed was hand-harvested at PM, HM, then again at 14 and 28 days post HM. Germination percentage, maximum at PM and ranging from 69.2 to 98.0 percent, did not significantly decline for up to 28 days post HM (Table 2). Across cultivars and years,

TABLE 1. Seed quality response of soybean to the timing of drought stress averaged over the 1985 and 1986 growing seasons at Ames, IA (Smiciklas et al., 1989).

Treatment	Germination	Dry-weight seedling^{-1}	Electrolytic conductivity seed^{-1}
	--- % ----	--- mg ---	--------μA --------
Nonstressed	95.6	61.8	141
R2 stress	94.4	61.8	153
R4 stress	92.4	60.3	154
R5 stress	85.4	58.0	159
R6 stress	91.0	58.1	153
LSD (0.05)	8.9	NS	NS

Treatment	Seed-Ca concentration	Seed Ca content
	---μg g^{-1}	-- mg --
Nonstressed	1648	15.3
R2 stress	1690	14.2
R4 stress	1662	13.8
R5 stress	1305	7.3
R6 stress	1614	9.5
LSD (0.05)	72	3.4

vigor (measured as accelerated aging germination percentage), however, dropped sharply from 82 percent by PM, to 66 percent by HM, and to 41 percent by 28 days post HM. In four of five production environments, seed vigor declined to a level of less than 50 percent accelerated aging germination within one month of harvest (TeKrony, Egli, and Balles, 1980).

At least four factors can contribute to soybean seed quality decline after PM. First, alternate drying and wetting in the field led to reduced seed quality because of the rapid and differential absorption of water by different seed tissues, then subsequent deterioration (Moore, 1971). Second, hot and dry weather after PM reduced seed vigor (Green et al., 1965; TeKrony, Egli, and Balles, 1980). Third, warm and humid weather that prolonged the desiccation period between PM and HM, when seed moisture declines from 50 to 15 percent, contributed to reduced seed quality (Howell, Collins, and Sedgwick, 1959; TeKrony, Egli, and Balles, 1980; TeKrony, Egli, and Phillips, 1980). Warm and humid conditions could reduce vigor in at least two ways: through the continued respiration and consumption of readily metabolized cell contents such as sugars, or through the infection by diseases such as pod and stem blight. Pod and stem blight infection of less than 25 percent promotes the possibility of maintaining germination percentages greater than 80 percent. Germination percentage was negatively correlated with infection by pod and stem blight ($r = -0.879$) for all cultivars tested (TeKrony, Egli, and Balles, 1980). Fourth, genetic variability also exists between cultivars such as Cutler 71 and Kent for seed quality.

Following HM, the seed is essentially stored on the maternal plant in the field until harvest. Because the maternal plant has senesced by this stage of development, it plays no active role in maintaining seed health and is unresponsive to environmental variables, except for structural support. As in the maturation environment, warm and humid weather conditions can promote the activity of plant pathogens, potentially reducing seed quality. Cultivars with thick pod walls and tolerance to the pod molds are advantageous in humid environments where harvest can occur late in the growing season.

TABLE 2. Trends in soybean seed germination and vigor for Kent (1973-1977; TeKrony, Egli, and Balles, 1980).

Year	PM[a]	HM	14	28
		Standard germination (%)		
1973	98.0	98.5	84.0	93.0
1974	93.5	90.5	92.0	89.5
1975	84.0	93.8	95.5	91.2
1976-May	77.7	84.3	77.3	81.8
1976-June	91.2	82.0	86.5	82.3
1977-May	69.2	77.3	80.2	76.3
1977-June	78.7	81.7	77.5	79.2
		Vigor (accelerated-aging germination) (%)		
1973	94.0	88.0	58.0*	65.0*
1974	84.5	82.5	84.5	40.5*
1975	89.0	84.5	65.8	56.8*
1976-May	82.8	28.0*	13.0*	7.3*
1976-June	96.7	58.3*	65.0*	52.2*
1977-May	71.0	61.8	39.5*	66.3
1977-June	78.7	69.0	66.0	5.3*

*Significantly lower (P = 0.05) than at physiological maturity
a. PM = physiological maturity; HM = harvest maturity;
14 and 28 = the number of days past harvest maturity

Genetic Effect

Genetic differences exist among cultivars for the ability to acquire and maintain good seed quality in stressful environments, but the differences appear to be small in comparison with the effect of stress itself. The environmental effect on seed quality is greater than the genetic effect, accounting for the majority of variation in seed quality among years (TeKrony, 1980). Nevertheless, differences among eight cultivars for seed quality were measured following irrigation (Korte et al., 1983). Elf, Will, and Woodworth maintained low seed quality scores (superior visual quality) in both two- and three-year averages, whereas Cutler 71, Harcor, Nebsoy and Amsoy 71 exhibited higher scores indicative of lower quality. Similarly, even though irrigation had no effect on seed quality, differences existed among the 16 cultivars tested (Kadhem, Specht, and Williams, 1985). As before, Elf, Will, and Hobbitt exhibited better visual seed quality than Nebsoy and Amcor. A significant genotype and environment interaction was detected following selection of 20 cultivars for improved seed germination ability (Unander, Lambert, and Orf, 1983). Genetic variability for seed quality exists but degree of potential improvement is small in comparison with the main effect of environment.

Substantial differences also exist among corn inbreds for the ability to tolerate adverse environmental conditions during production. While certain inbreds exhibit good stability by maintaining large yields of high-quality seed in various production environments, others fail to maintain productivity and satisfactory cold test results in adverse production environments (B. A. Martin, personal communication, Pioneer Hi-Bred International, Johnston, Iowa). A test of inbred stability in 18 locations for seed vigor, measured using the cold test germination, suggested that half the variability for seed quality was due to the genetic composition of the inbreds and half to the environment.

Relationship among Stress, Yield, the Yield Components, and Seed Quality

The effect of stress during seed development on seed yield and the yield components are well known, having received more attention than the effect on germination ability and vigor. Sizable reductions in soybean seed yield are associated with reduced seed quality in some cases, whereas seed quality was preserved in others.

Yield and Seed Quality

Irrigation during late pod development and early seedfill gave the largest yield advantage, but either had no effect on visual seed quality (Kadhem, Specht, and Williams, 1985) or improved visual seed quality only after early irrigation treatments (Korte et al., 1983). Severe drought imposed on potted soybean plants in the greenhouse or the field reduced yield between 32 and 42 percent but had little effect on seed germination and vigor (Vieira, TeKrony, and Egli, 1991; Vieira, TeKrony, and Egli, 1992). While yields were significantly reduced with little effect on seed quality, stress intensity was not quantified throughout the seedfill period.

Drought stress imposed and maintained throughout seedfill, then quantified by accumulating SDD, markedly reduced soybean seed yield and, to a lesser extent, seed quality (Smiciklas et al., 1989; Dornbos, Mullen, and Shibles, 1989; Dornbos and Mullen, 1991; Table 3,4). A 47 percent reduction in soybean seed yield because of severe drought during seedfill corresponded to 5 and 12 percent reductions in germination percentage and SADW, respectively, and a 19 percent increase in conductivity. In a similar study, high air temperature during seedfill reduced soybean yield, germination ability, and SADW by 30, 7, and 14 percent, respectively, whereas severe drought was associated with 48, 11, and 16 percent losses (Dornbos and Mullen, 1991). Not surprisingly, the largest reductions in yield and seed quality resulted when drought and heat stress occurred concomitantly. A 65 percent yield loss was associated with 26 and 30 percent losses in germination ability and SADW. Although the germination and vigor of the soybean seed was reduced substantially, seed quality was conserved to a large extent relative to the dramatic loss in yield.

Components of Yield and Seed Quality

In spite of extensive effort, no consistent relationship has been found between seed mass and viability or vigor (Edwards and Hartwig, 1971; Johnson and Leudders, 1974; Wetzel, 1975). Several factors may confound the interpretation of seed mass and vigor relationships by indirectly reducing the emergence ability of large seeds. Crusting soils may affect the emergence ability of seeds as a

TABLE 3. Yield, seed number, seed mass, and reproductive period duration of soybean at three drought stress levels (Dornbos, Mullen, and Shibles, 1989).

Stress degree days	Seed yield	Pod number	Seed number	Seed mass	Reproductive period duration
no.	g plant^{-1}	--no. plant^{-1}--		mg seed^{-1}	days
			1985		
60.6	42.6	91	203	0.210	35
91.7	29.7	71	154	0.193	34
113.4	26.6	61	147	0.182	29
LSD (0.05)					
14.7	5.7	11	21	0.018	4
			1986		
46.2	49.8	103	249	0.200	44
110.8	38.4	80	195	0.197	38
141.2	20.8	52	121	0.172	33
LSD (0.05)					
13.8	7.6	10	22	0.013	5

function of cotyledon dimension as opposed to mass. Seeds with larger mass may be subject to more damage during mechanical harvest. Finally, large differences exist between cultivars for seed mass, ranging from 2,200 to 2,900 seeds per pound on average with cultivars currently marketed in the Midwest. Seed mass differences due to genetically determined small seed mass, for example, have no bearing on potential seed vigor, as opposed to seed mass differences resulting from stress in the production environment.

In spite of these considerations, some investigations have identified a relationship between seed mass and quality, whereas others have not. Even when normally round seeds were reduced in size by up to 50 percent, germination and vigor levels were not compro-

TABLE 4. Yield components of soybean plants exposed to three water stress levels at an optimum and high air temperature.

Drought Stress	Experiment I			Experiment II		
	Seed Number Per Plant	Seed Mass (mg seed^{-1})	RPD+ (days)	Seed Number Per Plant	Seed Mass (mg seed^{-1})	RPD+ (days)
		29°C			27°C	
Control	166	208	39	204	188	35
Moderate	132	192	36	183	172	30
Severe	112	164	32	148	159	29
$LSD_{0.05}$++	14	12	7	18	8	4
		35°C			33°C	
Control	131	188	37	190	154	31
Moderate	108	135	29	174	150	28
Severe	87	99	22	140	120	22
$LSD_{0.05}$++	14	12	7	18	8	4
$LSD_{0.05}$+++	11	31	18	3	14	3

+ Reproductive period duration.
++ Comparison of water-stress means within a temperature.

mised (Vieira, TeKrony, and Egli, 1992). Extreme stress that gave rise to very small, shriveled, and misshapen seed negatively affected seed vigor, but not germination ability. In the absence of abnormally shaped seeds, the researchers concluded that drought stress is unlikely to affect seed germination or vigor. When separated according to mass and density, however, soybean seeds with large mass and low density from lots of intermediate vigor exhibited the lowest percentage and slowest rate of laboratory (Hoy and Gamble, 1985) and field (Hoy and Gamble, 1987) emergence.

Soybean is capable of strongly modulating the components of yield and possibly seed quality in response to stress during different stages of seed development. Severe drought throughout seedfill reduced seed number by 40 percent and single-seed mass by 14 percent (Dornbos, Mullen, and Shibles, 1989; Table 3). Single-seed mass declined from 205 to 177 mg $seed^{-1}$ and retained its normal shape while accounting for 25 percent of the yield reduction. Seed quality was reduced by a relatively small margin.

In a related study, when severe drought or high temperature occurred independently, seed number was again reduced by a greater proportion than seed mass. When drought and high air temperature occurred together, however, seed mass was reduced by a larger proportion. Thirty percent fewer seeds were produced by severely drought-stressed plants grown at both 29 and 35°C, but seed mass was reduced 18 percent by severe drought at the normal air temperature and 35 percent at the high air temperature. The incidence of abnormally shaped seeds was negligible when heat and drought stresses were applied independently, but increased to approximately 35 percent when heat and drought occurred together. Reductions in germination ability and SADW of 5 to 11 percent and 12 to 16 percent, respectively, were associated with 14 to 18 percent lower seed mass because of drought or high air temperature. But when drought and high temperature stress were imposed jointly, germination ability, SADW, and seed mass were reduced by 26, 30, and 35 percent, respectively.

Reductions in single-seed mass by stress were significantly correlated with germination percentage (0.70*) and conductivity (–0.95**) in 1985, and with SADW (0.80**) and conductivity (–0.90**) in 1986 (Dornbos, Mullen, and Shibles, 1989). Single-seed mass was correlated with germination percentage (0.89**) in experiment I, but not in experiment II, while SADW was correlated with mass in experiment I (0.87**) and experiment II (0.92**; Dornbos and Mullen, 1991). These data support the possibility that a relationship exists between seed mass following stress during seedfill and seed quality. Soybean plants sacrificed seed number while seed mass and quality was conserved when facing moderate stress that was maintained throughout seedfill, but when stress was sufficiently severe, mass and quality declined by a much greater margin.

Seed number was reduced by a greater extent than individual seed mass when drought stress occurred at the optimum air temperatures or when well-watered plants were exposed to high air temperatures, accounting for most of the yield reduction and supporting the maintenance of good seed germination and vigor. When moderate or severe drought stress and high air temperatures occurred concomitantly, average individual seed mass of the seedlot decreased, contributing to a larger loss in yield, seed germination, and vigor. Seed germination and vigor decreased proportionately with a decrease in the average mass of seed in the seedlot. When stress becomes sufficiently severe throughout seedfill, such that the capacity of the soybean plant to compensate by reducing seed number is exceeded, then the average mass of seeds is reduced and significant reductions in germination and vigor are possible.

Population Characteristics

The average seed mass, 100-seed weight, of a seedlot may not accurately reflect the predominant individual seed mass within the lot. The distribution of individual seed mass can be altered by source-sink manipulation, cultural selection (Egli, Wiralaga, and Ramseur, 1987), and environmental stress during reproductive development (Dornbos and Mullen, 1991). Soybean plants that developed under good environmental conditions produced seedlots comprised of a large proportion of uniformly heavy seeds (Dornbos and Mullen, 1991). These seedlots contained few seeds in each of the low-mass fractions collected. As air temperature during seedfill increased (Table 5) or drought stress became more severe (Table 6), the proportion of low-mass seeds in each seedlot increased progressively. Even severely stressed plants continued to produce some heavy seeds, but the distribution of seed mass was shifted toward a greater percentage of seeds with low mass exhibiting higher incidence of flattened and shriveled seed, thereby reducing the average individual mass of the seedlot. Seed size distribution was found not to change even though average seed mass of the seedlot decreased (Egli, Wiralaga, and Ramseur, 1987). Maintaining the production of at least a few heavy seeds in a population enhances the likelihood that annual plant species will produce at least a few seeds capable of germinating the subsequent year.

Role of Conditioning

Various seed cleaning devices can be utilized to remove seed of atypical shape or density from a seedlot, thereby potentially improving the quality of the lot. Spirals are often used by soybean seed processors to remove splits and flattened seeds. Removal of the lighter fraction of material separated from soybean seedlots using a gravity separator raised the average germination of the remaining seed by 2 percentage points (Misra, 1981). To the extent that environmental stress alters seed density or shape by increasing the proportion of small, flattened, or shrunken seeds, seed conditioning methods are particularly valuable tools for removing the poorly developed seed from the seedlot and thereby improving the quality of the remaining seed.

MODELS OF SEED VIABILITY AND VIGOR LOSS BECAUSE OF STRESS DURING SEEDFILL

Seed of low viability and vigor may be harvested for two reasons: stress during reproductive growth can limit the ability of seeds to develop properly, or conditions can cause seeds to deteriorate following PM (Figure 1). Mechanisms explaining seed quality loss after PM have been postulated, but mechanisms limiting the acquisition of adequate seed viability and vigor during development are incompletely understood.

Adequate seed quality may fail to be attained prior to PM in stressful environments for at least two reasons. First, insufficient quantities of chemical compounds (protein, lipid, and carbohydrate) may be stored, thereby failing to provide the energy and biochemical building blocks necessary to drive germination. Second, environmental stress during seedfill may promote the development of an impaired physiological apparatus with which to drive the germination process.

Physiological Function of Soybean Leaflets

Environmental stress dramatically alters the physiological functioning of the maternal plant and developing seed in ways directly

TABLE 5. Distribution of soybean seed number and seed mass after exposure to high temperature stress throughout seed fill.

Year	Air Temperature (°c)	Sieve size (diameter, mm) 4.8	4.4	4.0	3.6	3.2	<3.2
1985		Seed Number Per Plant					
	29	91	6	2	1	1	2
	35	42	10	6	5	4	9
	$LSD_{0.05}$	28	4	4	1	3	3
		Seed Mass (mg seed^{-1})					
	29	195	119	96	77	71	31
	35	172	132	112	94	81	50
	$LSD_{0.05}$	11	19	8	6	13	31
1986		Seed Number Per Plant					
	27	83	9	3	1	1	0
	33	53	16	8	6	3	4
	$LSD_{0.05}$	19	7	3	2	1	4
		Seed Mass (mg seed^{-1})					
	27	185	124	99	79	61	46
	33	166	127	107	90	73	50
	$LSD_{0.05}$	37	17	21	28	12	13

and indirectly related to yield and seed quality loss. The effect of environmental stress, particularly drought and heat, on the physiology of the maternal plant, and therefore yield, are well known. A 68 percent increase in leaf resistance, and 44 and 71 percent decreases in transpiration and photosynthetic rates, respectively, because of severe drought during seedfill corresponded to 58, 14, 6, and 9 percent reductions in yield, individual seed mass, germination ability, and SADW, respectively, and a 20 percent increase in conductivity (Dornbos, Mullen, and Shibles, 1989). Leaflet productivity, mass, and seed quality measurements were linearly related to increased stress intensity when regressed against SDD. Impairment of the physiological functioning of soybean leaflets by

TABLE 6. Distribution of soybean seed number and seed mass after exposure to drought stress throughout seed fill.

Year	Drought Stress	Sieve size (diameter, mm) 4.8	4.4	4.0	3.6	3.2	<3.2
1985		Seed Number Per Plant					
	Control	69	3	1	1	1	0
	Moderate	40	7	4	3	2	3
	Severe	23	5	3	3	3	7
	$LSD_{0.05}$	6	2	2	1	1	2
		Seed Mass (mg $seed^{-1}$)					
	Control	201	124	101	71	75	45
	Moderate	183	127	105	95	76	37
	Severe	166	125	106	90	77	40
	$LSD_{0.05}$	10	6	6	18	10	23
1986		Seed Number Per Plant					
	Control	54	7	2	1	1	1
	Moderate	54	10	4	2	1	1
	Severe	27	9	5	3	2	2
	$LSD_{0.05}$	7	1	1	1	1	1
		Seed Mass (mg $seed^{-1}$)					
	Control	180	127	105	85	69	48
	Moderate	183	124	99	80	65	49
	Severe	165	127	106	89	68	48
	$LSD_{0.05}$	12	8	10	13	10	9

environmental stress during seedfill may inhibit the ability of the plant to supply the developing seed with the necessary assimilate required to synthesize needed storage compounds and thereby drive the germination process. Protein, oil, and oligosaccharide content of developing soybean seeds increase rapidly between R5 and R7 (Dornbos and McDonald, 1986). Low single-seed mass because of environmental stress during seedfill may reflect inadequate levels of the stored compounds needed to drive the germination process.

Membrane Lipids of Seeds

Soybean seed membrane lipid composition is strongly modulated by environmental conditions during development. Larger quantities of total phospholipid were isolated from soybean seeds that developed under severe drought and high temperature stress (Dornbos, Mullen, and Hammond, 1989). Phospholipid class composition and the fatty acid composition of each phospholipid class was also altered by both drought and high air temperatures during seedfill. Drought increased the proportion of phosphatidylcholine (PC) at 27 and 33°C, but reduced phosphatidylethanolamine (PE) at 27°C and phosphatidylinositol (PI) at 33°C. Both drought and high temperature resulted in larger proportions of saturated (16:0 and 18:0) and monounsaturated (18:1) fatty acids, and smaller proportions of the polyunsaturated (18:2 and 18:3) fatty acids to be deposited in PC and PE. The activity of the enzyme oleic acid desaturase from flax is inhibited by high temperature during development and contributed to a reduction in the proportion of polyunsaturated fatty acids (Green, 1986). Similar enzymes may be operative in soybean seeds, facilitating the regulation of membrane lipid composition.

Changes in seed phospholipid class, and the fatty acid composition of each class, because of drought and high temperature stress during seedfill, were opposite to, and therefore consistent with, those reported for several plant types other than soybean following exposure to low temperature stress. When the temperature was reduced during vegetative growth, the phospholipid content of chilling-tolerant plant species doubled and the rate of fatty acid desaturation increased (Roughan, 1985; Thompson, 1990), thereby maintaining fluid membranes and growth (Lyons, Raison, and Steponkus, 1979; Pike, 1982). Changes in growth temperature are accompanied by changes in the composition of the phospholipid class and the fatty acid composition of each class, thereby maintaining optimum membrane fluidity and functionality at the new temperature (Sinensky, 1974). The composition of membrane lipids, the phospholipids, is dynamically controlled by environmental factors such as drought and temperature. A stressful production environment may limit the ability of the developing seed to synthesize and organize the appropriate membrane lipids needed to drive the germination process.

The deposition of a larger proportion of saturated fatty acids in PC and PE after exposure to a high temperature or drought stress during seedfill may represent an adjustment by developing soybean seeds to maintain optimum membrane fluidity. Although this lipid composition would represent an advantage for the developing seed at the high temperature or under severe drought stress, it may represent a disadvantage in a typical germination environment where the seedbed is typically cool and wet. During imbibition, the membranes of seeds adapted to the high air temperature may exhibit excessive permeability to cell solutes, contributing to reduced seed vigor relative to membranes with lipids adapted to less stressful environments during seed development.

Nutrient Imbalances

Elemental deficiencies and imbalances following development in stressful environments may be a causal factor of poor seed quality. Drought stress during R5 and R6 reduced the calcium content of soybean seed (Smiciklas et al., 1989; Table 1). Calcium content was positively correlated with soybean seed germination ability, while phosphorus, iron, and zinc were negatively correlated with germination percentage alone. Calcium-deficient seeds exhibited elevated electrolytic conductivity in this study, suggesting that membrane integrity was impaired. Calcium plays an important role in membrane integrity (Powell, 1986). The primary effect of calcium deficiency in plant cells is the loss of membrane integrity (Hecht-Buchholz, 1979). Potassium has also been implicated with the maintenance of good seed quality as significantly lower levels of diseased seed were produced when levels of potassium in the soil increased (Crittenden and Svec, 1974).

AGRONOMIC PRACTICES AND SEED PRODUCTION

Agronomic practices utilized to maximize grain production are essentially the same as those used to maximize seed production. It follows that practices selected to minimize stress intensity and spread production risk are similar for grain and seed production. Common climatological criteria that seed growers consider include: heat unit requirements until physiological maturity, day length, lack of extreme temperatures, irrigation availability, and quantities of irrigation water available.

Differences do exist regarding grain and seed production, however. Seed producers must consider more strongly those factors which deliver maximum viability and vigor by PM, and those which maintain seed quality after PM. Avoidance of environmental stress during the growing season cannot be planned because of the difficulty in forecasting long-term weather trends, but risk can be minimized by spreading seed production fields over a wide geographic region within their adapted area and by maintaining sound agronomic practices in seed production fields.

In addition to considering environmental characteristics alone, sound agronomic practices such as choice of suitable tillage practices, plant population, planting date, maintenance of high fertility, and good weed and insect control are important to minimize the impact of environmental stress on plant growth, should stress occur. The optimum plant population in a seed production field varies as a function of moisture availability (water-holding capacity of the soil), irrigation capability, genetic composition, and soil tilth. Growers strive to maximize the plant population and therefore yield potential while minimizing the risk of drought stress, should moisture become limiting during crop development. Many investigators have considered plant density and corn yield relationships (Craig, 1977). For corn seed production, a range of 22,000 to 26,000 plants per acre is typical. To maximize the amount of moisture and sunlight available to female plants in corn production fields, male plants are often destroyed after pollination is complete.

Phosphorus and potassium should be maintained at high levels and in balance with nitrogen levels to maximize efficiency of nutrient use. Phosphorus and potassium promote root development and therefore support the maintenance of good plant water relations. Efforts to maintain good soil tilth by minimizing compaction and preventing the formation of layers in the soil profile that limit downward root growth and water movement further serve to minimize the duration and severity of drought stress during the growing season. Because corn and sorghum inbred plants do not possess the vigor of hybrid plants, they are less able to forage for soil nutrients and water, and therefore show a greater response to high fertility levels and good soil tilth.

Weeds compete very aggressively with crop plants for moisture. Weeds may use substantial amounts of moisture during years in which it is limiting, thus exacerbating stress intensity and reducing productivity. Similarly, insects, particularly root-feeders such as rootworms and wireworms, and nematodes deter root growth, limiting the ability of the plant to forage for adequate moisture. Good management practices will effectively maintain good plant health and minimize the susceptibility of the plant to stress during crop growth and development.

Seed is harvested as soon as possible after PM. Care must be taken to harvest the seed as soon as feasible because environmental conditions that promote slow drying and the growth of pathogens frequently occur during the fall. Planting date strongly impacts seed quality because of the deleterious environmental conditions encountered during seedfill (Green et al., 1965). Early planted seed matured early in the growing season when the typically hot and dry weather during seed maturation caused seed of low quality to be produced. In contrast, late-planted seed matured late when cool weather conditions were conducive to better seed quality. Additional planning is also required when producing hybrid seed such as corn or sorghum because two inbreds must be interplanted within a field, possibly at different times if flowering dates do not coincide. Poor weather during the planting split can interrupt targeted planting dates.

Seed corn is harvested on the ear at 30 to 35 percent moisture. Soybean is combined conventionally as for grain, but between 15 and 17 percent moisture and at lower cylinder speeds. Mechanical damage to soybean seed becomes excessive if the moisture content of the seed is 12 percent or less (DeLouche and Andrews, 1964). Bruising damage occurred if seed was combined at moisture contents of 14 percent or more, suggesting that the optimum range for combining soybean seed to minimize mechanical damage varied between 12.5 and 14.5 percent (DeLouche, 1974). Strangely, drought stress enhanced the cold test results of inbred corn seed in a one-year study, presumably because stress caused the plant to mature prematurely and the seed to be drier at harvest (B. A. Martin, personal communication, Pioneer Hi-Bred International, Johnston, Iowa). Seed from unstressed control plants with a higher moisture content were mechanically bruised during harvest, resulting in a lower cold test.

Care must be taken to minimize mechanical damage to the seed to ensure that the maximum quality in the field is maintained during harvest operations and storage. Combined cylinder speeds either too fast or seed that is too dry will result in increased mechanical damage to the seed in the form of kernel breakage (Matlock, 1953; Green, Pinnell, and Cavanaugh, 1966). The vigor of wheat seed threshed at various moisture levels and three combine cylinder speeds (450, 850, and 1250 rpm) was lowest and highest for seed threshed at 1250 and 450 rpm, respectively (Moes, Entz, and Stobbe, 1990). Maximum yield of vigorous seed was obtained by threshing relatively low moisture content wheat at a medium cylinder speed. Breakage susceptibility in corn is an inherited trait that is also influenced by crop management (Vyn and Moes, 1988). Kernel breakage can be reduced by proper choice of hybrids, by not planting at too high of a plant density, drying at low air temperatures, and harvesting at a sufficiently low grain moisture content.

Factors in addition to seed moisture and cylinder speed may also be important to maintain a high level of seed quality. Seed with a higher incidence of seed coat etching, a phenomenon associated with environmental stress during seedfill (Keigley and Mullen, 1986), is more susceptible to mechanical damage during harvest and reduced seed viability (Burchett et al., 1985). Seedlots of etched seeds generally were found to have 18 percent lower germination and were more susceptible to damage from low drop heights than were non-etched seedlots.

REFERENCES

AOSA (Association of Official Seed Analysts). Seed vigor testing handbook. *Assoc. Official Seed Analysts* 32 (1983).

Blad, B. L., and N. J. Rosenberg. Measurement of crop temperature by leaf thermocouple, infrared thermometry, and remotely sensed thermal imagery. *Agron. J.* 68 (1976): 635-641.

Burchett, C. A., W. T. Schapaugh, Jr., C. B. Overly, and T. L. Walter. Influence of etched seed coats and environmental conditions on seed quality. *Crop Sci.* 25 (1985): 655-660.

Craig, W. F. Production of hybrid corn seed. In *Corn and Corn Improvement*, 2d ed., ed. G. F. Sprague (Madison: American Society of Agronomy, 1977), pp. 671-719.

Crittenden, H. W., and L. V. Svec. Effect of potassium on the incidence of *Diaporthe sojae* in soybeans. *Agron. J.* 66 (1974): 696-697.

DeLouche, J. C. Maintaining soybean seed quality. In *Soybean Production, Marketing, and Use. Bull.* Y-19 (Muscle Shoals: TVA, 1974), pp. 46-61.

DeLouche, J. C. Environmental effects on seed development and seed quality. *HortScience* 15 (1980): 775-779.

DeLouche, J. C., and C. H. Andrews. Tests show how injury lowers quality of seed. *Miss. Farm Res.* 27 (1964): 1-8.

Dornbos, D. L., Jr., and M. B. McDonald, Jr. Mass and composition of developing soybean seeds at five reproductive growth stages. *Crop Sci.* 26 (1986): 624-630.

Dornbos, D. L., Jr., and R. E. Mullen. Influence of stress during soybean seed fill on seed weight, germination, and seedling growth rate. *Can. J. Plant Sci.* 71 (1991): 373-383.

Dornbos, D. L., Jr., R. E. Mullen, and E. G. Hammond. Phospholipids of environmentally stressed soybean seeds. *J. Amer. Oil Chem. Soc.* 66 (1989): 1371-1373.

Dornbos, D. L., Jr., R. E. Mullen, and R. M. Shibles. Drought stress effects during seed fill on soybean seed germination and vigor. *Crop Sci.* 29 (1989): 476-480.

Drummond, E. A., J. L. Robb, and D. R. Melville. Effect of irrigation on soybean seed quality. *La. Agric.* 26 (1983): 9.

Edwards, C. J., and E. E. Hartwig. Effect of seed size upon rate of germination in soybean. *Agron. J.* 63 (1971): 429-430.

Egli, D. B., R. A. Wiralaga, and E. L. Ramseur. Variation in seed size of soybean. *Agron. J.* 79 (1987): 463-467.

Feaster, C. B. Influence of planting date on yield and other characteristics of soybean seeds grown in south-east Missouri. *Agron. J.* 41 (1942): 57-62.

Fehr, W. R., and C. E. Caviness. Stages of soybean development. *Iowa Agric. Home Econ. Exp. Stn., Iowa Coop. Ext. Serv. Spec. Rep.* 80 (1977).

Green, A. G. Effect of temperature during seed maturation on the oil composition of low-linolenic genotypes of flax. *Crop Sci.* 26 (1986): 961-965.

Green, D. E., E. L. Pinnell, and L. E. Cavanaugh. Effect of seed moisture content, field weathering, and combine cylinder speed on soybean quality. *Crop Sci.* 6 (1966): 7-10.

Green, D. E., E. L. Pinnell, L. E. Cavanaugh, and L. F. Williams. Effect of planting date and maturity date on soybean seed quality. *Agron. J.* 57 (1965): 165-168.

Harris, D. S., W. T. Schapaugh, Jr., and E. T. Kanemasu. Genetic diversity in soybeans for leaf canopy temperature and the association of leaf canopy temperature and yield. *Crop Sci.* 24 (1984): 839-842.

Harris, H. B., M. B. Parker, and B. J. Johnson. Influence of molybdenum content on soybean seed and other factors associated with seed source on progeny response to applied molybdenum. *Agron. J.* 57 (1965): 397-399.

Hecht-Buchholz, C. Calcium deficiency and plant ultrastructure. *Commun. Soil Sci. Plant Anal.* 10 (1979): 67-81.

Hesketh, J. D., D. L. Myhre, and C. R. Willey. Temperature control of time intervals between vegetative and reproductive events of soybeans. *Crop Sci.* 13 (1973): 250-254.

Howell, R. W., F. I. Collins, and V. E. Sedgwick. Respiration of soybean seeds as related to weathering losses during ripening. *Agron. J.* 51 (1959): 677-679.

Hoy, D. J., and E. E. Gamble. The effects of seed size and seed density on germination and vigor in soybean, *Glycine max*, (L.) Merr. *Can. J. Plant Sci.* 65 (1985): 1-8.

Hoy, D. J., and E. E. Gamble. Field performance in soybean with seeds of differing size and density. *Crop Sci.* 27 (1987): 121-127.

Idso, S. B., R. D. Jackson, and R. J. Reginato. Remote-sensing of crop yields. *Science* 196 (1977): 19-25.

Johnson, D. R., and V. D. Leudders. Effects of planted seed size on emergence and yield of soybeans (*Glycine max* (L.) Merr.). *Agron. J.* 66 (1974): 117-118.

Kadhem, F. A., J. E. Specht, and J. H. Williams. Soybean irrigation serially timed during stages R1 to R6. I. Agronomic responses. *Agron. J.* 77 (1985): 291-298.

Keigley, P. J., and R. E. Mullen. Changes in soybean seed quality from high temperature during seed fill and maturation. *Crop Sci.* 26 (1986): 1212-1216.

Korte, L. L., J. H. Williams, J. E. Specht, and R. C. Sorensen. Irrigation of soybean genotypes during reproductive ontogeny. I. Agronomic responses. *Crop Sci.* 23 (1983): 521-527.

Lyons, J. M., J. K. Raison, and P. L. Steponkus. The plant membrane in response to low temperature: An overview. In *Low Temperature Stress in Crop Plants*, eds. J. M. Lyons, D. Graham, and J. K. Raison (New York: Academic Press, Inc., 1979).

Mariga, I. K., and L. O. Copeland. Seed and pod development in dry beans (*Phaseolus vulgaris* L.) and their effect on seed quality. *Amer. Soc. Agron. Abstr.* (Madison: WI, 1989): 151.

Matlock, R. S. Influence of seed coat injury on the germination of soybean. *Amer. Soc. Agron. Abstr.* (Madison: WI, 1953): 117.

Meckel, L., D. B. Egli, R. E. Phillips, D. Radcliffe, and J. E. Leggett. Effect of moisture stress on seed growth in soybeans. *Agron. J.* 76 (1984): 647-650.

Miles, D. F. Effect of the stage of development and the desiccation environment on soybean seed quality and respiration during germination. PhD Dissertation, Univ. of Kentucky (1985), p. 98.

Miles, D. F., D. M. TeKrony, and D. B. Egli. Changes in viability, germination, and respiration of freshly harvested soybean seed during development. *Crop Sci.* 28 (1988): 700-704.

Misra, M. K. Seed quality loss to soybeans during conditioning. *Trans. ASAE* 28 (1981): 576-579.

Moes, J., M. H. Entz, and E. H. Stobbe. Wheat seed vigor in relation to harvest moisture content and combine cylinder speed. *Amer. Soc. Agron. Abstr.* (Madison: WI, 1990): 166-167.

Moore, R. P. Mechanisms of water damage to mature soybean seed. *Proc. AOSA* 61 (1971): 112-118.

Pike, C. S. Membrane lipid physical properties in annuals grown under contrasting thermal regimes. *Plant Physiol.* 70 (1982): 1764-1766.

Powell, A. A. Cell membranes and seed leachate conductivity in relation to the quality of seed for sowing. *J. Seed Technol.* 10 (1986): 81-100.

Reding, L. D., B. A. Martin, and C. F. Cerwick. Vigor, metabolism and composition of maize seed can be affected by environmental temperature during growth and development. *Amer. Soc. Agron. Abstr.* (Madison: WI, 1990): 167.

Roughan, P. G. Physphatidylglycerol and chilling sensitivity in plants. *Plant Physiol.* 77 (1985): 740-746.

Siddique, M. A., and P. B. Goodwin. Seed vigor in bean (*Phaseolus vulgaris* L. cv. Apollo) as influenced by temperature and water regime during development and maturation. *J. Exp. Bot.* 31 (1980): 313-323.

Sinensky, M. Homeoviscous adaptation: A homeostatic process that regulates the viscosity of membrane lipids in *Escherichia coli*. *Proc. Nat. Acad. Sci. USA* 72 (1974): 1649-1653.

Slayter, R. O. Physiological significance of internal water relations to crop yield. In *Physiological Aspects of Crop Yield*, ed. J. D. Eastin, (Madison: Amer. Soc. Agron., 1969), pp. 53-83.

Smiciklas, K. D., R. E. Mullen, R. E. Carlson, and A. D. Knapp. Drought-induced stress effect on soybean seed calcium and quality. *Crop Sci.* 29 (1989): 1519-1523.

TeKrony, D. M. Environmental influences on soybean seed quality during production. In *Proc. Fourth Annu. Seed Tech. Conf.*, Iowa St. Univ. (Ames: Iowa St. Univ., 1980), pp. 51-67.

TeKrony, D. M., D. B. Egli, and J. Balles. The effects of the field production environment on soybean seed quality. In *Seed Production*, ed. P. D. Hebblethwaite (London: Butterworth and Co., Ltd., 1980).

TeKrony, D. M., D. B. Egli, and A. D. Phillips. The effect of field weathering on the viability and vigor of soybean seed. *Agron. J.* 72 (1980): 749-753.

Thompson, G. A., Jr. The regulation of membrane lipid metabolism. (Boca Raton: CRC Press, Inc., 1990).

Unander, D. W., J. W. Lambert, and J. H. Orf. Cool temperature soybean germination: Genetic and environmental components. *Amer. Soc. Agron. Abstr.* (Madison: WI, 1983), pp. 83-84.

Vieira, R. D., D. M. TeKrony, and D. B. Egli. Effect of stress on soybean seed germination and vigor. *J. Seed Technol.* 16 (1991): 12-21.

Vieira, R. D., D. M. TeKrony, and D. B. Egli. Effect of drought and defoliation stress in the field on soybean seed germination and vigor. *Crop Sci.* 32 (1992): 471-475.

Vyn, T. J., and J. Moes. Breakage susceptibility of corn kernels in relation to crop management under long growing season conditions. *Agron. J.* 80 (1988): 915-920.

Wetzel, C. T. Soybean seed size and plant performance. *Proc. 23rd Short Course for Seedsmen, Mississippi State Univ.* 17 (1975): 95-102.

Yaklich, R. W. Moisture stress and soybean seed quality. *J. Seed Technol.* 9 (1984): 60-67.

Chapter 5

Seed Quality and Microorganisms

Martin M. Kulik

A high germination percentage and vigor level are two very important characteristics of high-quality seeds (Thomson, 1979). Unfortunately for agriculture and horticulture, seed germination and vigor can be markedly reduced by a number of factors, including the action of microorganisms (i.e., bacteria and fungi), viruses, and nematodes. The first part of this chapter deals with seed storage fungi, and includes information on the damage that they cause, the fungi involved, the means by which they are detected and identified, and methods for controlling them. The second part covers seedborne plant disease agents important in agriculture and horticulture, and provides examples of some of the economically important diseases caused by the various types of plant disease agents, the means by which these agents are detected and identified, and methods for controlling them. At the end of the chapter, there is an annotated bibliography of publications that provides the reader with additional information on all aspects of seed storage fungi and seedborne plant disease agents.

SEED STORAGE FUNGI

Etiology of Deterioration and the Causal Fungi

Seeds as they come from the production field may be carrying various types of microorganisms, including actinomycetes, bacteria, and fungi. The first two classes of microorganisms are usually not a problem in stored seeds because of their inability to grow on substrates with relatively low moisture levels.

Seedborne fungi belong basically to one of two categories, field fungi or storage fungi (Christensen and Kaufmann, 1965). The former can be either saprophytes that cause no injury to the seed, or parasites that can infect seeds during their development as well as attack seedlings and older plants. Field fungi usually remain quiescent during seed storage. The storage fungi, as their name implies, affect stored seeds because these fungi are able to grow under relatively dry conditions in which the field fungi cannot grow. In fact, many of the storage fungi grow best under relatively dry conditions. Some of them can even invade stored seeds whose moisture contents are in equilibrium with a relative humidity as low as 65 percent (Christensen, 1957; Kulik and Hanlin, 1968).

Storage fungi have been reported to invade and destroy seeds of virtually all crop plants, including cereals (Christensen and Kaufmann, 1969), cocoa beans (*Theobroma cacao* L. subsp. *cacao*) (Bunting, 1930), cotton (*Gossypium hirsutum* L.) (Arndt, 1946), grass (Kulik and Justice, 1967), peas (*Pisum sativum* L. subsp. *sativum*) (Fields and King, 1962), peanuts (*Arachis hypogaea* L.) (Diener, 1958), and soybeans (*Glycine max* [L.] Merr.) (Milner and Geddes, 1946). These fungi can attack almost any kind of seed under unfavorable storage conditions; in fact, they can grow on most organic materials. Invasion of seeds by storage fungi can result in embryo destruction and loss of viability, an increase in free fatty acids, a decrease in nonreducing sugars, development of musty odors, and discoloration. Deterioration can occur in a matter of a few days when seeds are stored under unfavorable conditions (Semeniuk, 1954).

Seed storage fungi are principally species of *Aspergillus* and *Penicillium*, which commonly occur throughout the world. Their spores are usually present in large numbers in the air and on surfaces in seed storage areas. However, they also may be present on a few seeds in a given lot as they come from the field. Sauer, Storey, and Walker (1984) tested over 2,500 grain samples from farm storage bins in 27 states and found storage fungi in 37 percent of the wheat (*Triticum* spp.) and 87 percent of the corn (*Zea mays* L. subsp. *mays*) samples. These fungi can invade and destroy seeds over a wide range of temperature (4 to 45°C) and relative humidity (65 to 100 percent) (Christensen and Kaufmann, 1969). Their activity is largely determined by the physical condition, vitality, and moisture content of the seed and the ambient temperature and relative humidity of the storage area. Consequently, the population of storage fungi reflects the kind of postharvest handling, conditioning, and storage environment of a given seedlot.

Much of our knowledge of seed storage fungi has come from studies of their activities in certain stored cereal grains. Of the many *Aspergillus* species that have been identified, only a small number have been reported to attack stored cereals (Christensen, 1982). *Aspergillus amstelodami* (Mangin) Thom & Church, *A. chevalieri* (Mangin) Thom & Church, *A. repens* de Bary, *A. restrictus* Smith, and *A. ruber* (Konig, Spieckermann, & Bremer) Thom & Church grow in seeds with a moisture content of 13.2 to 15 percent (wet weight basis) (Christensen, 1957). The major fungi found in cereal seeds above 15 percent moisture are *A. candidus* Link, *A. flavus* Link, *A. ochraceus* Wilhelm, *A. tamarii* Kita, and *A. versicolor* (Vuill.) Tiraboschi. *Penicillium* species, although not as common as *Aspergillus* species, are sometimes found in cereal seeds, particularly in lots with a moisture content above 16 percent and stored at relatively low temperatures (Christensen and Kaufmann, 1965).

The rapid growth of fungi in stored seeds can produce so-called hot spots caused by seed heating, which is a response to fungal infection. Using seeds with an 18 percent moisture content, Gilman and Barron (1930) found in laboratory tests that *A. flavus*, *A. fumigatus* Fresenius, and *A. niger* V. Tieghem raised the temperature of barley (*Hordeum vulgare* L.), oats (*Avena sativa* L.), and wheat from 17 to 43°C. These results were confirmed and extended by Christensen and Gordon (1948). Mead, Russell, and Ledingham (1942) showed that the rise in temperature can be even higher if the seedlot contains a high proportion of broken kernels and cracked seed coats. Similar hot spots can occur in bulked seeds containing pockets of moist seeds, as the natural insulating property of the surrounding drier seeds prevents rapid dissipation of the heat generated by the respiration of moist seeds and the associated storage fungi.

Even in seeds having a low moisture content, pockets of moist seeds can arise due to roof leaks, insect activity, and moisture translocation when temperature gradients within the seed mass are allowed to occur. The fungi associated with hot spots in seeds stored on the farm were studied by Wallace and Sinha (1962). They observed that hot spots may occur at any location in a storage bin and that temperatures can increase up to 53°C during the winter. Field fungi such as *Alternaria* species were not common in the heated seeds, although they were present in many of the viable seeds that had not undergone heating. The heated seeds were seen to be infected chiefly by storage fungi, mainly species of *Penicillium* and *Aspergillus*, especially *A. flavus*, *A. fumigatus*, and *A. versicolor*. Species of *Absidia* were also observed frequently. Christensen stated that "heating is likely to be the final and violent effect of mold invasion of the seed, not an indication of beginning deterioration" (1957), as was commonly thought at that time.

Sinha and Wallace (1965) investigated the succession of fungi and actinomycete populations associated with progressive stages in the heating process of an artificially caused fungus hot spot in stored wheat. The growth of *Penicillium cyclopium* Westling and *P. funiculosum* Thom in a moist pocket of grain stored at −5 to 8°C for four months began the heating process. A maximum temperature of 64°C was reached in this particular hot spot. The following ecological succession of microorganisms, which often overlapped, was observed: *P. cyclopium*, *P. funiculosum*, *Aspergillus flavus*, *A. versicolor*, *Absidia* species, and *Streptomyces* species.

Several investigators have studied the mycoflora of freshly harvested and stored peanuts (*Arachis hypogaea* L.). Hanlin (1966) examined the internally borne mycoflora of Georgia-grown peanuts. *Fusarium* species were found in 20.7 percent, *Penicillium* species in 18.5 percent, and *Aspergillus* species in 7.5 percent of the seeds at the time that they were harvested. Jackson (1965) used dilution methods to investigate the mycoflora of soil that had adhered to dried peanut pods. Relatively small numbers of *A. flavus*, *A. niger*, *A. terreus* Thom, *Rhizopus* species, and *Sclerotium bataticola* Taub. were found. *Penicillium citrinum* Thom, *P. funiculosum*, *P. rubrum* Stoll, and *Fusarium* species were found in large numbers. *Aspergillus flavus*, *A. niger*, *Rhizopus* species, and *S. bataticola* extensively penetrated pods and kernels when dry. The infested pods were hydrated for six days at 26, 32, or 38°C. As the temperature increased, infection by *A. flavus* and *A. niger* increased.

Garren (1966) studied the endogeocarpic floras of Virginia-grown peanuts and reported that "*Trichoderma viride* Pers. ex Fr. seems a dominant (fungus) and *Penicillium* species seem sub-dominants in the climax endogeocarpic community of sound and rotting peanut pods; and *Aspergillus flavus* and *A. niger*, which have a potential for causing trouble, are quantitatively minor but persistent species in this flora."

Diener (1960) investigated Georgia farmers' stock peanuts (uncleaned and unshelled) that had been stored for eight to 56 months. He found that the predominant fungi consisted of certain species of the *Aspergillus glaucus* group (*A. amstelodami*, *A. chevalieri*, *A. repens*, and *A. ruber*), *A. restrictus*, *A. tamarii*, *Cladosporium* species, *Penicillium citrinum*, and members of the Mucorales (water molds). Large numbers of fungi were directly associated with kernel moistures of 12.5 percent or greater at the time that the seeds were stored. Kernel moisture content decreased to about 7 percent, which is considered to be a safe level, in three to four weeks after the bins were filled, regardless of the initial moisture content of the kernels. During the five-year storage period, the moisture content of no sample ever exceeded 6.8 percent. There was no relationship between percentage of initial kernel damage, final damage, or increased damage during storage.

It should not be assumed that the effects of storage fungi on seeds of other field crops, flowers, or vegetables will necessarily always parallel those reported for stored cereals. Kulik (1973) reported that seeds of cabbage (*Brassica oleracea* var. *capitata* L.), cucumber (*Cucumis sativus* L.), pepper (*Capsicum annuum* L. var. *annuum*), radish, and turnip (*Brassica rapa* L. var. *rapa*,) inoculated with spores of either of two common storage fungi and stored at 85 percent relative humidity and 22 to 25°C for 30 days, remained largely free of invasion by these fungi. Conversely, Qasem and Christensen (1958) found that 84 percent of the corn seeds that they inoculated with a storage fungus and stored at 85 percent relative humidity and 20°C were invaded by the fungus after 30 days. Likewise, Tuite and Christensen (1957) found that 100 percent of the wheat seeds that they had inoculated with a storage fungus and stored at 80 percent relative humidity and 25°C, were invaded after 30 days.

For additional information on the effects of fungi on stored seeds, see Anonymous (1983), Christensen (1982), Christensen and Meronuck (1986), Halloin (1986), and Kulik (1978).

Detection and Identification of Seed Storage Fungi

Seed storage fungi, because of their ability or in some cases their requirement to grow on substrates having relatively low moisture levels, can usually be isolated by placing 100 to 400 seeds from a given lot on an agar medium containing malt extract or some other nutrient plus varying concentrations of sodium chloride or sucrose to raise the osmotic pressure of the culture medium (Christensen, 1957). The seeds may be given a pretreatment for one or more minutes in a solution of sodium hypochlorite (laundry bleach) to eradicate microorganisms that may be present on the surface of the seeds. This is done in order to determine the amount of internal fungal infection. However, it should be pointed out that the detection and particularly the identification of seed storage fungi can be difficult and is best left to the specialist who has the knowledge and laboratory equipment to do this type of work.

For additional information on the isolation and identification of *Aspergillus* and *Penicillium* species that attack stored seeds, see Kulik (1968) (*Penicillium*), Pitt (1979, 1985) (*Penicillium*), Raper and Fennell (1965) (*Aspergillus*), and Samson (1979) (*Aspergillus*).

Control of Seed Storage Fungi

Deterioration of stored seeds by fungi is controlled principally by drying seeds to a safe moisture content prior to storage in a dry, cool place and ensuring that the stored seeds remain dry (Christensen and Kaufmann, 1969). Most seed storage fungi cannot invade seeds that are in moisture equilibrium with a relative humidity of 65 percent or lower. However, as already mentioned, some seed storage fungi are xerophytic and can invade seeds that have a moisture content that is only slightly above the level that is considered safe. A seed storage area in which the temperature and humidity are controlled is ideal but is relatively costly to install and operate. Valuable seeds that are small in size can be dried and stored under an inert gas in sealed metal containers or over a desiccant such as phosphorus pentoxide in a glass container such as a laboratory desiccator.

Because the moisture content of a seedlot represents the average of many seeds, some individual seeds may possibly contain an unsafe amount of moisture even though the moisture content of the lot is at a safe level. Seeds containing too much moisture are of course susceptible to invasion by storage fungi, which may then spread throughout the seedlot. However, it is still extremely important to accurately measure the moisture content of a seedlot prior to storage. For best results, seeds should be dried as soon as possible after harvest, then thoroughly cleaned of extraneous material prior to storage in an environment where they will not absorb moisture.

Fumigation of stored seeds to control insects ordinarily has no detectable effect on storage fungi (Christensen and Hodson, 1960). Treating seeds with fungicides prior to storage can control storage fungi but many of these materials are hazardous to animals and humans and can pose a threat to the environment.

Kulik and Justice (1966) inoculated seeds of onion (*Allium cepa* L.), radish (*Raphanus sativus* L.), bluegrass (*Poa pratensis* L.), crimson clover (*Trifolium incarnatum* L.), sorghum (*Sorghum vulgare* Pers.), wheat (*Triticum aestivum* L.), and corn with spores of the common storage fungus *Aspergillus amstelodami*. Seeds of peanut were inoculated with spores of *A. flavus*, another common storage fungus. These seeds were exposed to doses of zero, five, ten, 20, 40, or 80 krad of gamma radiation from a cobalt 60 source. Some destruction of both fungi occurred at the highest dose but the seedling production potential of all the crop species, with the possible exception of radish and crimson clover, was greatly reduced. In general, these two storage fungi were found to be more resistant to radiation injury than were the seeds. Storage of inoculated, irradiated seeds under conditions that were favorable for the growth of storage fungi yielded no clear picture of the effect of gamma radiation on the subsequent behavior of the two test fungi on stored seeds.

For additional information on the control of seed storage fungi, see Christensen (1982), Christensen and Meronuck (1986), Kulik

(1978), and McLean (1989). See Justice and Bass (1978) for additional information on storing seeds, and Owen (1956) for safe moisture levels for seeds of various crop plants that are to be stored.

SEEDBORNE PLANT DISEASE AGENTS

Description and Importance

Seedborne plant disease agents include microorganisms, i.e., fungi and bacteria, viruses and viroids (not considered to be living organisms), and nematodes (also called eelworms).

Many of the fungi, some of the bacteria, and a smaller number of viruses, viroids, and nematodes that cause economically important plant diseases may be seedborne. Seeds that carry these plant disease agents are important in agriculture and horticulture because:

1. Infected seeds may not germinate or may be low in vigor. The resulting decrease in the seedling population can lead to fewer adult plants with a concomitant reduction in crop yield.
2. Infected seeds can be a source of inoculum which, under suitable environmental conditions, may introduce disease into a healthy crop and thus reduce the yield.
3. Seeds that are carrying plant disease agents can introduce these agents into geographic areas that are free of them.
4. Seeds that are infected with fungi or bacteria, even though they have been treated with a fungicide or bactericide, may still carry viable microorganisms and cause one or more of the results mentioned above (Kulik and Schoen, 1977).

Of the various kinds of seedborne plant disease agents, the fungi are responsible for the greatest number of plant diseases and thus it is not surprising that they are more commonly found associated with seeds than are any of the other plant disease agents (Neergaard, 1977). Seedborne pathogenic bacteria represent the second largest group of plant disease-causing agents. Plant pathogenic seedborne fungi and bacteria are usually host-specific and thus are found associated only with certain seeds. In addition to parasitic fungi and bacteria, saprophytic fungi and bacteria are often carried by seeds.

The latter are not host-specific and may be found on seeds of many different kinds of plants. Both parasitic and saprophytic fungi and bacteria may be superficially attached to the outer seed surface or lodged in cracks in the seed coat or under it, but the pathogens may also be present within the cotyledons and other parts of a seed, including the embryo.

The seeds of many agricultural and horticultural plants have been reported to harbor plant disease-causing fungi (Richardson, 1989). A lesser number have been reported to harbor pathogenic bacteria. However, for many of them, seed carriage of pathogenic fungi and bacteria is only of academic or quarantine interest since these agents usually do not cause a significant economic loss to the grower. Examples of economically important plant diseases in which seedborne fungi can play a major role are spring black stem of alfalfa (*Medicago sativa* L. subsp. *sativa*), caused by *Phoma medicaginis* Malbr. & Roum.; blackleg of cabbage and other crucifers caused by *Leptosphaeria maculans* (Desm. Ces. & de Not.); corn seedling blight (*Diplodia maydis* [Berk.] Sacc.); greenbean (*Phaseolus vulgaris* L.) anthracnose (*Colletotrichum lindemuthianum* [Sacc. & Magn.] Bri. & Cav.); rice (*Oryza sativa* L.) blast (*Pyricularia oryzae* Cav.); soybean pod and stem blight (*Phomopsis phaseoli* [Desm.] Sacc.); tomato (*Lycopersicon lycopersicum* [L.] Karsten var. *lycopersicum)* early blight (*Alternaria solani* Sorauer); wheat scab (*Gibberella zeae* [Schw.] Petch); and marigold (*Tagetes erecta* L. and T. *patula* L.) and zinnia (*Zinnia elegans* Jacq.) blight (*Alternaria zinniae* M. B. Ellis). In the case of seedborne pathogenic bacteria, barley leaf stripe (*Pyrenophora graminea* Ito & Kuribay), beet (*Beta vulgaris* L. subsp. *vulgaris*) bacterial blight (*Pseudomonas aptata* [Brown & Jamieson] Stevens), cucumber angular leaf spot (*Pseudomonas lachrymans* [E. F. Smith & Bryan] Carsner), greenbean fuscous blight (*Xanthomonas phaseoli* [E. F. Smith] Dowson var. *fuscans)*, and stock (*Matthiola incana* R. Br.) bacterial blight (*Xanthomonas incanae* [Kendrick & Baker] Starr & Weiss) may be mentioned (Richardson, 1989). See Neergaard (1977) for detailed information on seedborne fungi and bacteria.

Viruses are particles of nucleic acid surrounded by a protein coat. They can multiply only within living cells. Virus diseases of higher plants cause symptoms that include mottling, deformation, tumors, stunting, blighting, fasciation, and streaking. Seedborne viruses are much less common than seedborne fungi and bacteria but they can cause marked yield reductions in the crops in which they do occur (Agrios, 1988). In addition, germ plasm that is carrying plant pathogenic viruses poses a serious threat to agriculture and horticulture (Hampton et al., 1982). Examples of viruses that cause economically important plant diseases include mosaic of alfalfa, stripe mosaic of barley, common mosaic of greenbean, mosaics of lettuce (*Lactuca sativa* L.) and tomato, and ring spot of stone fruits (*Prunus* spp.). Viroids are somewhat similar to viruses but consist of low molecular weight, single-stranded ribonucleic acid molecules (Diener, 1971). A viroid that can be seedborne is the cause of the serious disease of Irish potato (*Solanum tuberosum* L.) known as potato spindle tuber. See Phatak (1974) and Neergaard (1977) for additional information on seedborne viruses.

Nematodes, or eelworms, are microscopic animals which can have a pronounced and harmful effect on plant growth. They mainly affect the underground portions of plants but they can also attack stems, leaves, flowers, and seeds (Agrios, 1988). Diseases caused by nematodes that can be seedborne include onion bloat (*Ditylenchus dipsaci* [Kuhn] Filipjev), rice white-tip disease (*Aphelenchoides besseyi* Christie), and wheat ear cockle (*Anguina tritici* [Steinbuch] Chitwood). Certain nematodes may sometimes be associated with seedlots in the form of galls or infested soil beds but are not found within seeds. An example of this is the soybean cyst nematode (*Heterodera glycines* Ichinohe). Symptoms of nematode attack include stunting, swelling, gall production, necrotic lesions, abnormal root growth, and generally poor growth (Agrios, 1988). In addition, some nematodes serve as vectors of certain viruses that cause plant diseases such as arabis mosaic virus, tomato black ring virus, and raspberry ring spot virus. See Neergaard (1977) for additional information on seedborne nematodes.

Isolation and Identification

Seedborne fungi are most often detected on a routine basis in seed testing laboratories by placing 200 to 400 seeds on a nutrient agar such as potato dextrose agar or on a moist, paper substrate such as blotting or filter paper (Kulik and Schoen, 1977). The seeds are incubated at a temperature that is favorable for fungal growth, often with 12 hours daily illumination from longwave ultraviolet or fluorescent lamps to stimulate fungal sporulation. After one or more weeks, the seeds are examined using a dissecting microscope to observe characteristic fruiting structures and colony growth forms (if on a nutrient agar). It is sometimes necessary to examine spores under a compound microscope before a correct identification can be made. There are special techniques for certain seedborne fungi (Kulik, 1981). See Kulik (1981), Kulik and Schoen (1977), Kulik and Stanwood (1984), and Neergaard (1977) for additional information on the routine detection and identification of seedborne fungi.

Similarly, seedborne bacteria are usually detected by placing the seeds on a suitable agar medium. After an incubation period that is usually shorter than for fungi, the bacteria will be seen to have formed a ring around a seed. It is then usually necessary to transfer the bacteria to special diagnostic media for identification (Schaad, 1982). See Schaad (1982) for additional information on the routine detection of seedborne bacteria, and Kulik (1984) and Van Vuurde (1987) for descriptions of newer techniques such as immunofluorescence.

Since viruses are incapable of growth outside of living cells, culturing seeds on agar would be fruitless. Instead, seeds are planted in soil or sand and allowed to germinate. Sometimes a virus can be detected by seedling symptoms but most of the time it is necessary to grow the plant to maturity (Phatak, 1974). However, positive identification of a virus may depend on the determination of one or more viral properties such as the thermal inactivation temperature or the dilution end point. Serological tests and examination of the isolated virus under the electron microscope may also be necessary to ensure accurate identification (Kulik, 1984). See Neergaard (1977) and Phatak (1974) for additional information on the routine detection of seedborne viruses. As is the case with viruses, the detection of seedborne viroids is rather complex and involves sophisticated instrumentation (Kulik, 1984).

Seedborne nematodes can be detected by crushing an infested seed to liberate the nematodes, and observing them under a compound microscope. See Neergaard (1977) for additional information on the routine detection of seedborne nematodes.

As is the case with seed storage fungi, the isolation and particularly the identification of seedborne plant disease agents is a task best left to a specialist such as a plant pathologist, mycologist, nematologist, or virologist. It should also be emphasized that the presence of a plant disease agent on a small number of seeds in a lot does not necessarily mean that economic losses will occur if these seeds are planted. For a discussion of the subject of inoculum thresholds for seedborne plant disease agents, see Gabrielson (1988) (fungi), Kuan (1988) (overview), Neergaard (1977) (all plant disease agents), Russell (1988) (sampling and statistics), Schaad (1988) (bacteria), and Stace-Smith and Hamilton (1988) (viruses).

Control of Seedborne Plant Disease Agents

Seedborne fungi and bacteria that cause plant diseases may be controlled by the use of fungicidal or bactericidal sprays to prevent seed infection in the production field, by treating seeds with fungicides or bactericides (antibiotics) prior to planting, or by subjecting them to treatment with hot water to eradicate the pathogen (Kulik, 1981). An effective method for obtaining seeds that are free of plant disease-causing agents is to produce the seeds in areas which are free of the pathogen or where the weather conditions do not favor infection of the developing seed crop. An example of the latter is the production of greenbeans for seed in arid areas of the western United States. Another way of controlling seedborne fungi, which is practiced mainly in Europe, consists of routinely testing seedlots for the presence of pathogens. Some countries prohibit the importation of any seeds that may be harboring any of a long list of pathogens, even in trace amounts. Sinclair (1988) described several innovative treatments for eradicating certain fungi and bacteria from the seeds of soybeans and other large-seeded legumes. These treatments include infusion of fungicides into dormant seeds using acetone or dichloromethane, and thermotherapy using hot oil.

Measures suggested for controlling seedborne viruses include using hot water-treated seeds, using seeds from lots that have been grown out and found to be virus- (or viroid-) free (i.e., indexing), and spraying seed production fields with insecticides to control insects that spread viruses (Kulik, 1981).

Control measures for seedborne nematodes include soil disinfestation using chemicals (soil drenches or fumigants) or heat, and the use of nematode-free seeds and nematode-resistant cultivars (Kulik, 1981).

For additional information on the control of seedborne plant disease agents see Chaube and Singh (1991), Kulik (1981), and Neergaard (1977).

REFERENCES

Agrios, G. N. *Plant Pathology*, third ed. (New York: Academic Press, 1988), 803 pp.

Anonymous. Symposium on deterioration mechanisms in seeds. *Phytopathology* 73(1983): 314-339.

Arndt, C. H. The internal infection of cotton seed and the loss of viability in storage. *Phytopathology* 36(1946): 30-37.

Bunting, R. H. Proceedings of the association of economic biologists. II. Mycological aspects. *Ann. Appl. Biol.* 17(1930): 402-407.

Chaube, H. S., and U. S. Singh. *Plant Disease Management* (Boca Raton, Florida: CRC Press, 1991), 319 pp.

Christensen, C. M. Deterioration of stored grains by fungi. *Bot. Rev.* 23(1957): 108-134.

_______ , ed. *Storage of Cereal Grains and Their Products*, third ed. (St. Paul, Minnesota: American Association of Cereal Chemists, 1982), 544 pp.

Christensen, C. M., and D. R. Gordon. The mold flora of stored wheat and corn and its relation to heating of moist grain. *Cereal Chem.* 25(1948): 40-51.

Christensen, C. M., and A. C. Hodson. Development of granary weevils and storage fungi in columns of wheat. II. *J. Econ. Entomol.* 53(1960): 375-380.

Christensen, C. M., and H. H. Kaufmann. Deterioration of stored grains by fungi. *Ann. Rev. Phytopathol.* 3(1965):69-84.

_______ . *Grain Storage: The Role of Fungi in Quality Loss*. (Minneapolis, Minnesota: University of Minnesota Press, 1969), 154 pp.

Christensen, C. M., and R. A. Meronuck. *Quality Maintenance in Stored Seeds and Grains*. (Minneapolis, Minnesota: University of Minnesota Press, 1986), 138 pp.

Diener, T. O. Potato spindle tuber "virus" IV. A replicating, low molecular weight RNA. *Virology* 45(1971): 411-428.

Diener, U. L. The mycoflora of stored peanuts. *Alabama Acad. Sci. J.* 30(1958): 5-6.

_______ . The mycoflora of peanuts in storage. *Phytopathology* 50(1960): 220-223.

Fields, R. W., and T. H. King. Influence of storage fungi on deterioration of stored pea seed. *Phytopathology* 52(1962): 336-339.

Gabrielson, R. L. Inoculum thresholds of seedborne pathogens. Fungi. *Phytopathology* 78(1988): 868-872.

Garren, K. H. Peanut (groundnut) microfloras and pathogenesis in peanut pod rot. *Phytopath. Z.* 55(1966): 359-367.

Gilman, J. C., and D. H. Barron. Effects of molds on temperature of stored grain. *Plant Physiol.* 5(1930): 565-573.

Halloin, J. M. Microorganisms and seed deterioration. In *Physiology of Seed Deterioration*, M. B. McDonald, Jr. and C. J. Nelson, eds. (Madison, Wisconsin: CSSA Special Publication No. 11, Crop Science Society of America, 1986), pp. 89-99.

Hampton, R. O., H. Waterworth, R. M. Goodman, and R. Lee. Importance of seedborne viruses in crop germplasm. *Plant Disease* 66 (1982): 977-978.

Hanlin, R. T. Current research on peanut fungi in Georgia. *Georgia Agric. Res.* 8(1966): 3-4.

Jackson, C. R. Peanut-pod mycoflora and kernel infection. *Plant & Soil* 23(1965): 203-212.

Justice, O. L., and L. N. Bass. *Principles and Practices of Seed Storage.* Agriculture Handbook No. 506, US Government Printing Office, Washington, DC, 1978, 289 pp.

Kuan, T. L. Inoculum thresholds of seedborne pathogens. Overview. *Phytopathology* 78(1988): 867-868.

Kulik, M. M. *A Compilation of Descriptions of New* Penicillium *Species.* Agriculture Handbook No. 351, US Government Printing Office, Washington, DC, 1968, 80 pp.

________ . Susceptibility of stored vegetable seeds to rapid invasion by *Aspergillus amstelodami* and *A. flavus* and effect on germinability. *Seed Sci. & Technol.* 1(1973): 799-803.

________ . Effects of pests and chemicals on seed deterioration in storage. In *Principles and Practices of Seed Storage*, O. L. Justice and L. N. Bass, eds. Agriculture Handbook No. 506, US Government Printing Office, Washington, DC, 1978, pp. 81-91.

________ . Identification and control of seed-borne pathogenic microorganisms. In *Handbook of Transportation and Marketing in Agriculture, Vol. II, Field Crops*, E. E. Finney, Jr., ed. (Boca Raton, Florida: CRC Press, 1981), pp. 207-237.

________ . New techniques for the detection of seed-borne pathogenic viruses, viroids, bacteria and fungi. *Seed Sci. & Technol.* 12(1984): 831-840.

Kulik, M. M., and R. T. Hanlin. Osmophilic strains of some *Aspergillus* species. *Mycologia* 60(1968): 961-964.

Kulik, M. M., and O. L. Justice. Survival of two storage fungi after gamma radiation of host seeds. *Rad. Bot.* 6(1966): 407-412.

________ . Some influences of storage fungi, temperature and relative humidity on the germinability of grass seeds. *J. Stored Prod. Res.* 3(1967): 335-343.

Kulik, M. M., and J. F. Schoen. Procedures for the routine detection of seed-borne pathogenic fungi in the seed-testing laboratory. *J. Seed Technol.* 2(1977): 29-39.

Kulik, M. M., and P. C. Stanwood. Horticultural seed pathology-An introduction. *J. Seed Technol.* 9(1984): 1-19.

McLean, K. A. *Drying and Storing Combinable Crops*, second ed. (Ipswich, UK: Farming Press Books, 1989), 257 pp.

Mead, H. W., R. C. Russell, and R. J. Ledingham. The examination of cereal seeds for disease and studies on the embryo exposure in wheat. *Scient. Agric.* 23(1942): 27-40.

Milner, M., and W. F. Geddes. Grain storage studies III. The relation between moisture content, mold growth, and respiration of soybeans. *Cereal Chem.* 23(1946): 225-247.

Neergaard, P. *Seed Pathology, Vol. I.* (London, UK: Macmillan Press, 1977), 839 pp.

Owen, E. B. The Storage of Seeds for Maintenance of Viability. Bulletin 43, Commonwealth Agricultural Bureaux, Bucks, England, 1956, 81 pp.

Phatak, H. C. Seed-borne plant viruses–Identification and diagnosis in seed health testing. *Seed Sci. & Technol.* 2(1974): 3-155.

Pitt, J. I. *The Genus* Penicillium *and its Teleomorphic States* Eupenicillium *and* Talaromyces (New York: Academic Press, 1979), 634 pp.

________ . *A Laboratory Guide to the Common* Penicillium *Species.* (North Ryde, NSW, Australia: Commonwealth Scientific and Industrial Research Organization, 1985), 182 pp.

Qasem, S. A., and C. M. Christensen. Influence of moisture content, temperature, and time on the deterioration of stored corn by fungi. *Phytopathology* 48(1958): 544-549.

Raper, K. B., and D. I. Fennell. *The Genus* Aspergillus. (Baltimore, Maryland: The Williams and Wilkins Company, 1965), 686 pp.

Richardson, M. J. *An Annotated List of Seed-Borne Diseases*, fourth ed. (Zurich, Switzerland: International Seed Testing Association, 1989), 320 pp.

Russell, T. S. Inoculum thresholds of seedborne pathogens. Some aspects of sampling and statistics in seed health testing and the establishment of threshold limits. *Phytopathology* 78(1988): 880-881.

Samson, R. A. A compilation of the aspergilli described since 1965. *Studies in Mycology* 18(1979): 1-38.

Sauer, D. B., C. L. Storey, and D. E. Walker. Fungal populations in U. S. farm-stored grain and their relationship to moisture, storage time, regions, and insect infestation. *Phytopathology* 74(1984): 1050-1053.

Schaad, N. W. Detection of seedborne bacterial plant pathogens. *Plant Disease* 66(1982): 885-890.

________ . Inoculum thresholds of seedborne pathogens. Bacteria. *Phytopathology* 78(1988): 872-875.

Semeniuk, G. Microflora. In *Storage of Cereal Grains and Their Products*, A. A.

Anderson and A. W. Alcock, eds. (St. Paul, Minnesota: American Association of Cereal Chemists, 1954), pp. 77-151.

Sinclair, J. B. Innovative treatments for large-seeded legumes. In *Experimental and Conceptual Plant Pathology, Vol. 1*, W. M. Hess, R. S. Singh, U. S. Singh, and D. J. Weber, eds. (New York: Gordon and Breach Publishers, 1988), pp. 121-136.

Sinha, R. N., and H. A. H. Wallace. Ecology of a fungus-induced hot spot in stored grain. *Can. J. Plant Sci.* 45(1965): 48-59.

Stace-Smith, R., and R. I. Hamilton. Inoculum thresholds of seedborne pathogens. Viruses. *Phytopathology* 78(1988): 875-880.

Thomson, J. R. *An Introduction to Seed Technology.* (New York: John Wiley and Sons, 1979), 252 pp.

Tuite, J. F., and C. M. Christensen. Grain storage studies. XXIV. Moisture content of wheat seed in relation to invasion of the seed by species of the *Aspergillus glaucus* group, and the effect of invasion upon germination of the seed. *Phytopathology* 47(1957): 323-327.

Van Vuurde, J. W. L. Detecting seedborne bacteria by immunofluorescence. In *Current Plant Science and Biotechnology in Agriculture–Plant Pathogenic Bacteria*, E. L. Civerolo, A. Collmer, R. E. Davis, and A. G. Gillaspie, eds. (Dordrecht, The Netherlands: Martinius Nijhoff Publishing Company, 1987), pp. 799-808.

Wallace, H. A. H., and R. N. Sinha. Fungi associated with hot spots in farm stored grains. *Can. J. Plant Sci.* 42(1962): 130-141.

AN ANNOTATED BIBLIOGRAPHY OF USEFUL REFERENCES ON SEED QUALITY AND MICROORGANISMS

Agrios, G. N. *Plant Pathology.* third ed. (New York: Academic Press), 1988, 803 pp. (Includes information on the various agents that cause plant disease.)

Anonymous. Symposium on deterioration mechanisms in seeds. *Phytopathology* 73(1983):314-339. (Includes papers on "Mechanisms of seed infection and pathogenesis," G. E. Harman, pp. 326-329, and "Deterioration resistance mechanisms in seeds," J. M. Halloin, pp. 335-339.)

Chaube, H. S., and U. S. Singh. *Plant Disease Management* (Boca Raton, Florida: CRC Press), 1991, p. 319. (Includes information on the control of seedborne diseases.)

Christensen, C. M., ed. *Storage of Cereal Grains and Their Products*, third ed. (St. Paul, Minnesota: American Association of Cereal Chemists, 1982), 544 pp. (Although principally concerned with cereal grains, there is a good deal of information that can be applied to stored seeds.)

Christensen, C. M., and R. A. Meronuck. *Quality Maintenance in Stored Seeds and Grains.* (Minneapolis, Minnesota: University of Minnesota Press, 1986), 138 pp. (Provides some useful information on proper methods of storing seeds.)

Dhingra, O. D., and J. B. Sinclair. *Basic Plant Pathology Methods*. (Boca Raton, Florida: CRC Press, 1985). pp. 355 (Provides information on methodology that is applicable to the detection of seed storage fungi and agents that cause plant disease.)

Geng, S., R. N. Campbell, M. Carter, and F. J. Hills. Quality control programs for seedborne pathogens. *Plant Disease* 67(1983): 236-242. (Provides a theoretical background and methods for determining the number of seeds to be used in testing for plant disease agents.)

Justice, O. L., and L. N. Bass. *Principles and Practices of Seed Storage*. Agriculture Handbook No. 506, U.S. Government Printing Office, Washington, DC, 1978, 289 pp. (Provides an in-depth treatment of this subject.)

Kulik, M. M. A Compilation of Descriptions of New *Penicillium* Species. Agriculture Handbook No. 351, US Government Printing Office, Washington, DC, 1968, 80 pp. (Contains a taxonomic key to 113 species of *Penicillium* described between 1949 and 1968, principally of use to plant pathologists and mycologists.)

_______ . Effects of pests and chemicals on seed deterioration in storage. In *Principles and Practices of Seed Storage*, O. L. Justice and L. N. Bass, eds. Agriculture Handbook No. 506, US Government Printing Office, Washington, DC, 1978, pp. 81-91. (Provides a concise account of seed storage fungi and their control.)

_______ . Identification and control of seedborne pathogenic microorganisms. In *Handbook of Transportation and Marketing in Agriculture, Vol. II, Field Crops*, E. E. Finney, Jr., ed. (Boca Raton, Florida: CRC Press, 1981), pp. 207-237. (Covers mainly in tabular fashion many important seedborne bacteria, fungi, nematodes, and viruses associated with a large number of crops.)

_______ . New techniques for the detection of seed-borne pathogenic viruses, viroids, bacteria, and fungi. *Seed Sci. & Technol.* 12(1984):831-840. (Discusses in some detail, newer, mostly serological procedures for detecting seedborne viruses, and covers some of the newer cultural, bioassay, and serological techniques for seedborne bacteria.)

Kulik, M. M., and J. F. Schoen. Procedures for the routine detection of seed-borne pathogenic fungi in the seed-testing laboratory. *J. Seed Technol.* 2(1977):29-39. (Presents detailed instructions for employing the commonly used agar and blotter techniques.)

Kulik, M. M., and P. C. Stanwood. Horticultural seed pathology–An introduction. *J. Seed Technol.* 9(1984):1-19. (Provides a concise account of this subject.)

Malone, J. P., and A. E. Muskett. Seed-borne fungi. *Proc. ISTA* 29(1964):179-384. (A useful collection of descriptions, illustrations, and cultural methods for 77 fungal species, both pathogens and saprophytes.)

McDonald, M. B., Jr., and C. J. Nelson, eds. *Physiology of Seed Deterioration*. (Madison, Wisconsin: CSSA Special Publication No. 11, Crop Science Society of America, 1986), 123 pp. (Includes sections dealing with "Precepts of successful seed storage," E. E. Roos, pp. 1-25; "Microorganisms and seed deterioration," J. M. Halloin, pp. 89-99; and "Quantifying seed deterioration," E. H. Roberts, pp. 101-123.)

McLean, K. A. *Drying and Storing Combinable Crops*, second ed. (Ipswich, UK: Farming Press Books, 1989), 257 pp. (Some of the information presented is applicable to seeds.)

Neergaard, P. *Seed Pathology. Vol. I.* 839 pp. (Includes seven chapters on methods used in testing seeds for plant disease-causing agents plus a wealth of information on all aspects of seed pathology.) *Vol. II.* 347 pp. (Consists of a glossary of seed pathology terms, references, and an index to volume I.) (London, UK: Macmillan Press, 1977).

Owen, E. B. The Storage of Seeds for Maintenance of Viability. Bulletin 43, Commonwealth Agricultural Bureaux, Bucks, England, 1956, 81 pp. (Although published 35 years ago, much of the material in this publication is still current and valuable.)

Phatak, H. C. Seed-borne plant viruses–Identification and diagnosis in seed health testing. *Seed Sci. & Technol.* 2(1974):3-155. (This is a comprehensive and useful treatment of the subject.)

Pitt, J. I. 1979. *The Genus* Penicillium *and its Teleomorphic States* Eupenicillium *and* Talaromyces. (New York: Academic Press, 1979), 634 pp. (A technical account of this genus principally of use to plant pathologists and mycologists.)

________ . *A Laboratory Guide to the Common* Penicillium *Species*. (North Ryde, NSW Australia: Commonwealth Scientific and Industrial Research Organization, 1985), 182 pp. (Provides a taxonomic key to 30 species of *Penicillium* plus descriptions and illustrations.)

Raper, K. B., and D. I. Fennell. *The Genus* Aspergillus. (Baltimore, Maryland: The Williams and Wilkins Company, 1965), 686 pp. (A technical account of this genus principally of use to plant pathologists and mycologists.)

Richardson, M. J. *An Annotated List of Seed-Borne Diseases*, Fourth Ed. (Zurich, Switzerland: International Seed Testing Association, 1989), 320 pp. (A useful reference that covers a large number of crops and some weeds, and the seed-borne fungi, bacteria, nematodes, and viruses associated with them.)

Samson, R. A. A compilation of the aspergilli described since 1965. *Studies in Mycology* 18(1979):1-38. (Provides a taxonomic key and descriptions principally of use to plant pathologists and mycologists.)

Schaad, N. W. Detection of seedborne bacterial plant pathogens. *Plant Disease* 66(1982):885-890. (Provides a detailed discussion of current methodology.)

Suryanarayana, D. *Seed Pathology.* (New Delhi, India: Vikas Publishing House, 1978), 111 pp. (This concise work includes useful information on diseases of 22 horticultural crops plus control measures.)

Thomson, J. R. *An Introduction to Seed Technology.* (New York: John Wiley and Sons, 1979), 252 pp. (Contains chapters on seed quality, harvesting and drying, storage, and testing for germination and vigor.)

Van Vuurde, J. W. L. Detecting seedborne bacteria by immunofluorescence. In *Current Plant Science and Biotechnology in Agriculture–Plant Pathogenic Bacteria*, E. L. Civerolo, A. Collmer, R. E. Davis, and A. G. Gillaspie, eds.,

(Dordrecht, The Netherlands: Martinius Nijhoff Publishing Company, 1987), pp. 799-808. (Presents a detailed discussion of this method.)

West, S. H., ed. *Physiological-Pathological Interactions Affecting Seed Deterioration*. (Madison, Wisconsin: CSSA Special Publication No. 12, Crop Science Society of America, 1986), 95 pp. (Includes sections on "Post-harvest insect-fungus associations affecting seed deterioration," J. T. Mills, pp. 39-51, and "Environmental factors associated with preharvest deterioration of seeds," D. C. McGee, pp. 53-63.)

Chapter 6

The Storage of Orthodox Seeds

Dale O. Wilson, Jr.

Adaptation of plants to the terrestrial environment was, in part, a reaction to low humidity. This adaptation consisted of an increase in size, development of a vascular system, desiccation barriers, and increased tolerance to desiccation in certain plant parts. Generations telescoped together and the developing sporophyte was sheltered and enclosed by the gametophyte, which itself was enclosed by the previous sporophyte generation. The young sporophyte became the dispersal unit, in contrast to spores of lower plants.

Two seed survival strategies were pursued by seed plants, resistance to, or avoidance of, desiccation (recalcitrant seeds), and tolerance to desiccation (orthodox seeds). The relatively large size of orthodox seeds compared to spores increased probability of survival by controlling the rate of drying and rehydration, by providing a barrier to predation, and by providing a larger food resource for establishment in a hostile environment. Different species differ in their inherent storability, and the reasons are not well understood. Justice and Bass (1978) have compiled information on the relative storability of a wide range of species.

Humans first exploited orthodox seeds to obtain the stored food, and in this way, the practice of seed (grain) storage was born. Repeated storage and planting of seeds resulted in domestication and development of food grains in their present form (Heiser, 1973).

In modern agriculture, plant breeding has become a fine art, and seed for planting has become a commodity. It is no longer simply grain, but a package for the genetic value added by the plant breeder. Yet the value of the seed goes beyond the genetic content. Growers plant *seed* rather than *grain* because doing so reduces their risk of crop failure. As the value of the genetic content increases, the value of the package increases. Careful storage to maximize seed vigor and viability is more important than ever. Seed storage is a fundamental component of systems and policies to preserve genetic diversity (Cohen et al., 1991).

The high value of seed justifies care in handling and storage. Many organisms are adapted to exploit seeds as a food resource. In this chapter I will describe these organisms and approaches for controlling them. Seed deteriorates physiologically even in the absence of pests. I present methods for quantification of this process, and approaches to minimize deterioration in storage. Finally, I describe current and emerging strategies for prolonging the storage life of seeds.

PESTS OF STORED SEED

Seed may be damaged or destroyed by a wide range of predators beginning at anthesis. Only those species which damage seed during storage will be discussed here, although infestation by several of these pests may begin before harvest. The pests are carried into the storage with the seeds, where they may greatly multiply.

Vertebrate Pests

Large amounts of stored grain and seed are consumed by rats and mice. Rats in particular are able to burrow and gnaw their way into grain and seed storage areas. Control is effected by a combination of sanitation, rodent-proofing, and trapping or poisoning. Reviews of rodent species, behavior, and control strategies have been published (Harris and Bauer, 1982; Mallis, 1982; Grolleau and Gramet, 1988).

Invertebrate Pests

Stored-product insects are the most insidious and difficult to control pests of stored seed. Only a few species are specialized and dangerous predators of sound seeds. These are called primary consumers.

The primary consumers breach the pericarp or testa, making feeding possible by a much wider range of insects, mites, and fungi. The primary consumers obtain water metabolically from seeds and release it into their immediate environment along with waste heat. At the same time the accumulating frass, grain dust, and carcasses restrict air movement, contributing to formation of a hot spot. A widening ecological niche may be established through which a whole succession of arthropod and fungal species may cascade (Mills, 1983). Meanwhile, the progeny of the primary consumers wander off and start new foci of infestation.

Important Species

Table 1 lists the most important primary consumer insect pests of stored seed. Those species which fly often lay their eggs in seeds before harvest. The generation time is often short, and in a few months the population can explode. Weevils complete their life cycle inside individual seeds and may become very numerous before their presence is noticed. The Angoumois grain moth also oviposits on seeds in the field, and completes its life cycle inside the seed.

The "other beetles" listed in Table 1 are less specialized than the weevils and will eat a wide variety of food products. Of these, the most dangerous to cereal seed is the lesser grain borer. It is able to multiply on sound kernels at relative humidities as low as 30 to 40 percent. Invasion by all insects is greatly facilitated by moisture.

Insects other than those listed in Table 1 may occasionally infest and damage seed. More comprehensive references are available (USDA, 1978; Cotton and Wilbur, 1982; Mallis, 1982; Fleurat-Lessard, 1988).

Insect Control

Constant vigilance and frequent inspection are the cornerstones of pest control in stored seed. If a bare arm is plunged full length into grain, warm spots may be detected, generally indicating foci of infestation. Small plastic traps, with or without pheromone bait, can be inserted in the seed bulk to scout for insects (Burkholder, 1984; Willson, 1987). Insects in stored grain or seed are controlled by avoidance, physical manipulation of the storage, and toxic materials.

TABLE 1. Common insect pests of sound stored seed and grain.

Scientific name Common name	Adult length	Seeds attacked	Life cycle	Can fly?	Comments
	mm		*Days*		
Weevils					
Sitophilus granarius Granary weevil	3-5	cereal grains	30-50	No	Eggs embedded in kernels, larvae feed and pupate inside kernels. Cold tolerant.
Sitophilus oryzae Rice weevil	2-3	cereal grains	25-30	Yes	Eggs embedded in kernels, larvae feed and pupate inside kernels.
Sitophilus zeamaise Maize weevil	2-4	cereal grains	30-40	Yes	Just like rice weevil, except larger.
Araecerus fasciculatus Coffee bean weevil	3	maize, coffee, many kinds of seed	30-70	Yes	Not very damaging on hard, well-dried seeds. No snout.
Acanthoscelides obtectus Bean weevil	3	beans, lentils peas, misc. seeds	21-80	Yes	Oviposits on pods in field or among seeds, larvae feed and pupate inside seed. No snout. Adapted to relatively low temp.
Zabrotes subfasciatus Mexican bean weevil	3?	beans and ??	??	Yes	Attaches eggs to testa. Larvae feed and pupate inside seed. Adapted to higher temperatures than *Acanthoscelides*.
Callosobruchus maculatus Cowpea weevil	3	many legume species	20-80	Yes	??
Other beetles					
Rhyzopertha dominica Lesser grain borer	3	cereal grains	30	Yes	Unlike the weevils, the adults feed as well as the larvae. Larvae crawl actively in grain.

Scientific name Common name	Adult length	Seeds attacked	Life cycle	Can fly?	Comments
	mm		*Days*		
Tenebroides mauritanicus Cadelle beetle	8-9	many kinds of seed. esp. cereals	65-410	Yes	Usually eats only the germ. Adults live up to 2 years.
Oryzaephilus surinamensis Sawtoothed grain beetle	2-3	Omnivorous	20-80	No	Larvae and adults actively feed and move about. Adults live up to 3 years.
Oryzaephilus mercator Merchant grain beetle	2-3	Like *o. surinamensis* but prefers oilseeds.	25-100	Yes	Just like *o. surinamensis.*
Faraxonotha kirschi Mexican grain beetle	5	Cereal grains	?	?	
Trogoderma granarium Khapra beetle	1.5-3	Omnivorous, but prefers cereal grains	30-1500	No	Only Dermested beetle which prefers seeds to animal products. Activity usually restricted to top 30 cm.
Moths					
Sitotroga cerealella Angoumois grain moth	13-17	Many kinds of seeds, prefers moist seed	35-65	Yes	Eggs are laid on seed. Larvae bore into seed and emerge as adults. Larvae active at relatively low temperatures.
Plodia interpunctella Indian meal moth	16	Omnivorous, but will attack many kinds of intact seed.	40-300	Yes	Larvae crawl among seeds producing abundant webbing and cocoons. Often resistant to malathion.
Corcyra cephalonica Rice moth	13-24	Rice and processed foods	40-140	Yes	Larvae crawl among seeds producing abundant silken tubes.

From: Mallis, 1982; Cotton and Wilbur, 1982; Fleurat-Lessard, 1988; USDA, 1978; Ebeling, 1975, Cardona, 1989.

Prompt harvest, drying, and thorough cleaning of the seed help prevent entry of insects into the storage. Good sanitation in the processing plant and harvest machinery is crucial, since seed storage pests often breed in small deposits of old seed in inaccessible places. Hard-to-reach spots should be disinfested by fumigation.

Stored product insects are tropical in origin, and a principal means of control is maintenance of low temperature and humidity. Minimum temperatures needed for population increase of species listed in Table 1 range from 15 to 24°C. Optimum temperatures are much higher than these, ranging from 24 to 37°C depending on the species. At temperatures below 15°C most stored product pests become inactive and eventually die (Mullen and Arbogast, 1984).

In large bulks, the temperature may remain high for weeks in the interior of the seed mass as ambient temperature declines in autumn. Temperature gradients cause migration of moisture, leading to hot-spot formation and exacerbation of pest problems. To effectively use low ambient temperatures for insect control in bulk seed, the seed must be aerated using forced air. This is most easily done with automated systems which take advantage of nighttime low temperatures.

Small lots of seed may be placed in a deep-freeze for a few days to kill the insects. At –20°C all storage insects are killed after ten hours. At –10°C, 62 hours are required. When large bags of seed are placed in a freezer, up to 100 hours may be needed to bring the interior of the bag to the freezer temperature. Insects previously exposed to cool weather are not killed by freezing as easily as insects reared under warm conditions (Mullen and Arbogast, 1984).

Sufficient drying to achieve complete control of insect pests is not feasible in conventional seed storage. Some insects will reproduce at relative humidities as low as 10 percent (O'Dowd and Dobie, 1983). However, maintaining the lowest practical moisture content increases the generation time, making management of the insects easier, and reduces the chance of hot spot formation.

Another physical method of insect control is the mixing with the seed of inert materials which damage insects or interfere with their development. Vegetable oil at rates as low as 1 ml kg^{-1} controls weevils in large-seeded legumes (Hill and Schoonhoven, 1981). Inert dusts such as diatomaceous earth or silica aerogel can be used to control insects in seed and grain by abrasion and desiccation (Ebeling, 1975; Mallis, 1982). The ash from rice hulls has been used to protect grain from insect injury (Peng, 1982).

An inhospitable atmosphere will control insects in seed or grain (Jay, 1984). Air can be replaced with an inert gas, or with air from which the oxygen has been removed by combustion. Carbon dioxide is a special case in that complete oxygen removal is not necessary. It kills insects by a metabolic disturbance. A CO_2 concentration of about 60 percent resulted in the best insect control (Jay, 1984). Northolt and Bullerman (1982) found that atmospheres of 20 to 40 percent CO_2 in air greatly inhibited growth of storage fungi, and this could not be accounted for simply by displacement of oxygen. Use of CO_2 for pest control may damage seeds (Moreno-Martinez, Menendez, Ramirez, 1987), so it should only be used cautiously.

A major cost of controlled atmosphere storage for insect control is the need to render the seed storage vessel airtight, since exclusion of air for a period of two to three weeks may be necessary. Even so, control using inert atmospheres may be economically competitive with fumigation (Jay, 1984).

Fumigation with a toxic gas or vapor is usually the most practical way to kill insects in seed. Most commercially available fumigants do not leave any residue, which may be important in case the seed must later be sold as food or feed. Two fumigants are commonly used, methyl bromide and phosphine.

Methyl bromide is very effective, but it can injure seeds, especially at high moisture content. Powell (1975) provides guidelines for fumigation of many kinds of seed with methyl bromide.

Phosphine is generally the best fumigant because its use does not require special equipment, it is handled and transported in a reasonably safe solid form (aluminum or magnesium phosphide), and it is less damaging to seeds than is methyl bromide. The disadvantages of phosphine compared to methyl bromide include the need for long exposure time (implying airtight storage) and warm temperatures (Bond, 1984), and emerging resistance in some insect populations (Price and Mills, 1988). The threat of insect resistance makes it imperative that phosphine not be used frequently and, when it is used, that exposure time is sufficient to kill all the insects.

Treatment with a residual insecticide is an important preventive measure for grain and seed. Products registered for use in the United States include malathion (Cythion), pirimiphos-methyl (Actellic), chlorpyrifos-methyl (Reldan), and *Bacillus thuringiensis* (Dipel). Even though other insecticides may be applied as seed treatments, and will control insects during storage, it is most practical to use insecticides allowed on grain, in case the seed must be diverted into food or feed channels.

Integrated management of insect pests in seed and grain is complex. Simulation may provide a way of optimizing the various management choices (Flinn and Hagstrum, 1990).

Fungal Pests

In the actively growing state in the field, the seed is subject to attack by fungi specialized to defeat the active defense mechanisms of the parent plant. Species which attack in the field are called field fungi (Christensen, 1973). At maturity, the structures enclosing the seed senesce. Senescent maternal tissue becomes subject to invasion by omnivorous facultative saprophytes such as species of *Alternaria*. As the seed dries, only xerotolerant fungi continue to grow. These are known as storage fungi (Christensen, 1973). *Penicillium* and *Aspergillus* are the two most important genera of storage fungi.

Fungal Species

Rigid classification into field versus storage fungi is an oversimplification (McLean and Berjak, 1987). For example, under hot, dry conditions, *Aspergillus* species often infect intact ears of corn as early as the milk stage by a specialized mechanism involving silk colonization (Lillehoj, 1983; Marsh and Payne, 1984; Zummo and Scott, 1990). *Fusarium* spp., normally classified as field fungi, occupy this same niche, and can persist and even increase during storage (McLean and Berjak, 1987). *Penicillium* spp. often invade ears of corn before harvest, through insect or bird injuries (Caldwell, Tuite, and Carlton, 1981). Many species of *Penicillium* and *Aspergillus* tolerate dry conditions, but most grow faster as the relative humidity (RH) approaches 100 percent.

Storage fungi are weak pathogens. They invade seeds because they grow at water activities too low for metabolism and defensive reactions by the seed. Infection of seeds in the field (Mclean and Berjak, 1987) and saprophytic activity on plant residue ensure that inoculum will always be present. Storage fungi initially attack only the nonliving portions of the seed, such as the pericarp or testa. In some kinds of seed, the embryo is never invaded, but hyphae ramify between the embryo and testa. In such cases, the embryo may be killed, apparently by diffusible toxins or enzymes, either before or during imbibition (Harman and Pfleger, 1974). In cereal grains, the embryo is typically colonized and darkened (Christensen, 1973; Harman and Pfleger, 1974).

The ecological determinants of fungal growth in stored seed are RH and temperature. Table 2 lists common species of storage fungi and their limits of growth. As seed dries and is placed in storage, a succession of species occurs as RH and temperature decline. Field fungi rapidly become inactive. If the seed is promptly dried and properly stored, storage fungi are active for only a short, insignificant time period. If RH is greater than 65 percent and the temperature above about 6°C, certain species of the *Aspergillus glaucus* group slowly grow. Even if the seed is too dry on average for fungal growth, invasion may occur in scattered spots which are more moist due to moisture migration or insect activity. Fungal growth is not always visible, and apparently normal, germinable seeds may be infected and reduced in vigor.

Once growth by the most xerotolerant species is underway, additional moisture is formed metabolically and the niche begins to change. Less xerotolerant, but faster-growing fungi such as those of the *Aspergillus flavus* group begin to dominate (McLean and Berjak, 1987). Temperature may rise from metabolic heat, leading to rapid decay. If the temperature remains lower than 5 to 10°C, *Penicillium* spp. may dominate, but these are not as xerotolerant as many *Aspergillus* spp. (Christensen, 1973). It should be emphasized that development of storage fungi to economically important levels may take years at temperatures below 15°C, even at relatively high moisture content (Christensen, 1973).

TABLE 2. Common species of storage fungi and their temperature and humidity limits.

Species	Min. RH	Min. Temp.	Comments
Aspergillus spp.			
halophilicus	0.65	?	Rare, slow growing, difficult to isolate.
restrictus	0.70		Slow growing, causes dark, purplish blue eye in cereals.
glaucus	0.73	?	Very common, causes incipient deterioration, slowly destroys germ in cereals (blue eye).
candidus	0.75	5-10	Fast growing, major cause of heating, germs turn black.
fumigatus	0.82	12	May dominate at high temperatures.
flavus	0.85	5-10	Can invade maize in the field, major cause of heating.
Penicillium spp.	>0.8	–2	Cause blue eye of high moisture maize at low temperatures, can initiate hot spots in seed at low temperatures.
Alternaria spp.	>0.9	?	Mainly field saprophytes.
Fusarium spp.	?	?	Invade in field, may grow after harvest, before and during drying.

From Christensen, 1973; Neergaard, 1977; Justice and Bass, 1978; Northolt and Bullerman, 1982.

Control of Storage Fungi

Storage fungi are controlled by prompt drying, maintenance of relative humidity below 65 percent, and prevention of insect infestation and moisture migration. Low temperature is effective if available. Fungicide treatment has been considered ineffective against storage fungi (Christensen, 1973), although certain fungicides apparently provide some control (Moreno-Martinez and Ramirez, 1985; White and Shove, 1988). Thiabendazole is currently labeled in Illinois for application to high-moisture maize grain at harvest. Such application delays fungal growth sufficiently to allow inexpensive, ambient air drying.

THE STORAGE ENVIRONMENT

Seed storage begins on the plant in the field, and overlaps with natural drying. It is not possible to draw a firm boundary between drying and storage. Indeed, the seed may dry several times in the field. Harvest at the point of physiological maturity is not feasible or beneficial, since rapid artificial drying often damages high-moisture seed, and fungi may multiply very rapidly before drying is possible. Yet the tendency in the seed industry is to harvest as early as possible in an effort to reduce risk, improve appearance, and increase yield. The optimum time to harvest seed crops is often a point of disagreement.

Seed Drying

Longevity of seed in storage depends on reducing seed moisture to a level sufficiently low to preclude physiological and pathological deterioration. But from the perspective of the seed, its period of quiescence has already begun, and deterioration has already started. It is important to slow down the process before seed quality becomes too low. Successful seed drying depends on achieving a balance between cost, damage caused by the act of drying itself, and destruction from storage at high moisture. This damage is caused by intrinsic physiological reactions of the seed to moisture and predation by fungi.

The Nature of Drying Injury

Drying can injure seed by physiological and physical mechanisms. Drying requires heat to vaporize water. The economic incentive in artificial drying is for increased throughput of seeds by increasing drying temperature. High temperature injures the seed by excessive drying rate or by interacting with moisture content to accelerate physiological deterioration.

Because seeds are not infinitely small, the act of drying establishes a moisture gradient inside the seed. Rapid drying of high-moisture seed causes a steep moisture gradient in seed tissue. Desiccation makes tissue shrink. In this way, moisture gradients in seeds are converted to mechanical stress, which can damage seed tissue (Eckstrom, Liljedahl, and Peart, 1965). Spaeth (1986) has provided a biophysical description of the process by which mechanical stress is established by moisture gradients in seeds. Large, brittle seeds with a large length to width ratio are most susceptible to this type of damage.

Moisture and temperature interact to cause physiological damage to seeds. The relationship of temperature, moisture content, and time to seed injury will be discussed in more detail below.

Practical Seed Drying

Artificial heat is expensive, and potentially damaging to seed. Solar drying is common, although it is often not thought of as such. The most efficient solar collector is the crop standing in the field. One reason for the comparative advantage and high seed vigor of arid regions is the ability to promptly dry the standing or windrowed crop.

Postharvest solar drying may be accomplished by spreading the commodity in a shallow layer on a floor or tarpaulin. This approach is probably more efficient than systems employing a discreet solar collector (Arinze, Schoenau, and Bigsby, 1987). Temperature of seed on the top may exceed 50°C (Wilson, unpublished data). Regular turning and covering at night (or during rain) is necessary to facilitate drying and to avoid damage. The higher the initial moisture, the shallower the layer must be to achieve sufficiently rapid drying.

Extensive modeling and optimization of grain drying has been performed (Hutchinson, 1944; Henderson and Perry, 1976; Nellist, 1980, 1981; Foster, 1982; Sokhansanj, 1987; Sokhansanj and Bruce, 1987; Lasseran, 1988; Sanderson et al., 1989). Caution is in order, since these authors consider only seed germination as a response, not more sensitive measures of seed vigor. The effect of drying injury on vigor of maize seed has been studied by Herter and Burris (1989a, 1989b).

Seeds may be rapidly and gently dried by storage over (Zhang and Tao, 1988), or intimate mixing with (Graham and Bilanski, 1986), dry solid absorbents. Silica gel granules work well, but cheaper materials may be more practical for large-scale use (Sadik and White, 1982; Raghavan et al., 1988; Watts, Bilanski, and Menzies, 1988). The desiccant is removed and regenerated after use. Cromarty (1982) provides a good review of approaches for drying seed for long-term storage.

Although it is probably possible to injure seed by drying it to an excessively low moisture content (Vertucci and Roos, 1990), indirect damage caused by rapid moisture uptake during imbibition is more common. Germination tests on such seed often yield anomalously low results if the seed is not slowly rehydrated before testing.

Effect of Moisture on Seed Longevity

Seed moisture content is the most important determinant of longevity in storage. Harrington's (1960) rule of thumb was that for each 1 percent decrease in seed moisture content, storage life doubled. Within a seedlot, if the log of the average seed lifespan is plotted against the log of the moisture content, a straight line results. This relationship exists in the range of water activities from 0.08 to 0.90 (Roberts and Ellis, 1989). Other work suggests that the lower limit for effect of moisture on longevity occurs at a water activity of 0.10 (Ellis, Hong, and Roberts, 1989); or 0.19 (Vertucci and Roos, 1990). The position of the lower limit is important, since drying seed beyond the limit does not increase longevity.

Water Activity and Percent Moisture

As seeds lose water to the atmosphere, water that remains is bound more and more tightly to seed macromolecules. Each increment of water removed requires a progressively lower RH to pull it out of the seed. This gives rise to characteristic curves relating RH to moisture content on a weight basis. Every substance has its own unique moisture desorption relationship, the shape of which depends on its composition. A similar curve can be plotted as the substance absorbs moisture. The two curves are usually not exactly the same because once moisture is absorbed at a given RH, a slightly lower RH is needed to release it. This phenomenon is called hysteresis.

Water activity (A_w) is the chemical potential of water in the system. It indicates the degree to which water is chemically available. Water activity equals RH divided by 100, assuming that water vapor is an ideal gas, an assumption which yields errors of less than 0.2 percent (Multon, 1988a). This chapter will use A_w, and percent moisture (wet wt. basis) to express moisture content of seeds.

The important thing from the perspective of chemical reactions in the seed or organisms preying on the seed is A_w, not percent by weight. Storage of one kind of seed at 14 percent moisture may be feasible, while for a different species 14 percent moisture might be too high. Oilseeds equilibrate to a higher RH at a given moisture content by weight than do non-oily seeds. Figure 1 shows moisture absorption curves (sorption isotherms) for common bean (*Phaseolus vulgaris* L.), soybean (*Glycine max* Merr.), and rape (*Brassica napus* L. var. *napus*). A safe storage moisture content from the perspective of deterioration by fungi would be 13 to 14 percent for bean, 9 to 10 percent for soybean, and 6 to 7 percent for rape. Decisions concerning safe moisture content are risky without some information about the sorption isotherm for that kind of seed.

Holding percent moisture constant, increasing seed temperature increases A_w. Sorption isotherms for seeds exhibit a characteristic shape, and the family of curves resulting from equilibration at a range of temperatures has been described as a two-parameter model (Henderson and Perry, 1976; Aguerre, Suarez, and Viollaz, 1986). Information about equilibrium moisture contents or sorption isotherms for a wide range of seeds has been published (Justice and Bass, 1978; Multon, 1988a).

To construct sorption isotherms, samples of seed are equilibrated in atmospheres of known relative humidity, usually over slushes made by mixing water with various salts (Winston and Bates, 1960). When the seed samples no longer change weight, the seed is removed and moisture content determined (ISTA, 1985; Grabe, 1989). It is easiest if the trial is arranged in such a way that all the samples are either absorbing or desorbing moisture. Otherwise, discontinuities in the curve may arise from hysteresis.

In the case of species for which little information is available, an alternative to estimation of A_w from percent moisture via the sorption isotherm is direct measurement of RH in the atmosphere surrounding the seeds. The atmosphere must be in equilibrium with the seed, and this could take several days, depending on the required

FIGURE 1. Moisture absorption curves (sorption isotherms) for common bean, soybean, and rape.

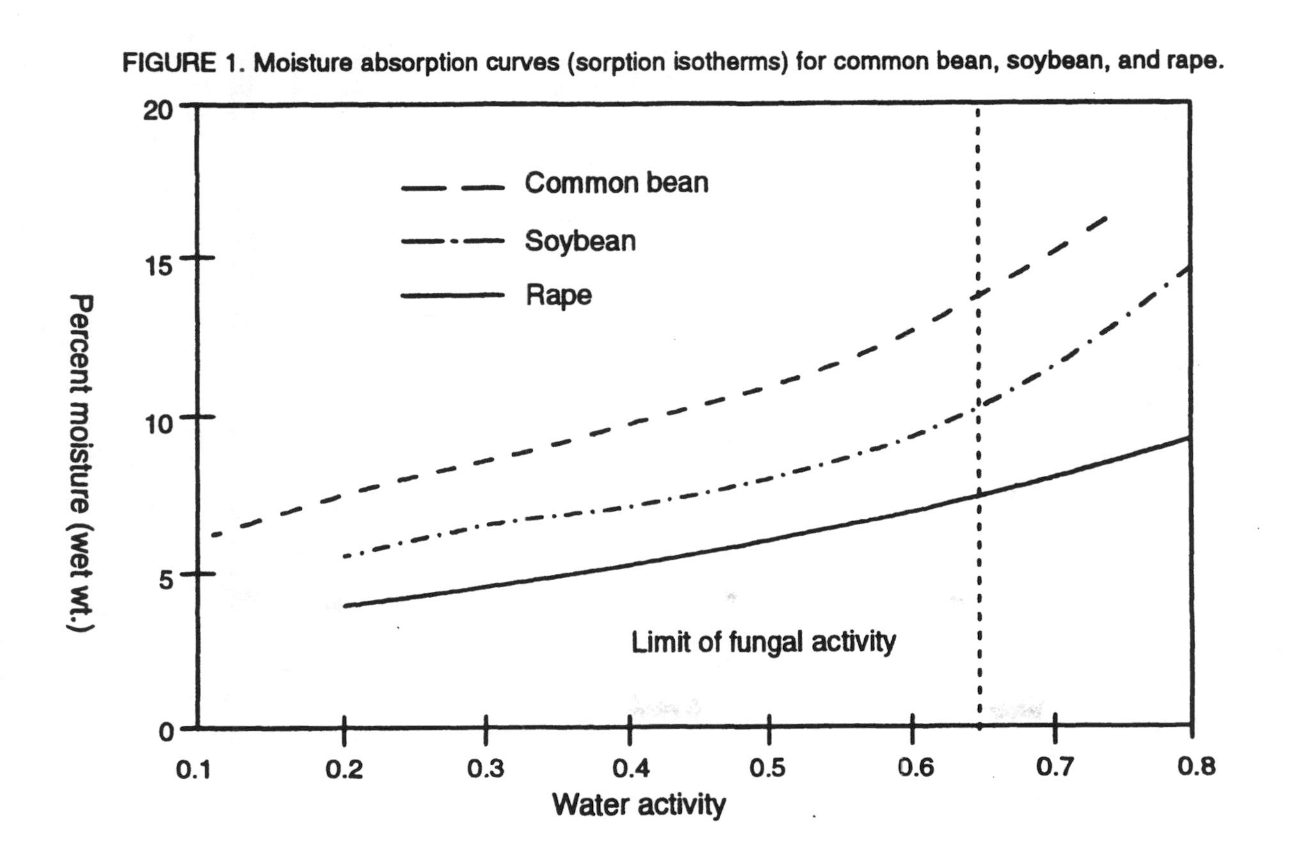

precision. Inexpensive (100 to 500 dollars) hand-held electronic hygrometers are available, and would probably serve well. For example, a relative humidity probe could be plunged deeply into stored seed and a reading taken to give an indication of whether the moisture content is sufficiently low for safe storage.

Control of Moisture During Storage

Seed must be kept dry to maintain quality in storage. If the seed is well dried and handled in a large bulk, it will absorb water only very slowly from the outside environment. This assumes that liquid water is not permitted to leak into the storage. Moisture normally moves very slowly by diffusion among air spaces in seed. It can move significantly only if a net movement of air occurs in the grain mass, usually associated with temperature gradients.

The moisture content of seed can be maintained by controlling the RH of the warehouse atmosphere. Various dehumidifier options and construction details are discussed elsewhere (Delouche et al., 1973; Justice and Bass, 1978; Burrell, 1982; O'Dowd and Dobie, 1983). If subambient temperature is maintained in the seed storage room using refrigeration equipment, dehumidification becomes crucial. As air is cooled by the refrigerator, its RH increases. The net effect of refrigeration without dehumidification is usually decreased seed storability compared to ambient storage. Of course, if the seed is hermetically sealed in moisture-proof containers, it will not absorb moisture from the air. In this case, refrigeration alone will greatly extend seed longevity.

If seed is packed in cloth or paper bags and loosely stacked under ambient conditions, equilibration with the environment occurs fairly quickly, and seed viability and vigor may be lost in a few months in a warm, humid environment (O'Dowd and Dobie, 1983). For distribution and small-scale storage under these conditions, the well-dried seed should be packed in moisture-proof containers. Metal or plastic drums work well. Not all apparently moisture-proof, flexible materials are equally resistant to the passage of water vapor. Justice and Bass (1978) present information on the performance of a wide range of moisture barrier packing materials for seeds.

Temperature and Seed Longevity

The rate of chemical reactions typically increases by a factor of two as temperature is increased by 10°C. Longevity of seed has been shown to be a negative exponential function of temperature (Crocker and Groves, 1915; Hutchinson, 1944; Hukill, 1963). Ellis and Roberts (1981) found that the applicable temperature range could be extended by adding a temperature squared term to the exponent. The bulk of this work concerns high temperatures (30 to 70°C) and is most relevant to drying injury. Uncertainty remains about applicability of these relationships to subambient temperature, since the time periods involved are too long for convenient study.

Temperature Gradients

In bulk seed, temperature gradients cause moisture to migrate. At a point of heat input, the RH of interstitial air is lowered by the increase in temperature. This causes water to move from adjacent seed into the air. Warm air rises, and when it encounters cold seed or bin walls, its temperature drops, increasing the RH. Water then moves from the moist air into the newly adjacent grain. Moisture migration is most commonly observed when autumn weather cools the exterior of a bin loaded with warm grain. It may also occur when only one side of an outdoor storage bin is exposed to the sun.

The possibility of moisture migration necessitates a margin of safety in allowable moisture level. Seed must be dry enough so that a few percent increase in local moisture content will still allow safe storage. Moisture migration is controlled by aeration, which brings the seed close to ambient temperature.

Self-Heating

Stored seed or grain represents a thermodynamic disequilibrium in the presence of oxygen. Seed or grain is said to "heat" when energy generated by respiration of the seeds and associated organisms increases the temperature. Grain exhibits negligible heat conductivity and low specific heat (Multon, 1988b). This is probably true for most kinds of seed. Relatively small heat inputs significantly increase temperature. Heat production is autocatalytic, since the heat-producing reactions go faster at higher temperature. Heating may not involve the whole mass of seed, but occur in scattered foci of insect infestation, high-moisture pockets, or areas containing fine debris. Local hot spots grow as warm moist air rises, increasing moisture content of cooler seeds above. Heating of seeds is mainly a problem in the early postharvest period before the product has been dried.

The portion of the bin directly under the spout through which grain is added is particularly subject to heating and spoilage. As grain is added to the bin, a cone forms. Larger, smoother, intact seeds tend to roll down and away from the cone, but broken grain, debris, and insects remain in the center of the developing pile. This core is subject to reduced air movement and higher moisture content, and may form a hot spot. Other areas subject to spoilage include the upper and outer portions which may be increased in moisture content due to moisture migration.

Atmospheric Composition

Storage of grain under reduced oxygen or high CO_2 is seen as a way to kill animal pests (Jay, 1984; Shejbal and DeBoislambert, 1988). In fairly moist (approx. 15 percent H_2O) grain, this can be accomplished by respiration of pests in hermetic storage (Shejbal and De Boislambert, 1988). More often, the atmosphere is actively modified by flushing with nitrogen, CO_2, or combusted air (Jay, 1984). Fungi cannot be controlled in practice by reducing oxygen concentration, but are inhibited by high CO_2 levels (Northolt and Bullerman, 1982).

Since deterioration of dry seed is thought to result from attack by oxygen (Senaratna, Gusse, and McKersie, 1988; Wilson and McDonald, 1986), attempts have been made to increase seed longevity by storage under inert gas. In high-moisture soybeans (20 percent H_2O, 25°C, 16 days), loss of germinability was linearly related to oxygen pressure (0 to 7.7 atm) (Ohlrogge and Kernan, 1982). Other results have been variable (Justice and Bass, 1978). Variability of response probably stems from multiple mechanisms of deterioration in stored seed and inconsistent and unpredictable change in atmospheric composition during sealed storage.

Dry maize (less than 12.6 percent H_2O) produces CO_2 and consumes O_2, but this seems to have nothing to do with respiration (Bartholomew and Loomis, 1967). There is no reason to think that respiration either occurs or is necessary for survival in very dry seeds. Other, mainly deleterious, chemical reactions are sufficient to account for oxygen uptake. Seeds may exist in an "ametabolic state" (Stewart, 1989). On the basis of oxygen uptake, respiratory quotients, ATP synthesis, and action of respiratory inhibitors, it is thought that mitochondrial electron transport only occurs at water activities above about 0.91 (Vertucci and Roos, 1990). As moisture content of seeds is increased, longevity decreases, but a point is reached at which increasing moisture content results in increasing longevity (Ibrahim and Roberts, 1983; Ward and Powell, 1983). This transition suggests a change in the mechanism of survival, from a state of suspended animation to active metabolism. Below this point, longevity may be unaffected or enhanced by hermetic or inert atmosphere storage (Rao and Roberts, 1990). Above this point, however, oxygen is clearly beneficial. In most storage experiments at these high moisture levels, the seed is subject to attack by fungi. When seed of intermediate or high moisture content is placed in hermetic storage, CO_2 may accumulate, inhibiting metabolism of both fungi and seeds. The resulting low O_2 content might also injure the seeds if they are sufficiently moist for metabolic reactions.

QUANTIFICATION OF LONGEVITY IN STORAGE

By the 1950s it had become accepted that longevity decreased exponentially with increasing temperature (Crocker and Groves, 1915; Hutchinson, 1944; Hukill, 1963). Relationship between seed moisture content and longevity was either exponential (Hukill, 1963) or a power curve (Hutchinson, 1944). A practical problem that remained was how to define longevity. Hutchinson's (1944) approach of noting just when reduction in viability began, or when all the seeds were dead, was not practical because the longevity depended on how many seeds were tested. Testing large numbers of seeds always revealed outliers at both ends.

The Roberts/Ellis Model

A significant advance was made by Roberts (1960), who recognized that seed mortality events were normally distributed over time in storage. In other words, percent (or probability) of germination over time approximated a negative cumulative normal distribution. Roberts borrowed a statistical technique from dose-response bioassay called probit analysis (Finney, 1971). A similar technique, which does not depend on normality of the underlying distribution, using the Weibull curve has been described (Moore and Jolliffe, 1987).

Estimation of Median Viability Period

In probit analysis the population of seeds is represented by the area under a normal frequency distribution curve. The standard deviation is arbitrarily set to one unit (one probit). By projecting a line downward from the frequency distribution curve to cut off various proportions of the population (mortality), the standard units on the horizontal axis can be used to express the proportion of surviving seeds. It turns out that the standard units are roughly proportional to time in storage. Probit analysis is a convenient way to fit cumulative normal distributions, since the relationship between probits and the independent variable (time, in the case of seed storage) becomes linear. A good in-depth presentation of the rationale of probit analysis in this context is given by Ellis (1982).

Roberts used probit analysis to calculate average or median longevity (P_{50}) just like toxicologists calculate LD_{50}. Then he formulated models to relate P_{50} to storage temperature and seed moisture such as:

$$\text{Log } T_{50} = K_v - C_1M - C_2T \qquad (1)$$

or:

$$\text{Log } T_{50} = K_v - C_1\text{Log}M - C_2T \qquad (2)$$

Where M = seed moisture content, T = storage temperature, and K_v, C_1, and C_2 are constants (Roberts, 1973).

The New Improved Viability Equation

The seed survival curve in the form of probits is given as:

$$v = K_i - p(1/\sigma) \quad (3)$$

where v is the viability after probit transformation, K_i is the intercept, or initial viability on the probit scale, and p is a period of time in storage. The reciprocal of the slope of the line described by equation 3 is the standard deviation of the frequency distribution of seed deaths in time. The slope of the probit line increases as storage conditions become more severe.

The slope of the probit viability curve can be predicted from the environmental conditions of storage (M, T) just like P_{50}. An improved viability equation (Ellis and Roberts, 1981) combines equation 3 with an improved version of equation 3:

$$v = K_i - p/10^{(K_e - C_w \text{Log } M - C_h t - C_q t^2)} \quad (4)$$

Ellis and Roberts (1981) found the parameters to this model by a two-step process. First, probit analysis was conducted on an environment-by-environment basis. Then they determined the parameters K_e, C_w, C_h, and C_q by regressing the log of $1/\sigma$ over the exponent of equation 4. Within species, these parameters are reported to be stable. With the parameter values, and some knowledge of initial viability (K_i) of the seedlot, it is possible to estimate the viability after any combination of time, temperature, and moisture content. A more computationally intensive approach to estimating the parameters involves specifying equation 4 as a nonlinear regression model (Wilson, McDonald, and St. Martin, 1989). This may be more effective for sparse data sets.

Critique of the Roberts/Ellis Model

Extrapolation

Ellis and Roberts (1980) stress "... the unity of the process of seed deterioration." But seeds probably die by different mechanisms under different storage environments, even if a simple model fits the data. Not enough is known about the mechanisms of seed aging to simulate it mathematically. The Roberts/Ellis model is purely empirical, so it should not be used to make predictions outside the data space from which the parameters were calculated. For example, using parameter values from a storage experiment with bean (*Phaseolus vulgaris* L.) seed stored at seed moisture contents ranging from 12 to 24 percent and temperatures from 20 to 60°C up to one year, good predictions of germinability were obtained within the range of environments used (Wilson, McDonald, and St. Martin, 1989). But the model also predicted that if the lot were stored at 5°C and 5 percent moisture, germination would still be 60 percent after 1,000 years.

Application to Typical Seedlots

Most of the seedlots employed by Roberts and Ellis exhibited extremely high initial germination, so high, in fact, that they were indistinguishable from 100 percent in germination tests with reasonable sample size (less than 1,000 seeds). These may have been examples of very high-quality lots, but the germinations were also high because radicle protrusion was used as the criterion of germinability rather than the usual criteria employed by seed technologists.

When Roberts/Ellis style probit analysis is applied to seedlots of lower initial germination (less than 98 percent or so), curves always result instead of the straight line implied by equation 3. It does not matter whether the initial germination was low because the lot was bad or because only normal seedlings were counted. The model used by Roberts and Ellis (equation 3) for probit analysis is technically incorrect (Finney, 1971; Wilson, McDonald, and St. Martin, 1989). The portion of the seed population which fails to germinate initially should not be considered part of the stored population being studied. The initial proportion of nongerminable seeds should be adjusted out of the model or jointly estimated using the maximum-likelihood method (Finney, 1971). This correction is important even when initial germination is above 90 percent (Hoekstra, 1987). If correction for initial germination is not made, programs such as SAS PROC PROBIT (SAS Institute, 1985) often report that the normal distribution does not fit the data, and the slope is underestimated (Wilson, McDonald, and St. Martin, 1989).

Nevertheless, the Roberts/Ellis approach is probably still useful if the initial germination is not lower than about 95 percent. Practical advice on using the viability equation (equation 4) is given by Ellis (1988). If correction for initial germination must be made, models such as the new improved viability equation (equation 4) probably cannot be formulated. But probit analysis can still be used to estimate P_{50}.

Meaning of K_i

When no correction is made for initial germination, and it is significantly lower than 100 percent, the probit analysis intercept (K_i) is an estimate of initial germination. The meaning of K_i is less clear when initial germination is very high. The difference between, say, 99.9 and 99.999 percent is large on the probit scale, but biologically nebulous. Yet, K_i does convey information about the storability of a seedlot. It may be that the controlled deterioration vigor test (Ellis and Roberts, 1980), which provides an estimate of K_i, is useful as an index of storability or resistance to drying damage.

EXTENDING THE STORAGE LIFE OF SEEDS

Cryogenic Storage

Theoretically, seed should remain viable for an extraordinarily long time at liquid nitrogen (LN_2) temperature (–196°C), perhaps hundreds of years. Repositories for storing genetic material as seeds spend a large fraction of their resources testing seed as it deteriorates in conventional storage, and growing out accessions to produce new lots of seed. Cryogenic storage is estimated to cost only one fourth as much as conventional low-temperature (–18°C) storage when averaged over 100 years (Stanwood and Bass, 1981). Stanwood (1985) provides a good review of cryopreservation methods for seed.

The majority of orthodox seeds, dried sufficiently for storage under conventional regimes, can be stored over LN_2 and rewarmed uninjured, without any specific precautions (Stanwood, 1985).

Damage, when it occurs, results from either phase transition or thermal shock. Normally, cryopreserved materials are not immersed in LN_2 for a variety of practical reasons, but kept in the vapor over LN_2 at about –150°C (Stanwood, 1985).

Moist seeds are injured or killed when placed over LN_2 because free water in the tissue freezes. The critical moisture level at which this occurs is called the high moisture freezing limit (HMFL). Stanwood (1985) provides a list of HMFLs for common species. Many oily seeds, even when very dry, are damaged by rapid cooling as a result of a phase transition of the seed oil. Damage is minimized by reducing the cooling rate to 8°C min^{-1} or less, by sheathing the cryovials with insulation (Vertucci, 1989). Rate of warming after storage is apparently not critical.

Some seeds are mechanically damaged by plunging into LN_2 (Stanwood, 1985; Pritchard, Manger, and Prendergast, 1988). This probably results from a thermal shock phenomenon, as tissue contracts in response to low temperature.

Hydration-Dehydration Treatments

Treatment with water is used in several ways to enhance seed quality. Seed priming involves restriction of growth of imbibed seeds by regulation of water potential, or temperature, for a period of four to 15 days followed by redrying. Hydration-dehydration treatments are less extreme than priming, and are defined here as increasing A_w to about 0.9 or higher for about one day, followed by redrying.

Hydration-dehydration treatments improve survival in field planting, response to vigor tests, repair damage from previous aging treatments (Dey and Mukherjee, 1986), and reduce lipid oxidation products (Saha, Mandal, and Basu, 1990). A peculiar effect of these treatments is that they mitigate the effects of aging treatments applied after the hydration-dehydration event (Savino, Haigh, and De Leo, 1979; Basu and Pal, 1980; Dearman, Brocklehurst, and Drew, 1985; Dey and Mukherjee, 1986; Georghiou, Thanos, and Passam, 1987; Chourhuri and Basu, 1988; Thanos, Georghiou, and Passam, 1989). It is almost as if the hydration treatment solved a problem that had not yet occurred. Yet, the amelioration of future aging is consistent with the view of seed aging as a cumulative phenomenon

(Wilson and McDonald, 1986), and an understanding of aging as beginning in the field before harvest.

Anyone familiar with the agronomic performance of seeds understands that decline in seed viability during storage is not a stochastic process. Each seed gradually weakens and eventually slips past the point of yielding a normal seedling in the germination test. Certain factors or damage begin to accumulate, even before harvest, and each additional increment of these factors reduces the probability of survival when the seed is later planted or tested. Apparently, hydration-dehydration treatments reset the clock in some way, either by repairing, quenching, detoxifying, or releasing some age-related deleterious factor. Neither protein synthesis (Pan and Basu, 1985) nor oxygen (Goldsworthy, Fielding, and Dove, 1982) appear to be necessary for the hydration effect, so there is some doubt this phenomenon results from macromolecular turnover and repair. The effect on subsequent aging by these treatments is most dramatic in seeds that have already been aged for a time (Basu and Pal, 1980; Saha and Basu, 1984; Chourhuri and Basu, 1988).

Hydration-dehydration treatments work across a wide variety of species. The treatments have been performed by dipping, soaking, uptake from osmotic media, and by vapor-phase hydration. It may be possible to extend the life of seeds by periodic hydration-dehydration cycles during storage.

Antioxidant Treatment

If damage to seeds results from free radical attack (Wilson and McDonald, 1986; Senaratna, Gusse, and McKersie, 1988), antioxidants might be expected to prolong the life of seeds. Evidence has been equivocal. Attempts to infuse water soluble reducing agents into seeds have revealed the effects of the agents to be minor compared to the beneficial effect of water alone (Basu and Dasgupta, 1978; Barnes and Berjak, 1978). Infusion of lipid soluble antioxidants into seeds using organic solvents has occasionally reduced the rate of aging (Bahler and Parrish, 1981; Woodstock et al., 1983). However, there is some question about the depth of penetration of organic solvents in seeds. Studies using antioxidants have failed to include post-aging antioxidant treatment as a control to separate possible effects on the germination process from effects during aging. Treatment of soybeans with a volatile antioxidant resulted in a fairly dramatic 30 percent gain in germination after aging, possibly because the antioxidant was removed before germination (Yang and Yu, 1982).

Treatment of seeds with halogens has resulted in increased storability (Basu and Rudrapal, 1980; Rudrapal and Nakamura, 1988). The halogen adds across the double bond of unsaturated fatty acids rendering them less susceptible to hydrogen abstraction and free radical formation.

A phenomenon probably related to free radical reactions is reduced aging rate of seeds in an electric field (Barnes and Berjak, 1978; Berjak, 1978). The mechanism is not understood.

Vigor and Storability

Assuming a good storage environment, the most practical way to further extend the storage life of seeds is to begin with seed of the highest possible vigor. Deterioration of seed in storage is a cumulative phenomenon, and harvest before optimum maturity, poor weather near harvest, severe drying conditions, or temporary storage at high moisture can take years off the storage life of seeds. The focus should be on vigor rather than germinability.

Seed storage reduces vigor more rapidly than viability. Under dry conditions, seeds gradually become very weak, yet retain the ability to germinate. At high moisture levels, both vigor and viability drop off suddenly (Wilson, 1986). Figure 2 shows the estimated relative frequency distributions of germinability loss events in storage of bean seed. Counts were made at either four days (vigor test) or seven days (standard germination test). Under favorable storage conditions (Figure 2, top graph), germination at four days was reduced much sooner than germination at seven days. Under severe conditions (bottom graph), seven- and four-day germination was lost at about the same time. Rao, Roberts, and Ellis (1988) found that under high-moisture storage of lettuce seed, loss of viability occurred before much chromosome damage was observed. Under low-moisture storage, chromosome damage occurred at high levels before viability was lost.

The danger in long-term storage is that subtle damage can occur to the seed which is undetectable by germination under favorable

FIGURE 2. Relative frequency distribution of germination reduction in storage of bean seed.

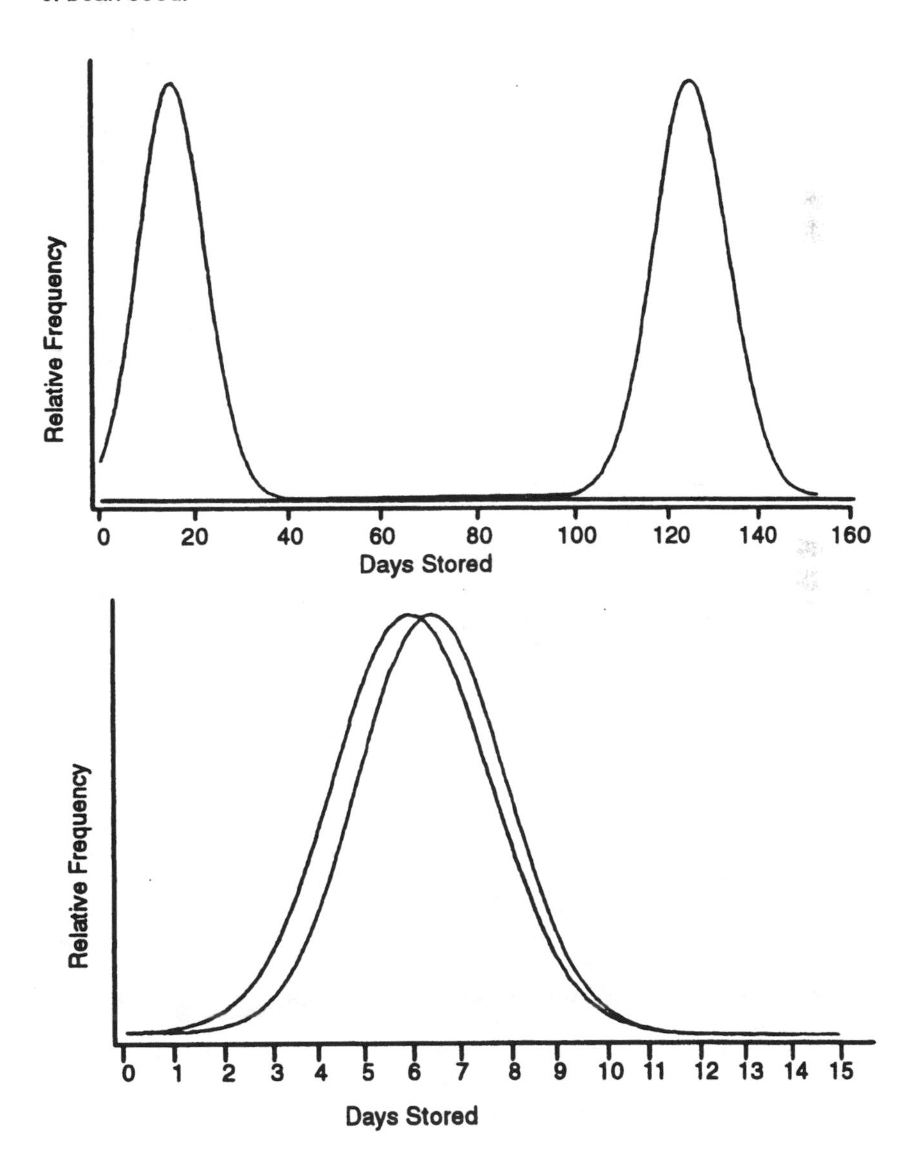

conditions. This is dangerous for germplasm conservation because it might lead to genetic changes. It is dangerous for sowing purposes since the ability of the seed to resist stress is impaired.

CONCLUSION

The aspect of seed storage that distinguishes it from the storage of other agricultural commodities is the fact that seeds are alive and the value of the seed lies in this life. Seed storage means protection of this life. Recognition of this must begin as the seed matures in the production field, and continue until it is planted. Orthodox seeds represent an astounding adaptation to storage and dispersal in dry environments. But seeds are defenseless against predation and environmental injury. Seeds cannot fight off or compensate for damage, and this damage accumulates.

Often, seed deterioration during storage is manifested not as a significant decline in germinability, but as a loss of seed vigor. This results in increased risk and ultimately cost, to growers in the case of commercial seed, and to germ plasm storage facilities in the case of genetic conservation. Design and implementation of seed storage systems should focus on maintenance of seed vigor.

REFERENCES

Aguerre, R. J., C. Suarez, and P. E,. Viollaz. Enthaly-entropy compensation in sorption phenomena: Application to the prediction of the effect of temperature on food isotherms. *J. Food Sci.* 51 (1986): 1547-1549.

Arinze, E.A., G. J. Schoenau, and F. W. Bigsby. Determination of solar energy absorption and thermal radiative properties of some agricultural products. *Trans. ASAE* 30 (1987): 259-265.

Bahler, C. C., and D. J. Parrish. Use of lipid antioxidant to prolong storage life of soybean seeds. *Agronomy Abstracts, Am. Soc. Agron.* (1981): 117.

Barnes, G., and P. Berjak. The effect of some antioxidants on the viability of stored seeds. *Proc. Electron Microsc. Soc. S. Afr.* 8 (1978): 95-96.

Bartholomew, D. P., and W. E. Loomis. Carbon dioxide production by dry grain of *Zea mays*. *Plant Physiol.* 42 (1967): 120-124.

Basu R. N., and M. Dasgupta. Control of seed deterioration by free radical controlling agents. *Ind. J. Exp. Biol.* 16 (1978): 1070-1073.

Basu, R. N., and P. Pal. Control of rice seed deterioration by hydration-dehydration pretreatments. *Seed Sci. & Technol.* 8 (1980): 151.

Basu, R. N., and A. B. Rudrapal. Iodination of mustard seed for the maintenance of vigor and viability. *Ind. J. Exp. Biol.* 18 (1980): 492-494.

Berjak, P. Viability extension and improvement of stored seeds. *S. Afr. J. Sci.* 74 (1978): 365-368.

Bond, E. J. Fumigation of raw and processed commodities. In *Insect Management for Food Storage and Processing*, ed. F. J. Baur (St. Paul, MN: Am. Assoc. of Cereal Chem., 1984), pp. 143-160.

Burkholder, W. E. Use of pheromones and food attractants for monitoring and trapping stored-product insects. In *Insect Management for Food Storage and Processing*, ed. F. J. Baur (St. Paul, MN: Am. Assoc. of Cereal Chem., 1984), pp. 69-86.

Burrell, N. J. Refrigeration. In *Storage of Cereal Grains and Their Products*, ed. C. M. Christensen (St. Paul, MN: Am. Assoc. of Cereal Chem., 1982), pp. 407-442.

Caldwell, R. W., J. Tuite, and W. W. Carlton. Pathogenicity of *Penicillia* to corn ears. *Phytopathology*, 71 (1981): 175-180.

Cardona, C. Insects and other invertebrate bean pests in Latin America. In *Bean Production Problems in the Tropics*, eds. H. F. Schwartz and M. A. Pastor-Corrales (Cali, Colombia: Centro Internacional de Agricultura Tropical CIAT, 1989).

Chourhuri, N., and R. N. Basua. Maintenance of seed vigour and viability of onion (*Allium cepa* L.). *Seed Sci. & Technol.* 16 (1988): 51-63.

Christensen, C. M. Loss of viability in storage: Microflora. *Seed Sci. & Technol.* 1 (1973): 547-562.

Cohen, J. I., J. T. Williams, D. L. Plunknett, and H. Shands. *Ex situ* conservation of plant genetic resources: Global development and environmental concerns. *Science* 253(1991): 866-872.

Cotton, R. T., and D. A. Wilbur. Insects. In *Storage of Cereal Grains and Their Products*, ed. C. M. Christensen (St. Paul, MN: Am. Assoc. of Cereal Chem., 1982), pp. 281-318.

Crocker, W., and J. F. Groves. A method of prophesying the life duration of seeds. *Proc. Nat. Acad. Sci. USA* 1 (1915): 152-155.

Cromarty, A. Techniques for drying seeds. In *Seed Management Techniques for Genebanks*, (Kew, UK: Report of a Workshop held July 6-9 at the Royal Botanical Gardens, 1982), pp. 88-130.

Dearman, J., P. A. Brocklehurst, and R. L. K. Drew. Effects of osmotic priming and ageing on onion seed germination. *Ann. Appl. Biol.* 108 (1985): 639-648.

Delouche, J. C., R. K. Matthes, G. M. Dougherty, and A. H. Boyd. Storage of seed in sub-tropical and tropical regions. *Seed Sci. & Technol.* 1 (1973): 671-700.

Dey, G., and I. I. T. Mukherjee. Deteriorative change in seeds during storage and its control by hydration-dehydration pretreatments. *Seed Res.* 14 (1986): 49-59.

Ebeling, W. *Urban Entomology* (Los Angeles, CA: Univ. of California Div. of Agric. Sci., 1975).

Eckstrom, G. A., J. B. Liljedahl, and R. M. Peart. Thermal expansion and tensile properties of corn kernels and their relationship to cracking during drying. Paper: American Society of Agricultural Engineers Winter Meeting–Chicago, II No. 65-809 (1965): 1-17.

Ellis, R. H. The meaning of viability. In *Seed Management Techniques for Genebanks* (Kew, UK: Report of a Workshop held July 6-9 at the Royal Botanical Gardens, 1982), pp. 146-180.

Ellis, R. H. The viability equation, seed viability nomographs, and practical advice on seed storage. *Seed Sci. & Technol.* 16 (1988): 29-50.

Ellis, R. H., and E. H. Roberts, Toward a rational basis for testing seed quality. In *Seed Production*, ed. P. D. Hebblethwaite (London: Butterworths, 1980), pp. 605-636.

Ellis, R. H., and E. H. Roberts. The quantification of ageing and survival in orthodox seeds. *Seed Sci. & Technol.* 9 (1981): 373-409.

Ellis, R.H., T. D. Hong, and E. H. Roberts. A comparison of the low moisture-content limit to the logarithmic relation between seed moisture and longevity in twelve species. *Ann. Bot.* 63 (1989): 601-611.

Finney, D. J. *Probit Analysis*, third edition (Cambridge: Cambridge Press, 1971).

Fleurat-Lessard, F. *Insects, Preservation and Storage of Grains, Seeds and Their By-Products*, trans. D. March, and A. J. Eydt (New York: Lavoisier Publishing, 1988), pp. 367-408. Conservation et Stockage des Grains et Graines at Produits Derives, ed. J. L. Multon (Paris: Technique et Documentation–Lavoisier, 1982).

Flinn, P. W., and D. W. Hagstrum. Simulations comparing the effectiveness of various stored-grain management practices used to control *Rhyzopertha dominica* (*Coleoptera:Bostrichidae*). *Env. Entom.* 19 (1990): 725-729.

Foster, G. H. Drying cereal grains. In *Storage of Cereal Grains and Their Products*, ed. C. M. Christensen (St. Paul, MN: Am. Assoc. of Cereal Chem., 1982), pp. 79-116.

Georghiou, K., C. A. Thanos, and H. C. Passam. Osmoconditioning as a means of counteracting the ageing of pepper seeds during high-temperature storage. *Ann. Bot.* 60 (1987): 279-285.

Goldsworthy, A., J. L. Fielding, and M. B. J. Dove. Flash imbibition: A method for the re-invigoration of aged wheat seed. *Seed Sci. & Technol.* 10 (1982): 55-67.

Grabe, D. F. Measurement of Seed Moisture. In *Seed Moisture*, eds. P. C. Stanwood and M. B. McDonald, Jr. (Madison, WI: Crop Sci. Soc. Am. Special Publication Number 14, 1989), pp. 69-92.

Graham, V. A., and W. K. Bilanski. Simulation of grain drying in intimate contact with adsorbents. *Trans. ASAE* 29 (1986): 1776-1783.

Grolleau, G., and P. Gramet. Vertebrate pests of stored grains and seeds. In *Preservation and Storage of Grains, Seeds and Their By-Products*, trans. D. March, and A. J. Eydt (New York: Lavoisier Publishing, 1988), pp. 417-434. Conservation et Stockage des Grains et Graines et Produits Derives, ed. J. L. Multon (Paris: Technique et Documentation–Lavoisier, 1982).

Harman, G. E., and F. L. Pfleger. Pathogenicity and infection sites of *Aspergillus* species in stored seeds. *Phytopathology*, 64 (1974): 1339-1344.

Harrington, J. F. Thumb rules of drying seed. *Crops & Soils* 13 (1960): 16-17.

Harris, K. L., and F. J. Bauer. Rodents. In *Storage of Cereal Grains and Their Products*, ed. C. M. Christensen (St. Paul, MN: Am. Assoc. of Cereal Chem., 1982), pp. 363-406.

Heiser, J. C. B. *Seed To Civilization–The Story of Man's Food.* (San Francisco: W. H. Freeman and Co., 1973).

Henderson, S. M., and R. L. Perry. *Agricultural Process Engineering* (Westport, CN: AVI Publishing, 1976), pp. 303-339.

Herter, U., and J. S. Burris. Effect of drying rate and temperature on drying of corn seed. *Can. J. Plant Sci.* 69 (1989a) 763-774.

Herter, U., and J. S. Burris. Preconditioning reduces the susceptibility to drying injury in corn seed. *Can. J. Plant Sci.* 69 (1989b): 775-789.

Hill, J., and A. V. Schoonhoven. Effectiveness of vegetable oil fractions in controlling the Mexican bean weevil (Coleoptera, Bruchidae) on stored beans. *J. Econ. Entom.* 74 (1981): 478-479.

Hoekstra, J. A. Acute bioassays with control mortality. *Water, Air, and Soil Pollution* 35 (1987): 311-317.

Hukill, W. V. Storage of Seeds. *Proc. ISTA* 28 (1963): 871-873.

Hutchinson, J. B. The drying of wheat III. The effect of temperature on germination capacity. *J. Soc. Chem. Ind.* 63 (1944): 104-107.

Ibrahim, A. E., and E. H. Roberts. Viability of lettuce seeds 1. Survival in hermetic storage. *J. Exp. Bot.* 34 (1983): 620-630.

ISTA. International Rules for Seed Testing. *Seed Sci. & Technol.* 13 (1985): 299-355.

Jay, E. Recent advances in the use of modified atmospheres for the control of stored-product insects. In *Insect Management for Food Storage and Processing*, ed. F. J. Baur (St. Paul, MN: Am. Assoc. of Cereal Chem., 1984), pp. 239-254.

Justice, O. L., and L. N. Bass. Principles and Practices of Seed Storage. *USDA Agriculture Handbook* 506 (1978).

Lasseran, J. C. The Aeration of Grains and the Measurement of Grain Temperature in Storage Bins (Silo-Thermometry). In *Preservation and Storage of Grains, Seeds and Their By-Products*, trans. D. Marsh, and A. J. Eydt (New York: Lavoisier Publishing, 1988), pp. 664-748. Conservation et Stockage des Grains et Graines et Produits Derives, ed. J. L. Multon (Paris: Technique et Documentation-Lavoisier, 1982).

Lillehoj, E. B. Effect of environmental and cultural factors on aflatoxin contamination of developing corn kernels. In *Aflatoxin and* Aspergillus flavus *in Corn*, eds. U. L. Diener, R. L. Asquith, and J. W. Dickens, *Southern Cooperative Series Bulletin* 279 (Auburn University, Alabama: Alabama Agric. Exp. Stat. 1983), pp. 27-34.

Mallis, A. *Handbook of Pest Control* (Cleveland, OH: Franzak & Foster, 1982), pp. 507-592.

Marsh S. F., and G. A. Payne. Preharvest infection of corn silks and kernels by *Aspergillus flavus*. *Phytopathology* 74 (1984): 1284-1289.

McLean, M., and P. Berjak. Maize grains and their associated mycoflora–A microecological consideration. *Seed Sci. & Technol.* 15 (1987): 831-850.

Mills, J. T. Insect-fungus associations influencing seed deterioration. *Phytopathology* 73 (1983): 330-335.

Moore, F. D., and P. A. Jolliffe. Mathematical characterization of seed mortality in storage. *J. Amer. Soc. Hort. Sci.* 112 (1987): 681-686.

Moreno-Martinez, E., and J. Ramirez. Protective effects of fungicides on corn seed, stored with low and high moisture contents. *Seed Sci. & Technol.* 13 (1985): 285-290.

Moreno-Martinez, E., A. Menendez, and J. Ramirez. Behavior of maize (*Zea mays* L.) seed under different storage systems. *Turrialba* 37 (1987): 267-274.

Mullen, M. A., and R. T. Arbogast. Low temperatures to control stored product insects. In *Insect Management for Food Storage and Processing*, ed. F. J. Baur (St. Paul, MN: Am. Assoc. of Cereal Chem., 1984), pp. 255-264.

Multon, J. L. Interactions between water and the constituents of grains, seeds and by-products. In *Preservation and Storage of Grains, Seeds and Their By-Products*, trans. D. Marsh, and A. J. Eydt (New York: Lavoisier Publishing, 1988a), pp. 89-159. Conservation et Stockage des Grains et Graines et Produits Derives, ed. J. L. Multon (Paris: Technique et Documentation–Lavoisier, 1982).

Multon, J. L. Spoilage mechanisms of grains and seeds in the postharvest ecosystem, the resulting losses and strategies for the defense of stocks. In *Preservation and Storage of Grains, Seeds and Their By-Products*, trans. D. Marsh and A. J. Eydt (New York: Lavoisier Publishing, 1988b), pp. 3-63. Conservation et Stockage des Grains et Graines et Produits derives, ed. J. L. Multon (Paris: Technique et Documentation–Lavoisier, 1982).

Neergaard, P. *Seed Pathology. Vol. 1* (New York: John Wiley & Sons, 1977), pp. 282-297.

Nellist, M. E. Safe drying temperatures for seed grain. In *Seed Production*, ed. P. D. Hebblethwaite (London: Butterworths, 1980), pp. 371-388.

Nellist, M. E. Predicting the viability of seeds dried with heated air. *Seed Sci. & Technol.* 9 (1981): 439-455.

Northolt, M. D., and L. B. Bullerman. Prevention of mold growth and toxin production through control of environmental conditions. *J. Food Protect.* 45 (1982): 519-526.

O'Dowd, T., and P. Dobie. Reducing viability losses in open seed stores in tropical climates. *Seed Sci. & Technol.* 11 (1983): 57-75.

Ohlrogge, J. B., and T. P. Kernan. Oxygen dependent aging of seeds. *Plant Physiol.* 70 (1982): 791-794.

Pan, D., and R. N. Basu. Absence of *de novo* protein synthesis in moisture equilibration-drying treatment for maintenance of lettuce seed viability. *Ind. J. Exp. Biol.* 23 (1985): 375-379.

Peng, T. S. Padi husk ash as a grain protectant against beetles. *Mardi* 10 (1982): 323-326.

Powell, D. F. The fumigation of seeds with methyl bromide. *Ann. Appl. Biol.* 81 (1975): 425-431.

Price, L. A., and K. A. Mills. The toxicity of phosphine to the immature stages of resistant and susceptible strains of some common stored product beetles, and implications for their control. *J. Stored Prod. Res.* 24 (1988): 51-59.

Pritchard, H. W., K. R. Manger, and F. G. Prendergast. Changes in *Trifolium arvense* seed quality following alternating temperature treatment using liquid nitrogen. *Ann. Bot.* 62 (1988): 1-11.

Raghavan, G. S. V., Z. Alikhani, M. Fanous, and E. Block. Enhanced grain drying by conduction heating using molecular sieves. *Trans. ASAE* 3 (1988): 1289-1294.

Rao, N. K., and E. H. Roberts. The effect of oxygen on seed survival and accumulation of chromosome damage in lettuce (*Lactuca sativa* L.). *Seed Sci. & Technol.* 18 (1990): 229-238.

Rao, N. K., E. H. Roberts, and R. H. Ellis. A comparison of the quantitative effects of seed moisture content and temperatures on the accumulation of chromosome damage and loss of seed viability in lettuce. *Ann. Bot.* 62 (1988): 245-248.

Roberts, E. H. The viability of cereal seed in relation to temperature and moisture. *Ann. Bot.* 24 (1960): 12-31.

Roberts, E. H. Storage environment and the control of viability. In *Viability of Seeds*, ed. E. H. Roberts (London: Chapman and Hall, 1973), pp. 14-58.

Roberts, E. H., and R. H. Ellis. Water and seed survival. *Ann. Bot.* 63 (1989): 39-52.

Rudrapal, D., and S. Nakamura. Use of halogens in controlling eggplant and radish seed deterioration. *Seed Sci. & Technol.* 16 (1988): 115-121.

Sadik, S., and J. W. White. True potato seed drying over rice. *Potato Res.* 25 (1982): 269-272.

Saha, R., and R. N. Basu. Invigoration of soybean seed for the alleviation of soaking injury and ageing damage on germinability. *Seed Sci. & Technol.* 12 (1984): 613-622.

Saha, R., A. K. Mandal, and R. N. Basu. Physiology of seed invigoration treatments in soybean (*Glycine max.* L.). *Seed Sci. & Technol.* 18 (1990): 269-276.

Sanderson, D. B., W. E. Muir, R. N. Sinha, D. Tuma, and C. I. Kitson. Evaluation of a model of drying and deterioration of stored wheat at near-ambient conditions. *J. Agric. Eng. Res.* 42 (1989): 219-233.

SAS Institute. *SAS User's Guide. Statistics* (Cary, NC: SAS Institute Inc., 1985): pp. 639-653.

Savino, G., P. M. Haigh, and P. De Leo. Effects of presoaking upon seed vigour and viability during storage. *Seed Sci. & Technol.* 7 (1979): 57-64.

Senaratna, T., J. F. Gusse, and B. D. McKersie. Age-induced changes in cellular membranes of imbibed soybean seed axes. *Physiol. Plant.* 73 (1988): 85-91.

Shejbal, J., and J. N. De Boislambert. Modified atmosphere storage of grains. In *Preservation and Storage of Grains, Seeds and Their By-Products*, trans. D. Marsh and A. J. Eydt (New York: Lavoisier Publishing, 1988), pp. 749-777.

Conservation et Stockage des Grains et Graines et Produits Derives, ed. J. L. Multon (Paris: Technique et Documentation–Lavoisier, 1982).
Sokhansanj, S. Improved heat and mass transfer models to predict grain quality. *Drying Technol.* 5 (1987): 511-525.
Sokhansanj, S., and D. M. Bruce. A conduction model to predict grain temperatures in grain drying simulation. *Trans. ASAE* 30 (1987): 1181-1184.
Spaeth, S. C. Imbibitional stress and transverse cracking of bean, pea, and chickpea cotyledons. *HortScience* 21 (1986): 110-111.
Stanwood, P. C. Cryopreservation of seed germplasm for genetic conservation. In *Cryopreservation of Plant Cells and Organs*, ed. K. K. Kartha (Boca Raton, FL: CRC Press, 1985) pp. 199-226.
Stanwood, P. C., and L. N. Bass. Seed germplasm preservation using liquid nitrogen. *Seed Sci. & Technol.* 9 (1981): 423-437.
Stewart G. R. Desiccation injury, anhydrobiosis and survival. In *Plants Under Stress*, eds. H. G. Jones, T. J. Flowers, and M. B. Jones (Cambridge: Cambridge Univ. Press, 1989), pp. 115-130.
Thanos, C. A., K. Georghiou, and H. C. Passam. Osmoconditioning and ageing of pepper seeds during storage. *Ann. Bot.* 63 (1989): 65-69.
USDA. *Stored-Grain Insects.* Agricultural Research Service Agriculture Handbook No. 500 (1978).
Vertucci, C. W. Effects of cooling rate on seeds exposed to liquid nitrogen temperatures. *Plant Physiol.* 90 (1989): 1478-1485.
Vertucci, C. W., and E. E. Roos. Theoretical basis of protocols for seed storage. *Plant Physiol.* 94 (1990): 1019-1023.
Ward, F. H., and A. A. Powell. Evidence for repair processes in onion seeds during storage at high seed moisture contents. *J. Exp. Bot.* 34 (1983): 277-282.
Watts, K. C., W. K. Bilanski, and D. R. Menzies. Design curves for drying grains with bentonite. *Can. Agric. Eng.* 30 (1988): 237-242.
White, D. G., and G. C. Shove. Fungicides reduce corn drying and storage risks. Paper presented at the 1988 Summer Meeting of the American Society of Agricultural Engineers (St. Joseph, MI: ASAE, 1988).
Willson, H. R. *Stored Grain Insect Control. Bulletin* 153 (Columbus, OH: Ohio State Univ. Cooperative Extension Service, 1987).
Wilson, D. O., Jr. *Threshing Injury and Mathematical Modeling of Storage Deterioration in Field Bean Seed* (Phaseolus vulgaris L.) (Columbus: The Ohio State Univ., 1986), pp. 178-181.
Wilson, D. O., Jr. Unpublished data.
Wilson, D. O., Jr., and M. B. McDonald, Jr. The lipid peroxidation model of seed aging. *Seed Sci. & Technol.* 14 (1986): 269-300.
Wilson, D. O., Jr., M. B. McDonald, Jr., and S. K. St. Martin. A probit planes method for analyzing seed deterioration data. *Crop Sci.* 29 (1989): 471-476.
Winston, P. W., and D. H. Bates. Saturated salt solutions for control of humidity in biological research. *Ecology* 41 (1960): 232-237.
Woodstock, L. W., S. Maxon, K. Faul, and L. Bass. Use of freeze-drying and acetone impregnation with natural and synthetic anti-oxidants to improve stor-

ability of onion, pepper, and parsley seeds. *J. Amer. Soc. Hort. Sci.* 108 (1983): 692-696.

Yang, S. F., and Y. B. Yu. Lipid peroxidation in relation to aging and loss of seed viability. *Search* (American Seed Research Foundation) 16 (1982): 2-7.

Zhang, X. Y., and K. L. Tao. Silica gel seed drying for germplasm conservation–Practical guidelines. *Plant Genetic Resources Newsletter* 75/76 (1988): 1-5.

Zummo, N., and G. E. Scott. Cob and kernel infection by *Aspergillus flavus* and *Fusarium moniliforme* in inoculated, field-grown maize ears. *Plant Disease* 74 (1990): 627-631.

Chapter 7

Storage of Recalcitrant Seeds

H. F. Chin

The Indo-Malaysian rainforest is renowned for its great diversity of tropical flora. It has over 25,000 species of flowering plants and is one of the World's major centers of origin of many species of fruit trees, timber, spices, and medicinal plants. Conservation of these important genetic resources is of vital importance for the survival of future generations. More and more people are becoming aware of the loss of plant genetic resources and the need for environmental protection, hence great efforts are being made to conserve plant genetic resources as crop plants or their wild relatives. National and international programs are implemented for their conservation by various methods and techniques. At present, besides *in situ* conservation such as national parks and forest reserves, *ex situ* live collections of these species are also found in arboreta and botanical gardens. Gene banks such as seed banks and *in vitro* gene banks have been established fairly rapidly all over the world, but the seed banks are limited to orthodox seeds while there is none for recalcitrant seeds, i.e., seeds which cannot tolerate desiccation and freezing temperatures.

Seeds may be stored for long periods in seed banks or they may be kept for the short-term or mid-term or as carry-over seeds for the next planting season. In recalcitrant seeds it is very difficult to maintain quality during storage as the seeds themselves are very variable in moisture, size, and viability. Furthermore, there is a lack of understanding of basic mechanisms and behavior of this group of seeds. Therefore, finding suitable methods for storage of recalcitrant seeds for the short- or long-term poses a great challenge to seed scientists and technologists. Alternatives such as using excised embryos and cryogenic storage may be the answer for conserving the seed, as genetic resources in the seed banks.

CLASSIFICATION AND CHARACTERISTICS OF RECALCITRANT SEEDS

In the past, seeds have been classified in many ways according to their usage, families, structure, longevity, etc. For example, classifications such as cover crop seeds, pasture seeds, leguminous seeds, and monocotyledonous endospermic seeds, and such terms as microbiotic, mesobiotic, and macrobiotic have been used. They are still used today but still are not clearly defined or based on the physiological behavior of seeds. In 1973, when the journal *Seed Science and Technology* was launched, Roberts (1973) introduced the term "recalcitrant" for the group of seeds whose viability decreases with a reduction in moisture content below some relatively high value (between 12 and 31 percent) and which cannot tolerate freezing temperatures, as opposed to orthodox seeds which can be dried and can tolerate freezing temperatures.

The definition of the term recalcitrant has been debated at various conferences. Literally, recalcitrant means obstinate, hard to control, obstinately disobedient, does not obey the normal rules. This definition exactly explains the behavior of recalcitrant seeds which is opposite to the "normal" behavior of orthodox seeds, but even then the term is not too clear and descriptive. A few other terms have been proposed to overcome this problem. Hanson (1984) suggested that the terms "desiccation sensitive" and "non-desiccation sensitive" more accurately describe recalcitrant and orthodox seeds, respectively. Lately Berjak (1989) suggested the use of another term, "homoiohydrous," which is concerned with desiccation sensitivity. However, the term recalcitrant seeds is now already well known throughout the world among seed technologists, and I think it should be retained as such because it has become acceptable and is widely used and understood by seed scientists.

Recalcitrant seeds are mainly found in the tropics, coming from trees growing in the humid tropical forests, but relatively few temperate tree species also produce recalcitrant seeds, such as acorns of the English oak *Quercus robur* (Suszka and Tylkowski, 1980) and *Araucaria* (Tompsett, 1982). In the tropics recalcitrant seeds come from many important tropical timbers, crops, and fruits. Most recalcitrant seeds are very large and heavy with 1,000 seed weight often exceeding 500g (Chin, Hor, and Mohd Lassim, 1984). The moisture content of these seeds ranges from 30 to 70 percent at maturity. Recalcitrant seeds, besides having very high moisture content, are very variable. The coefficient of variation between individuals may be as high as 13 percent as against 2 to 3 percent in orthodox seeds such as maize and bean. Because of their large size and high moisture content they are very difficult to store. Many examples illustrating the characteristics of recalcitrant seeds are listed in King and Roberts (1979) and Chin and Roberts (1980).

The most important characteristic is their sensitivity to desiccation and freezing. Some perish even at 26 percent moisture content or at temperatures of 15°C which is well above freezing. Therefore, their storage methods are directly opposite to that of orthodox seeds. Orthodox seeds can be dried to a very low moisture content. In fact, the lower the better: subzero temperature is often recommended for their storage.

Desiccation sensitivity is the main criteria for the identification of recalcitrant seeds. One way to test and identify recalcitrant seeds is by a series of drying studies followed by germination tests to find out the critical moisture level at which seed viability drops or is totally lost. In orthodox seeds, drying to a very low moisture enhances their storability as well as longevity.

Recalcitrant seeds are well known for their sensitivity to desiccation, especially the large-seeded species generally found in the tropics. Their degree of sensitivity varies according to the species, the critical moisture content ranging from 12 to 35 percent. For example, the critical moisture content below which seeds are killed is 17 percent for *Shorea talura* (Sasaki, 1976), 15 to 20 percent for *Hevea brasiliensis* (Chin et al., 1981), 20 percent for *Nephelium lappaceum* (Chin, 1975), 26 percent for *Theobroma cacao* (Hor, Chin, and Karim, 1984), 35 percent for *Hopea helferi* (Tamari, 1976), and 38 percent for *Quercus robur* (Vlase, 1970). From these few examples, it is obvious that recalcitrant seeds are very sensitive to desiccation. As a result, storage techniques in current use are limited to imbibed or moist storage. It is interesting to note that lemon (*Citrus limon*) was formerly identified as a recalcitrant seed, but Mumford and Grout (1979) found that if testa is removed from lemon seed, it exhibits orthodox behavior and is tolerant to desiccation and low temperatures. It is therefore important to apply the correct technique of dehydration or desiccation, as faulty drying can lead to a seed being wrongly identified as recalcitrant.

The degree of sensitivity to desiccation in the seed itself varies between species, but generally the minimum moisture content is 20 to 35 percent. Seeds can be killed if their moisture content falls below this range. The critical moisture content which kills all seeds varies from species to species, but is a relatively high value, usually within the range of 12 to 31 percent (Roberts, 1973). The critical moisture content referred to here is probably best described as the "lowest safe moisture content" (LSMC) (Tompsett, 1987). Variation in susceptibility to desiccation is found not only between species, but also within the same species or seedlot. The high coefficient of variation in the minimum tolerated moisture content of recalcitrant seeds may possibly account for their variation in susceptibility to drying injury.

The reason dehydration causes the death of recalcitrant seeds is still not clear. King and Roberts (1979) suggested two alternatives: either death occurs rapidly at or below some critical moisture content (critical moisture content hypothesis); or loss of viability occurs at a rate which is negatively correlated to moisture content over a wide range of moisture contents (non-critical moisture content hypothesis). It is also possible that a reduction in moisture content causes a loss in membrane integrity and nuclear disintegration, as has been shown to occur with rubber seeds when they are sun dried (Chin et al., 1981). The seeds of many tropical plants contain high concentrations of phenolic compounds and phenolic oxidases. These compounds are normally compartmentalized within the cells. On desiccation, the cell membranes are damaged and the phenolic compounds are released. They are then oxidized and protein/phenol complexes are formed, with a consequent loss of enzyme activity (Loomis and Battaile, 1966).

Farrant, Pammenter, and Berjak (1986) have proposed a new hypothesis based on work with the recalcitrant seeds of *Avicennia marina*. These seeds behave as imbibed orthodox seeds when they are first shed, and can withstand the loss of at least 18 percent of their initial water content and still remain viable. The seeds normally start to germinate immediately after they are shed. They become more sensitive to desiccation with the onset of cell division and vacuolation during the early stages of germination (Bewley, 1979). Dehydration in seeds stored for more than four days caused a marked decline in germination. As germination proceeded, the lower limit below which seeds died also fell.

Most recalcitrant seeds are found in the humid tropics, and are ecologically more adapted to high temperature, hence they do not tolerate freezing temperatures. In nature, many trees and shrubs can survive freezing temperatures, but many others are prone to injury. Most recalcitrant seeds belonging to timber, plantation, and fruit species originate in moist tropical forest habitat. Thus, it is not surprising that they do not tolerate freezing temperatures, although the failure of the seeds of some species to survive at 15°C is difficult to understand. In contrast, many orthodox seeds can survive even at temperatures of −196°C. Among the recalcitrant seeds, there are varying degrees of tolerance to low temperatures, and temperate species are of course more tolerant. Seeds of *Quercus* spp., for example, can germinate at 2°C after eight months in cold storage, whereas the seeds of many tropical species are killed at subambient temperatures or suffer chilling injury. Examples of the latter are *Theobroma cacao* (Hor, Chin, and Karim, 1984), *Nephelium lappaceum* (Chin, 1975), *Dryobalanops aromatica* (Jensen, 1971), *Hopea odorata* (Tang and Tamari, 1973), *Shorea ovalis* (Sasaki, 1976), and *Garcinia mangostana* (Winters, 1963).

The literature on chilling injury in seeds covers both orthodox and recalcitrant seeds. It was already known 50 to 60 years ago that if orthodox seeds are not dried to a sufficiently low moisture content they can be killed by temperatures below freezing (Becquerel, 1925; Lipman and Lewis, 1934). Stanwood (1983) has stated that there is a high moisture freezing limit (HMFL) which is the threshold, and if it is exceeded the viability of a seed sample will be reduced during liquid nitrogen cooling.

This threshold is normally a relatively narrow range in seed moisture content for any one particular species, but it can vary between species. The cause of seed death in such cases is similar to that of recalcitrant seeds which have a high moisture content when they are frozen. Freezing damage in moist seeds is presumably associated with the formation of ice crystals, and usually occurs when the moisture content is higher than 14 to 20 percent (Roberts, 1972).

The reason recalcitrant seeds may be killed at subambient temperatures is not known. Some of the results on the deleterious effects of subambient temperatures on orthodox seeds may be applicable to recalcitrant seeds. Simon et al. (1976) attributed the low temperature susceptibility of the orthodox seeds of cucumber (*Cucumis sativa* cv. Long Green Improved) and mungbean (*Vigna radiata*) to protein denaturation. Wolfe (1978) suggested that the declining fluidity of membranal lipids which occurs during chilling can result in changes in membrane thickness and permeability, which may in turn affect membrane-bound enzymes.

Studies on cocoa seeds have shown that the fall in viability with declining temperature can be extremely abrupt. Possible reasons given by Boroughs and Hunter (1963) were: (1) the presence of some temperature-dependent, rate-limiting reaction, the cessation of which causes lethal metabolic disruption; (2) the absence of some protective substance which is present in those seeds not susceptible to chilling; and (3) the liberation of some toxic material owing to cold-induced changes in membrane permeability.

Hor (1984) noticed a very sharp reduction in the storability of cocoa seeds at 15°C compared to 17°C, indicating that they are very sensitive to slight temperature changes around a critical value. This temperature difference of only 2°C causes seeds to die in less than two weeks, which suggests that only a few interrelated reactions may be involved in seed death. A number of physiological, biochemical, and ultrastructural changes have been detected, such as a threefold increase in leachate conductivity and lower (^{14}C) leucine incorporation, while major ultrastructural changes have been observed in the cell membrane system.

METHODS OF STORAGE

Many economically important crop, timber, and fruit species of the tropics are the backbone of developing countries. Crops like rubber (*Hevea brasiliensis*), cocoa (*Theobroma cacao*), and coconut (*Cocos nucifera*), and fruits such as mango (*Mangifera indica*), durian (*Durio zibethinus*), jack fruit (*Artocarpus theterophyllus*), and rambutan (*Nephelium lappaceum*) are some of the major plant species which produce recalcitrant seeds. The conservation of recalcitrant seeds is hindered by the lack of expertise in their storage as compared to orthodox seeds, for which the technology is very well developed. There are over 100 seed banks throughout the world, while there are just a few *in vitro* gene banks for recalcitrant species.

The typical recalcitrant seeds are short-lived, with a life span of a few days to a few weeks, and are very sensitive to desiccation and low temperatures. As such it is very difficult, or rather impossible, to store them for germplasm conservation. Over the last two decades, the storage technology has improved, thus enabling storage of recalcitrant seeds for several months or a year, but not on a long-term basis. Improvement even in short term storage can be very useful as it will ease the problem of field collection and transportation of these species to gene banks. At present, the storage has been limited to imbibed storage, as seeds do not tolerate desiccation, although recently partial desiccation techniques have been developed. Roberts, King, and Ellis (1984) are of the opinion that it is unlikely that the conventional methods of short-term storage of recalcitrant seeds can be used for germplasm conservation. According to them, the most promising method is storage in liquid nitrogen, although its problems are formidable. After nearly two decades, there is slight improvement in storage period, from a few months to a year for some recalcitrant seeds. As an alternative to seed, attempts have been made to store embryonic axes. It has been found that the embryonic axes are less sensitive to desiccation and hence may be dried to lower moisture content (to 12 percent) and subsequently cryopreserved in liquid nitrogen. These cryopreserved embryos are then thawed and cultured by *in vitro* techniques to regenerate plants. This technique has the potential to become a practical method for germplasm conservation.

The storage techniques currently used for recalcitrant seeds can be grouped into four main types, (1) moist or imbibed storage, (2) partial desiccation technique, (3) controlled atmosphere storage, and (4) cryogenic storage.

Moist or Imbibed Storage

Unlike orthodox seeds, recalcitrant species have to be kept moist or at a high level of moisture content. As such they present more problems, especially with pests and diseases. Recalcitrant seeds have been known to survive when kept in moist conditions, such as in moist media and, in extreme cases, they can be submerged or stored in water. Ong and Lauw (1963) demonstrated that *Hevea* seeds immersed in water for a month resulted in 60 percent germination. King and Roberts (1982) also used the same technique for cocoa seeds for three months, with similar results. The media commonly used are sawdust, charcoal, and sand, moistened with water. The period of storage using this method ranges from a few weeks to a few months. To prevent early germination, a lower temperature of 7°C has been used. Inhibitors of germination such as the natural juice from the aril or pulp of the fruits have also been used. This method of moist storage has been employed for a number of crops including rubber (Ang, 1977), rambutan (Chin, 1975) and cocoa (Evans, 1953).

Partial Desiccation Technique

The moisture content of recalcitrant seeds at maturity is in the range of 30 to 70 percent. Some species are killed when their moisture content is lowered to 25 percent while others can be dried to a lower moisture level. Hence instead of keeping them moist, attempts have been made to partially dry them, i.e., surface drying. This can be done in association with a fungicidal treatment. For example, a technique has been developed for cocoa and *Hevea* seeds in which the seeds were treated and then partially dried before storage. It involves partial drying in air at a temperature of 20°C. In the case of rubber, the seeds are processed, cleaned, soaked in 0.3 percent benlate, drained, and then surface dried to a moisture content of 20 percent before storage at 20°C ± 3°C. Seeds stored in this manner in perforated plastic bags at ambient temperature remained viable with over 50 percent germination after one year in storage.

For cocoa seeds, Hor (1984) processed the seeds by partial drying to 35 percent, after which the seeds were dusted with 0.2 percent w/w benlate-thiram mixture and packed in batches of 500 seeds in thin (0.15 mm), perforated plastic bags. These bags should be stored in a loosely closed box in an air-conditioned room. Aeration in perforated bags and a loosely packed box is important. When stored under such conditions they remain viable for over six months with over 50 percent germination. This shows an improvement in the imbibed storage by about three months.

Controlled Atmosphere Storage

Over the years experts have experimented with storing seeds in various gases but there is not as yet a single successful method for practical application. Other methods have been tried, but with little success. Seeds of cocoa have been sealed in their pods by waxing the entire pod (Pyke, Leonard, and Wardlaw, 1934). Another important tropical fruit seed is the durian. After processing and cleaning, it was sealed in a container, but remained viable for barely a month (Soepadmo and Eow, 1976). Seeds of *Shorea talura*, a timber species, sealed in an inflated polythene bag containing a fungicide, remained viable for one month (Sasaki, 1976). Similarly, Villa (1962), using carbon dioxide, has reported a slight improvement in the storage life of cocoa seeds to 45 days. Ong and Lauw (1963) stored rubber seeds in a carbon dioxide atmosphere with similar results. From all these reports, it is unlikely that controlled atmosphere storage will be of practical application in the storage of recalcitrant seeds.

Cryogenic Storage

To survive at very low storage temperature, reduced moisture in seeds is a prerequisite. Usually when seeds have a very low moisture content, they will not suffer from freezing injury even at –196°C (Stanwood and Bass, 1978). Roberts, King, and Ellis (1984) are of the opinion that the most promising method of germplasm conservation for recalcitrant species is storage in liquid nitrogen. So far, there is no report of successful storage of recalcitrant seeds using this method. It is not very practical to store such large seeds in cryogenic tanks or containers. My experience with fairly large seeds such as rubber is that they tend to explode, leading to physical damage to the seeds, and many other types of large seeds tend to crack and suffer from mechanical damage. Other methods of conservation involving other parts of the plant such as excised embryos have been recommended as alternatives.

FUTURE TECHNIQUES AND PROSPECTS

Withers (1980) and Bajaj (1985) have suggested that the germ plasm of recalcitrant seeds could possibly be conserved through cryopreservation of excised embryos or their segments. It has been found that plant parts on dehydration in an oven (Sun, 1958) or under a vacuum (Withers, 1979) exhibit remarkable resistance to cryogenic damage. It was further confirmed by Chin, Hor, and Krishnapillay (1989) that excised embryos of recalcitrant species can withstand desiccation better than the whole seeds. Drying technique was further refined by Vertucci (1991). She proposed the use of "flash drying" followed by storage of the embryos in freezers at −112°F.

A clearer understanding of the high moisture freezing limit (HMFL), the seed moisture threshold that will result in a decrease in viability of a seed sample when it is cooled in liquid nitrogen, is needed. Stanwood (1983) has published a wide range of HMFL values from 9.6 percent for sesame to as high as 28.5 percent for bean. It is possible that some embryos of recalcitrant seeds can be dried to below their HMFL, and then cooled and stored in liquid nitrogen. Grout (1979) has used the low-temperature storage of imbibed tomato seeds as a model for recalcitrant seed storage. Normah, Chin, and Hor (1986) successfully cryopreserved excised embryos of rubber, a very typical recalcitrant seed. Similarly, Chin, Hor, and Krishnapillay (1989) have reported success in cryopreservation of a number of excised embryos of both orthodox and recalcitrant species and have recommended this technique as an alternative to seed storage. Pence (1990) studied the cryostorage of embryo axes of several large-seeded temperate tree species. The embryo axes of 18 species in seven genera of temperate trees with large recalcitrant or suborthodox seeds survived desiccation and dry freezing. Temperate tree species with recalcitrant seeds include species of *Aesculus, Castanea,* and *Quercus*. Results of Pence (1990) and many other reports of cryopreservation of zygotic and somatic embryos indicate that cryostorage should be possible for a number of difficult to store temperate as well as tropical seeds. This technique could also significantly reduce the amount of space needed for germplasm preservation in these species.

After 50 years of research on recalcitrant seeds, there is not a single method of long-term storage available for conservation of germplasm of important tropical crops, such as rubber, cocoa, and coconuts. With the recent developments in biotechnology and cryobiology there are indications that cryopreservation of excised embryos may be the suitable technique for long-term storage of such seeds. In the near future, seed banks will not only store orthodox seeds but will have a section devoted to *in vitro* seed banks which store cyropreserved embryos and culture them for use when the need arises.

SUMMARY

Many tropical tree species, and some temperate ones, produce recalcitrant seeds. This group of seeds classified as recalcitrant are usually very large, high in moisture content, limited in viability, and very sensitive to desiccation and low temperatures. Examples of recalcitrant seeds are found in tropical timber, fruits, and plantation crops such as rubber, cocoa, and coconuts. They play a significant role in the economy of most developing countries in the tropics. Conservation of these important species as genetic resources is vital for survival of the future generation. At present, there is not a single method available for storage of recalcitrant seeds on a long-term basis. Unlike orthodox seeds, few seed banks are established for their conservation.

The available technology only permits storage of recalcitrant seeds for about one year. In the past, seeds have been stored for a few weeks to a few months, either in imbibed storage or gaseous storage. Lately a technique of partial desiccation and surface dressing with fungicide has been introduced with prolongation of viability, to a year. There is no method for long-term storage for germplasm conservation. The technique of cryopreservation of excised embryos seem to be a possibility for their conservation. This technique, with further research and refinement, may become the answer to long-term conservation of recalcitrant seeds.

REFERENCES

Ang, B. B. Problems of rubber seed storage. In *Seed Technology in the Tropics*, eds. H. F. Chin, I. C. Enoch, and R. M. Raja Harun (Kuala Lumpur, Malaysia: University Pertanian Malaysia, 1977), pp. 117-122.

Bajaj, Y. P. S. Cryopreservation of embryos. In *Cryopreservation of Plant Cells, and Organs*, ed. K. K. Kartha (Boca Raton, FL: CRC Press, 1985), pp. 228-242.

Becquerel, P. La suspension de la vie des graines dana la vide a la temperature de l'helium liquide. *C. R. Acad. Sci., Paris* 181 (1925): 805-807.

Berjak, P. Storage behaviour of seed of *Hevea brasiliensis*. *J. Nat. Rubb. Res.* 4 (1989): 195-203.

Bewley, J. D. Physiological aspects of desiccation tolerance. *Annu. Rev. Plant Physiol.* 30 (1979): 195-238.

Boroughs, H., and J. R. Hunter. The effect of temperature on the germination of cocoa seeds. *Proc. Amer. Soc. Hort. Sci.* 82 (1963): 222.

Chin, H. F. Germination and storage of rambutan (*Nephelium lappaceum*) seeds. *Malay. Agric. Res.* 4 (1975): 173-180.

Chin, H. F., and E. H. Roberts, eds. *Recalcitrant Crop Seeds* (Kuala Lumpur, Malaysia: Tropical Press, 1980).

Chin, H. F., Y. L. Hor, and B. Krishnapillay. Excised embryos–An alternative to seed storage (Edinburgh, Scotland: *ISTA Congress*, 1989), p. I.

Chin, H. F., Y. L. Hor, and M. B. Mohd Lassim. Identification of recalcitrant seeds. *Seed Sci. & Technol.* 12 (1984): 429-436.

Chin, H. F., M. Aziz, B. B. Ang, and S. Hamzah. The effect of moisture and temperature on the ultrastructure and viability of seeds of *Hevea brasiliensis*. *Seed Sci. & Technol.* 9 (1981): 411-422.

Evans, H. The preservation of cacao seeds for transport purposes. In *A Report on Cacao Research, 1945-1951*. (St. Augustine, Trinidad: Imperial College of Tropical Agriculture, 1953), p. 79.

Farrant, J. M., N. W. Pammenter, and P. Berjak. The increasing desiccation sensitivity of recalcitrant *Avicennia marina* seeds with storage time. *Physiol. Plant.* 67 (1986): 291-298.

Grout, B. W. W. Low temperature storage of imbibed tomato seeds: A model for recalcitrant seed storage. *Cryo-Lett.* 1 (1979): 71-76.

Hanson, J. The storage of seeds of tropical tree fruits. In *Crop Genetic Resources: Conservation and Evaluation*, eds. J. H. W. Holden and J. T. Williams (London: Allen and Unwin, 1984), pp. 53-62.

Hor, Y. L. Storage of cocoa (*Theobroma cacao*) seeds and changes associated with their deterioration. PhD Thesis, Universiti Pertanian Malaysia, Kuala Lumpur, 1984.

Hor, Y. L., H. F. Chin, and M. Z. Karim. The effect of seed moisture and storage temperature on the storability of cocoa (*Theobroma cacao*) seeds. *Seed Sci. & Technol.* 12 (1984): 415-420.

Jensen, L. A. Observations on the viability of Borneo camphor (*Dryobalanops aromatica* Gaertn.). *Proc. ISTA* 36 (1971): 141-146.

King, M. W., and E. H. Roberts. *The Storage of Recalcitrant seeds. Achievements and Possible Approaches* (Rome: International Board of Plant Genetic Resources, 1979).

King, M. W., and E. H. Roberts. The imbibed storage of cocoa (*Theobroma cacao*) seeds. *Seed Sci. & Technol.* 10 (1982): 535-540.

Lipman, C. B., and G. N. Lewis. The tolerance of liquid air temperature by seeds of higher plants for 60 days. *Plant Physiol.* 9 (1934): 392-394.

Loomis, W. D., and J. Battaile. Plant phenolic compounds and the isolation of plant enzymes. *Phytochemistry* 5 (1966): 423-438.

Mumford, P. M., and B. W. W. Grout. Desiccation and low temperature (−196°C) tolerance of *Citrus limon* seed. *Seed Sci. & Technol.* 7 (1979): 407-410.

Normah, M. N., H. F. Chin, and Y. L. Hor. Desiccation and cryo-preservation of embryonic axes of *Hevea brasiliensis. Muell-Arg. Pertanika* 9 (1986): 299-303.

Ong, T. P., and I. K. Lauw. Results on storage test with seed of *Hevea brasiliensis. Menara Perkebunan* 32 (1963): 183-192.

Pence, V. C. Cryostorage of embryos axes of several large seeded temperate tree species. *Cryobiology* 27 (1990): 212-218.

Pyke, E. E., E. R. Leonard, and C. W. Wardlaw. On the viability of cacao seeds after storage. *Trop. Agric.* 11 (1934): 303-307.

Roberts, E. H. Storage environment and the control of viability. In *Viability of Seeds*, ed. E. H. Roberts (London: Chapman and Hall, 1972), pp. 14-58.

Roberts, E. H. Predicting the storage life of seeds. *Seed Sci. & Technol.* 1 (1973): 499-514.

Roberts, E. H., W. W. King, and R. H. Ellis. Recalcitrant seeds, their recognition and storage. In *Crop Genetic Resources: Conservation and Evaluation*, eds. J. H. W. Holden and J. T. Williams (London: Allen and Unwin, 1984), pp. 38-52.

Sasaki, S. The physiology, storage, and germination of timber seeds. In *Seed Technology in the Tropics*, eds. H. F. Chin, I. C. Enoch, and R. M. Raja Harun (Kuala Lumpur: Universiti Pertanian Malaysia, 1976), pp. 11-15.

Simon, E. W., A. Minchin, M. M. McMenamin, and J. M. Smith. The low temperature limit for seed germination. *New Phytol.* 77 (1976): 301-311.

Soepadmo, E., and B. K. Eow. The reproductive biology of *Durio zibethinus* Murr. *Garden's Bull.* 29 (1976): 25-33.

Stanwood, P. C. Cryopreservation of seed germplasm for genetic conservation. In *Cryopreservation of Plant Cells and Organs*, ed. K. K. Kartha (Boca Raton, FL: CRC Press, 1983), pp. 200-226.

Stanwood, P. C., and L. N. Bass. Ultra cold preservation of seed germplasm. In *Plant Cold Hardiness and Freezing Stress. Mechanisms and Crop Implica-*

tions, eds. P. H. Li and A. Sakai (New York: Academic Press, 1978), pp. 361-372.

Sun, C. N. The survival of excised pea seedlings after drying and freezing in liquid nitrogen. *Bot. Gaz.* 119 (1958): 234-236.

Suszka, B., and T. Tylkowski. Storage of acorns of the English oak (*Quercus robur* L.) over 1-5 winters. *Arboretum Kornickie* 25 (1980): 199-229.

Tamari, C. Phenology and Seed Storage Trials of Dipterocarps. Research Pamphlet No. 69, Forest Research Institute, Kepong, Malaysia, 1976.

Tang, H. T., and C. Tamari. Seed description and storage tests of some dipterocarps. *Malay. For.* 36 (1973): 38.

Tompsett, P. B. The effect of desiccation on the longevity of seeds of *Araucaria hunstecini* and *A. cunninghamii*. *Ann. Bot.* 50 (1982): 693-704.

Tompsett, P. B. Desiccation and storage studies on Dipterocarp seeds. *Ann. Appl. Biol.* 110 (1987): 371-379.

Vertucci, C. Flash drying saves seeds. *Agric. Res.* 39 (1991): 23.

Villa, C. L. Pruebas para ampliar el periodo de viability d'el cacao. *Agricultura Tecnica en Mexico* 2 (1962): 133-136.

Vlase, I. Contribution to the development of a method of storing acorns of, *Quercus robur* L. for long periods. *Revista Padurilor* 85 (1970): 616-619.

Winters, H. F. The mangosteen. *Fruits Var. Hort. Digest* 8 (1963): 57-58.

Withers, L. A. Freeze preservation of somatic embryos and clonal plantlets of carrot (*Daucus carota* L.). *Plant Physiol.* 63 (1979): 460-467.

Withers, L. A. Cryopreservation of plant and tissue cultures. In *Tissue Culture Methods for Plant Pathologists*, eds. D. S. Ingram and J. P. Helgeson (Oxford: Blackwell Scientific Publication, 1980), pp. 63-70.

Wolfe, J. Chilling injury in plants–The role of membrance lipid fluidity. *Plant Cell & Environ.* 1 (1978): 241-247.

Chapter 8

Mechanisms of Seed Deterioration

Peter Coolbear

Seed deterioration or seed aging is probably best described as the loss of seed quality with time. In this way we can make a distinction from seed damage, the loss of seed quality through mechanical injury. Of course, these two causes of reduced seed quality are not entirely exclusive: it is well known, for example, that damaged seeds deteriorate more rapidly than undamaged ones.

Over the last 25 years there has been considerable research into the mechanisms of seed deterioration, but while it is fair to say that we now have a reasonable overview of the range of events which occur as seeds age, the literature in some areas is, at best, confusing. It is clear we still have a great deal to learn about the relative importance of each of these events, how they interact with each other, and how they are affected by the prestorage history of the seed and different ambient storage conditions.

A major step toward the advancement of research in this area has already been taken with the publication of two excellent critical texts (McDonald and Nelson, 1986; Priestley, 1986). In both of these, considerable attempts have been made to make sense of many of the contradictions in the literature and to discuss the limitations of research accomplished to date. Given the existence of these

---

I would like to thank my PhD students, Adèle Francis, Nit Sakunnarak, and Suresh Nath for their contributions to the literature research in this review. In particular, thanks to Nit for compiling the data which comprise Tables 4 and 5 and to Suresh for permission to reproduce Figures 1 and 5. Most of all, however, I would like to thank them all for their friendship and the many hours of stimulating and thought-provoking discussion they have engendered in the years we have been conducting research on this topic.

two publications, my objective in this chapter is to develop an overview of current ideas about seed deterioration, revisit some of the possible reasons for contradictions in the literature, and then begin to develop some ideas for potential ways forward in this area of research, an area which, I believe, comprises an essential component of our understanding of seed quality. While most of the chapter will be concerned with the mechanisms of deterioration of seeds in storage, it would be incomplete without consideration of deteriorative events which may take place on the parent plant prior to harvesting and processing. Accordingly, the first main section comprises a brief overview of this area of study.

Before continuing, however, I shall discuss two widespread misconceptions still current among many seed technologists about the process of seed deterioration. These ideas have arisen from earlier writings in this area and, like many other fallacies, any fault tends to lie not with the original authors, but rather subsequent workers who have taken tentative ideas or qualified statements and raised them to the status of general dogma.

The first fallacy is that seed deterioration is always irreversible. Seed deterioration may be an inevitable and inexorable process, but statements such as:

> [It is] an irreversible change in the quality of a seed after it reaches its maximum quality level (Abdul-Baki and Anderson, 1972, p. 284)

or

> Seed Deterioration is an irreversible process. Once seed deterioration has occurred, this catabolic process cannot be reversed (Copeland and McDonald, 1985, p. 146)

overlook the fact that seeds not only have inbuilt systems to counter the impact of deteriorative events, but also, given the right conditions, have active mechanisms for self-repair. From an ecological point of view this is only to be expected of an extremely successful unit of propagation and dispersal: it is part of a seed's survival strategy.

Connected with these considerations is the question of when seeds actually die. How many deteriorated seeds actually lose viability under the stresses of attempting to germinate after sowing rather than during storage itself? An increasing number of studies (e.g., Matthews, Powell, and Rogerson, 1980; Woodstock and Tao, 1981; Saha and Basu, 1984) have shown that pre-sowing treatment or restriction of water uptake during early germination can allow deteriorated seeds to show dramatic improvements in germination compared to their performance in the normally optimum conditions of the standard germination test. If there is adequate time for the effective operation of repair processes, severely deteriorated material may regain the ability to produce a normal plant.

The second fallacy is that seed deterioration is a sequence of events. In the early 1970s both Heydecker (1972) and Delouche and Baskin (1973) proposed tentative models of deterioration which described the process as a sequence starting with a series of biochemical events, in particular membrane damage and the impairment of biosynthetic reactions, and then the resulting losses of various seed performance attributes, starting with reduced germination rate and culminating in reduced field emergence, increased numbers of abnormal seedlings, and, finally, seed death. While this model is attractive for its simplicity, we now know that several types of cellular changes occur during deterioration and, although interrelated, they do not necessarily follow a predetermined sequence. More than this, their relative importance may vary with different seeds and with different storage conditions (Priestley, 1986).

A sequential model implies that all aspects of loss of seed quality (decreased rate of germination, decreased storability, decreased stress tolerance, etc.) are in the same continuum which ultimately leads to loss of seed viability. There is an increasing amount of evidence suggesting that this is not the case. For instance, Coolbear, Francis, and Grierson (1984) suggested from their work on artificially aged tomato seeds that the factors limiting rate of germination may not be the same as those crucial to whether or not the seed remains viable. Indeed, in his pioneering work on self-repair in seeds, Villiers (1973) ventured to suggest that one of the reasons why partially deteriorated seeds germinate more slowly is that they take additional time to undergo the necessary processes of self-repair. Equally, from the vigor testing point of view, the fact that a single standardizable vigor test has yet to be developed after many decades of research (Hampton and Coolbear, 1990) would strongly suggest that all these deteriorative changes are not part of a single, inevitable continuum.

DETERIORATION ON THE PARENT PLANT

It is almost axiomatic in the seed industry that maximum seed quality is attained from the time the developing seed reaches its maximum dry weight, normally referred to as physiological maturity (Harrington, 1972; Justice and Bass, 1978). It is now clear, however, that this term is a little unfortunate in that most seeds continue to undergo physiological changes after maximum dry weight has been achieved. For instance, many seeds continue to accumulate specific reserves after this time (e.g., Aldana, Fites, and Pattee, 1972) or require loss of moisture to induce germination ability (Kermode et al., 1986). Most recently, Filho and Ellis (1991) have demonstrated that maximum seed quality in spring barley is attained some considerable time (up to nearly four weeks in some cases) after maximum dry weight is attained. Nevertheless, whenever maximum seed quality is reached, seeds will deteriorate if left for prolonged periods on the parent plant after this time. Many species show little evidence of this simply because ripe seeds are quickly shed from the parent plant; in other species, however, this weathering damage can be a major problem.

Weathering

Weathering is a problem for seed quality in many crops and may often occur in the period between the attainment of maximum dry weight and harvest ripeness (TeKrony, Egli, and Balles, 1980). In general, high temperature and high humidity (either singly or in combination) seem to be the major environmental factors favoring weathering damage (see review by Delouche, 1980).

Very little work has been done on the actual mechanism of this type of damage in the field. Invasion of seed by field microflora (which will be discussed briefly later) is clearly a significant factor in loss of quality (Christensen and Kaufmann, 1974; Delouche,

1980; TeKrony, Egli, and Balles, 1980), while, as mentioned in the introduction, it is sometimes difficult to separate increased susceptibility to mechanical damage from actual physiological deterioration. This was illustrated, for example, by Biddle (1980), who undertook a study on the effects of seed moisture content (SMC) at harvest. His data show clearly that increased susceptibility to mechanical damage, while a contributing factor to seed deterioration, only accounts for just over 25 percent of the variation in seed quality measured by tetrazolium staining. Late harvesting also induces physiological deterioration.

One of the very few studies on the physiology of seed deterioration during weathering has been undertaken by Woodstock, Furman, and Leffler (1985) on a range of cotton cultivars. Proportions of these seed crops were held in the field for five weeks after normal harvesting, and field emergence of weathered seed were considerably reduced (by up to 50 percent) over unweathered controls. Both electron microscopy and measurements of leakage of cations indicated that weathering causes disruption to membranes, while there was also evidence of changes in both lipid and protein bodies as well as loss of ribosomes and impaired respiratory capacity.

It is quite likely that the physiological component of deterioration in the field is very similar to those mechanisms involved in seeds stored at higher moisture contents. Clearly more work needs to be done in this area before we can understand this and thus exploit evident cultivar variations within crop species in their ability to resist this kind of stress.

Sprouting Damage

Another category of loss of seed quality which occurs in a range of species is not general deterioration *per se*, but the loss of specific primary dormancy mechanisms which are designed to prevent germination of developing seed on the parent plant. In extreme cases this may result in vivipary (sprouting). Often, however, it is not associated with visible germination, but rather increases in hydrolytic enzymes such as α-amylase which are normally not released until germination of the mature seed is underway. This kind of problem can seriously affect the food value of many commercial grains and can be of major economic concern, particularly in climates where there is a high probability of wet weather during seed ripening and at harvest time. For example, Humphrey-Taylor and Larsen (1990) have surveyed the extent of this problem in New Zealand wheats over the 16-year period from 1971 to 1986. During this time. 12.5 percent of New Zealand wheat suffered from sprouting damage due to increased α-amylase activity, with the problem being especially acute in cooler regimes, occurring in well over 20 percent of the total crop.

Walker-Simmons (1987) demonstrated that sensitivity to the plant growth regulator (PGR) abscisic acid, an inhibitor of both germination and α-amylase production, was associated with susceptibility to sprouting damage in wheat. Alternatively, susceptibility to sprouting damage may be induced by increased sensitivity to gibberellic acid, the PGR which promotes α-amylase production (Mitchell et al., 1980). There is some doubt however, that all sprouting damage effects can be attributed to environmentally induced responses to PGRs. In a major project evaluating the hereditability of sprouting damage in wheat conducted over several years at Massey University, the conclusion has been reached that there are at least three independently inheritable physiological traits which can make this crop susceptible to sprouting damage. (Gordon, Smith, and Sereepraseri, 1990).

THE SYMPTOMS OF SEED DETERIORATION IN STORAGE

Losses of Vigor Preceding Losses of Seed Viability

As described in Chapter 6, unless the storage conditions are particularly severe or the seeds are of very poor vigor, there will be a period in storage when there is little change in seed viability. When the viability of a homogenous seed population does decline, it will follow a negative sigmoidal pattern, reflecting the normal distribution of seed deaths with time. Well before any detectable loss in viability, other deteriorative changes will become evident. Seeds will begin to lose vigor, manifested, for example, by marked decreases in rates of germination (Figure 1B).

Similarly, there may be marked increases in the conductivity of leachates from deteriorating seeds as they are placed in contact with

FIGURE 1. Changes in germination and vigor of wheat grains, cv Karamu held at 35°C and a constant 15 percent seed moisture content. A. Percentage seed viability (data have been subjected to an arcsin $\sqrt{\%}$ transformation for statistical analysis); B. Time to 50 percent radicle emergence (T50); C. Conductivity of seed leachate after 24h soaking; D. Mean root dry weight of normal seedlings after 15d growth in the germination test conducted at 10°C. Data from Nath (1991).

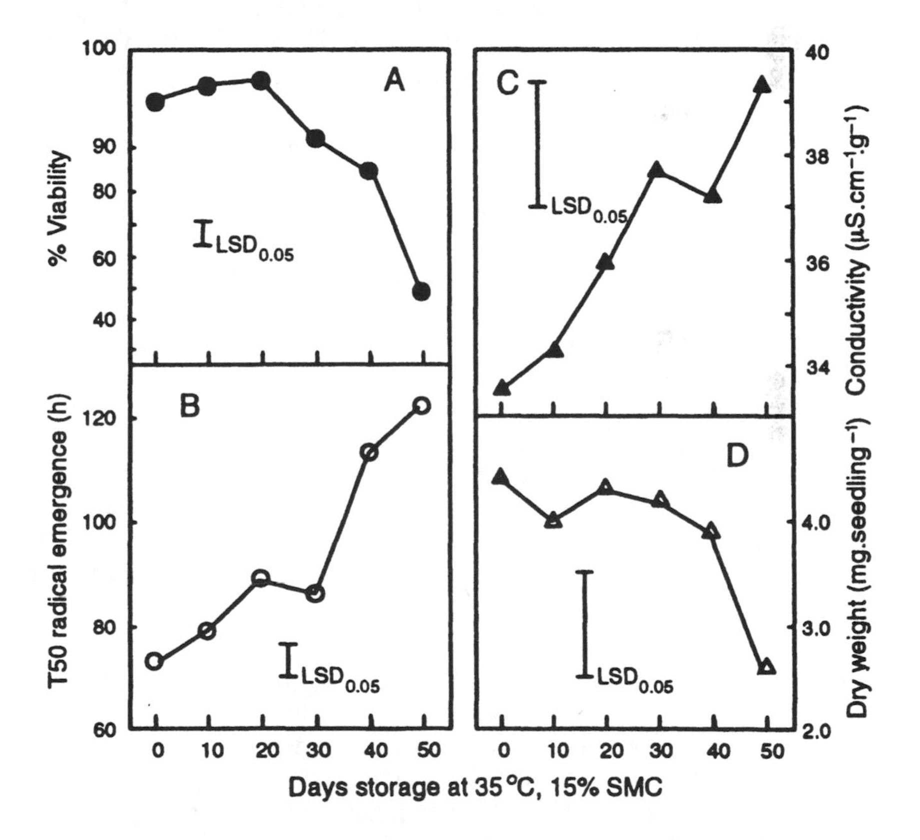

water. This has been well documented in large-seeded legumes (e.g., Powell and Matthews, 1977; McDonald and Wilson, 1980) and also recently in red clover (Wang, 1989). Such changes may indicate a reduction in the ability of seeds to reorganize membranes upon imbibition and can be used routinely as vigor tests (see Chapter 3). Significant conductivity changes for wheat seed during storage at a constant 35°C and 15 percent SMC are shown in Figure 1C. Interestingly, in this study, the patterns of conductivity change differed considerably with storage conditions, while no significant differences in conductivity were observed when seeds were held under severe accelerated aging conditions (100 percent relative humidity, 40°C, for up to six days) which caused equivalent losses in viability (Nath, Coolbear, and Hampton, 1991). Clearly we must avoid overgeneralizing about the kinds of changes in vigor found in deteriorating seeds.

It is assumed that many of these vigor losses would be reflected in decreased field performance. Indeed the basic problem of germination testing is that seeds performing well in germination tests do not necessarily perform well in the field (see Chapter 3). Much of these quality differences no doubt reflect seed deterioration, although surprisingly, direct evidence for this is hard to come by (see Priestley, 1986 for a good discussion of this problem). The key experimental difficulties are twofold. Firstly, it is very difficult to design an experiment to unequivocally test the hypothesis in question. Either studies use seeds which have been harvested in sequential years, and thus production conditions in each growing season become a confounding variable, or field trials are done at different times for the same harvest, in which case care has to be taken to allow for differences in field conditions at each planting. The only obvious way around this is to use rapid-aging techniques to induce rapid deterioration of a seedlot, but then the assumption must be made that any results can be extrapolated to the much longer time frames of deterioration during commercial practice.

A second problem is that vigor differences may not show themselves in the field under optimal conditions where seeds may be able to recover from some of the damage incurred during storage. Figure 1D shows changes with increasing storage times of root dry weight of surviving normal wheat seedlings grown under the relatively benign conditions of a laboratory germination test. It can be seen that significant losses in dry weights do not occur until after there has been a significant loss in seed viability. This is quite a common finding in studies of this type and, of course, does not confirm general ideas about losses in seed vigor always preceding losses in viability. What it does show is that, given the right subsequent growing conditions, a proportion of seeds in the population may be irreversibly damaged by storage before surviving seeds have appreciably lost the ability to recover from any deterioration in store.

Other Symptoms of Seed Aging

An increase in the numbers of abnormal seedlings appearing in the germination test is another common symptom of seed deterioration. Often this kind of problem results from the specific failure of one area of meristematic tissue within the seedling. The most commonly quoted example of this is in deteriorated onion seeds where the cotyledonary knee which breaks through the soil surface fails to develop (Harrington, 1973). Other abnormalities produced in plants are not identifiable at the germination testing stage. Priestley (1986) reviews many examples of commercially important crops where pollen viability is decreased in plants grown from aged seed material. Other evidence has shown that floral morphology or the proportions of male to female flowers in monoecious crops may be affected by the age of seed planted.

Two other common symptoms of seed deterioration may be color changes or increases in free fatty acid levels within seeds (Harrington, 1973; Priestley, 1986). Darkening of seed coats with age is very likely due to the oxidation of phenolics or similar compounds within the seed coat. Fatty acid production is usually the result of lipase action on seed reserves. There is considerable discussion in the literature about how much of this is caused by fungal enzymes and how much is due to activity of lipases within the seed itself. As seed moisture contents increase during storage, infection by storage fungi, in particular *Aspergillus* and *Penicillium* species, is almost inevitable. When these organisms take hold, symptoms of rapid decay become evident and include musty odors, raised temperatures, and increased moisture contents within areas of the stored seed. These latter increases, due to respiratory activity of fungi (and also seeds as moisture levels rise above 15 percent) will promote growth of more aggressive storage fungi and eventually allow bacterial invasion as well. When the moisture content is this high, seeds will show obvious signs of rotting. Such effects are quite likely to be localized within a bulk seed store, but usually when such symptoms as these (even simply visible signs of fungal growth) are observed, the planting value of the entire seed stock may already have been seriously compromised (e.g., Christensen, 1973).

MECHANISMS OF SEED DETERIORATION IN STORAGE

Table 1 summarizes the types of physiological events which have been recorded in the literature during seed deterioration. I will discuss these briefly, summarizing current ideas in each area. As previously mentioned, for more detailed reviews see McDonald and Nelson (1986) and Priestley (1986). Before we embark on a discussion of the items on the list in Table 1, however, there are a few general ideas which require consideration.

One of the crucial issues in current research into seed deterioration is the relative importance of each type of event, i.e., which are primary causes of deterioration and which are secondary ones. For example, it might be suggested (as has often been the case) that membrane damage is the fundamental cause of seed deterioration. Very likely this and genetic damage are the prime candidates. Nevertheless, a vigorous seed has major capacities for (a) resisting membrane damage and (b) membrane repair. Possibly the inability to dispose of toxic metabolites (e.g., short chain fatty acids) may be the ultimate reason why an individual seed fails to survive or germinates with considerably impaired vigor. It will become clear, too, in the ensuing discussion, that generalizations are dangerous. Different factors may be of importance for different species or even for the same seedlot under different storage conditions.

Similarly, it must be borne in mind that deteriorative events do not occur in isolation from each other. Damage to mitochondrial membranes, for instance, may have direct effects on respiratory activity, while damage to the plasmalemma-endoplasmic reticulum-golgi membrane system may have a major impact on a cell's capacity for the synthesis and processing of protein molecules. Damage to DNA will, of course, have a direct consequence for subsequent transcriptional activity.

Conceptually, then, aging should be visualized as an integrated matrix of deteriorative events, superimposed on which is the seed's capacity for detoxification and repair. A preliminary model of this

TABLE 1. Mechanisms of seed deterioration in storage

Category of event	Typical evidence	Likely causes
Membrane damage	Changes in seed conductivity	Free radical peroxidation via
	Ultrastructural changes	a) autoxidation
	Changes in membrane composition	b) lipoxygenase
	Methods designed to detect free radical action	Hydrolytic damage
Genetic damage	Point mutations	Autolysis
	Chromosomal aberrations during cell division	Hydrolytic damage
	Ultrastructural changes	
	Direct analysis of extracted DNA	
Changes in respiratory activity	Gas exchange	Damage to mitochondrial membranes
	Ultrastructural changes	Changes in availability of respiratory substrate
	ATP production	Damage to respiratory enzymes
	Analysis of components of respiratory metabolism	Toxic metabolites
Enzyme and protein changes	Measurements of enzyme activities	Free radical damage, e.g., cross-linking
	Protein analysis	Hydrolytic damage
	Protein synthesis changes	Changes in cofactors
	RNA metabolism changes	Impairment of protein synthesis
		Loss of compartmentalization
		Loss of membrane-based organization in cells

TABLE 1 (continued)

Hormonal changes	PGR treatments of aged seeds Demonstration of changes of sensitivity to PGRs	Disruption of metabolism Loss of sensitivity to PGRs via membrane damage or decreased capacity of cell to respond
Toxic metabolites	Direct detection	Primary or secondary products of deteriorative reactions Failure of detoxification systems
Microorganisims	Direct detection Inoculation experiments Mycotoxin measurements	Elevation of seed moisture content and/or temperature[1]

[1]Storage fungi are essentially ubiquitous in seed stores, it is almost inevitable in commercial practice that hulled seeds will be contaminated with spores (e.g., Christensen, 1967).

type was proposed by Osborne (1980) and represents a working scheme for the kind of events which might occur in a dry seed with little or no "free" water (Figure 2). This type of matrix model is likely to be a much more realistic version of what is going on in seed deterioration than the simplistic sequential models mentioned earlier. A truly comprehensive model would have to allow for all possible events at a range of likely seed moisture contents and also incorporate the effect of any prestorage factors plus the influences of microfloral infections within the store. As will soon become clear in the ensuing discussion, our knowledge about individual events in the matrix is much too incomplete to make such a model an achievable proposition at the present time.

Membrane Damage

Increased Conductivity of Seed Leachate

Disruption of membrane integrity is a major physiological symptom of seed deterioration and we have already discussed increases in conductivity of seed leachate which are regarded as a symptom of this deterioration. It must, however, be remembered that conductivity changes should be treated with caution. An increase in conductivity can arise from

1. flushing out the contents of broken cells under areas of mechanical damage
2. an increase in the amount of specific solutes available to leak out
3. loss of membrane integrity

Aware of these problems of interpretation, Powell and Matthews (1977) were able to demonstrate that increased leakage occurred from pea seeds before there was any incidence of cell death as determined by tetrazolium chloride staining. This, then, must be regarded as strong evidence that membrane damage may be an early event in seed deterioration. Ferguson, TeKrony, and Egli (1990a) were similarly able to show that increases in conductivity of leachate from isolated axes were a very early symptom of soybean deterioration. Special precautions were taken in this study to pre-

FIGURE 2. A suggested scheme for the matrix of possible events occurring during the deterioration of dry stored seeds. Reprinted with permission from Osborne (1980). Copyright CRC Press, Inc., Boca Raton, Florida.

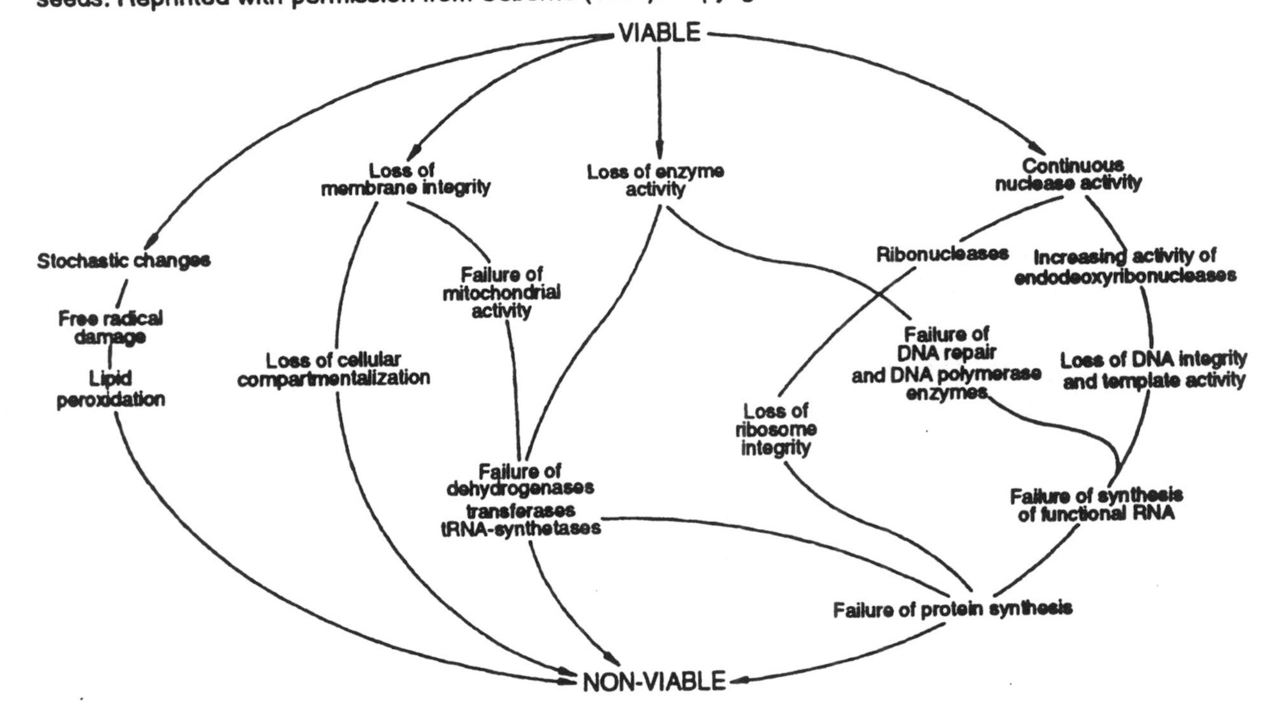

equilibrate the excised tissues to around 30 percent SMC to avoid any imbibition injury before axes were placed in the soak water used to measure conductivity. However, as already mentioned, data like that of Nath, Coolbear, and Hampton (1991) suggests that such changes are not found in all seeds under all deterioration conditions.

Ultrastructural Changes

Another major category of evidence for membrane damage in deteriorating seeds is ultrastructural data gained by comparison of fresh and aged seed tissues under the transmission electron microscope. Despite problems of interpretation which can occur due to the state of hydration of seed tissue prior to fixing (Priestley, 1986), it is clear that membrane damage does occur. In particular, the plasmalemma may withdraw from the cell walls as seeds age and cytoplasmic contents may be seen to have leaked through the membrane (e.g., Hallam, 1973; Villiers, 1980; Sakunnarak, 1992; in rye, lettuce, and soybean respectively). Other changes include fusion of lipid bodies (Harman and Granett, 1972), changes in the appearance of mitochondria, plastids, and golgi stacks (Berjak and Villiers, 1972a), and damage to the nuclear envelope (Vishnyakova et al., 1976; Villiers, 1980).

Of key interest are the findings of Berjak and Villiers (1972a) who noted that changes to cell organelles were observable in tissue from seeds which had not yet deteriorated sufficiently to show much decrease in germination and were reversible on imbibition.

Losses of Membrane Phospholipid

As with conductivity increases, changes in membrane composition have been demonstrated to occur at very early stages of deterioration when storage conditions are severe. For example, Powell and Matthews (1981) were able to demonstrate that losses in phospholipid (PL), particularly phosphatidylcholine, were very early events during aging in pea seeds (Figure 3). Losses of phospholipid from cotyledons were associated with a decreased percentage of cotyledons showing full tetrazolium staining, but occurred before there was any evidence of decreased embryo axis viability.

Bewley (1986) summarized a range of available data for phospholipid changes in a variety of species during aging. It is quite clear from this that, under high moisture content aging conditions, seeds lose phospholipids as they lose viability. It is less clear whether such changes occur under low moisture content storage. Priestley and Leopold (1983) found little change in PL during nearly four years storage at 8 to 10 percent SMC, 4°C. These storage conditions did, however, cause only a small loss in germinability during this time, so it is possible that deterioration was not allowed to proceed far enough for clear changes to emerge. It is evident from the data in Figure 3 that lipid determinations from aged seed can be highly variable. This is hardly surprising in view of the number of different confounding factors involved. Not the least of these is the general observation that a major symptom of deterioration is a loss of seed uniformity (e.g., Delouche and Baskin, 1973). Certainly, both Pearce and Abdel Samad (1980) and Francis and Coolbear (1987) found marked declines in phospholipid composition in slowly aged (low SMC) peanut and tomato seeds, respectively.

Few workers have undertaken careful comparative studies on the effects of different aging conditions in this area, but in their study in wheat, Petruzzelli and Taranto (1984) compared PL changes in seeds stored under a range of conditions (including two years dry storage) which resulted in equivalent losses in germinability. While there was a small but significant loss of phospholipid from embryos of dry stored seeds, this decrease was much smaller than that found in seeds stored at higher moisture contents.

A further complication is that correlative evidence of this type, demonstrating a relationship between phospholipid and germination losses, does not demonstrate cause and effect. Both Powell and Matthews (1981) and Petruzzelli and Taranto (1984) showed that PL changes occurred before appreciable changes in germinability. Nevertheless, as the Powell and Matthews data in Figure 3 show, changes in PL in peas could not be divorced from cell death in cotyledonary tissues. In this context a most interesting finding was reported by Francis and Coolbear (1987) in aged tomato seeds, cv. Kingley Cross. In long-term storage, PL losses paralleled loss in germinability, but under high moisture content deterioration conditions seeds lost germinability before any significant loss in PL was

FIGURE 3. Changes in total extractable cotyledon phospholipid with seed aging in peas. Open and closed symbols refer to data from different experiments. Data plotted from Powell and Matthews (1981). Losses in total phospholipid were largely due to losses of phosphatidylcholine.

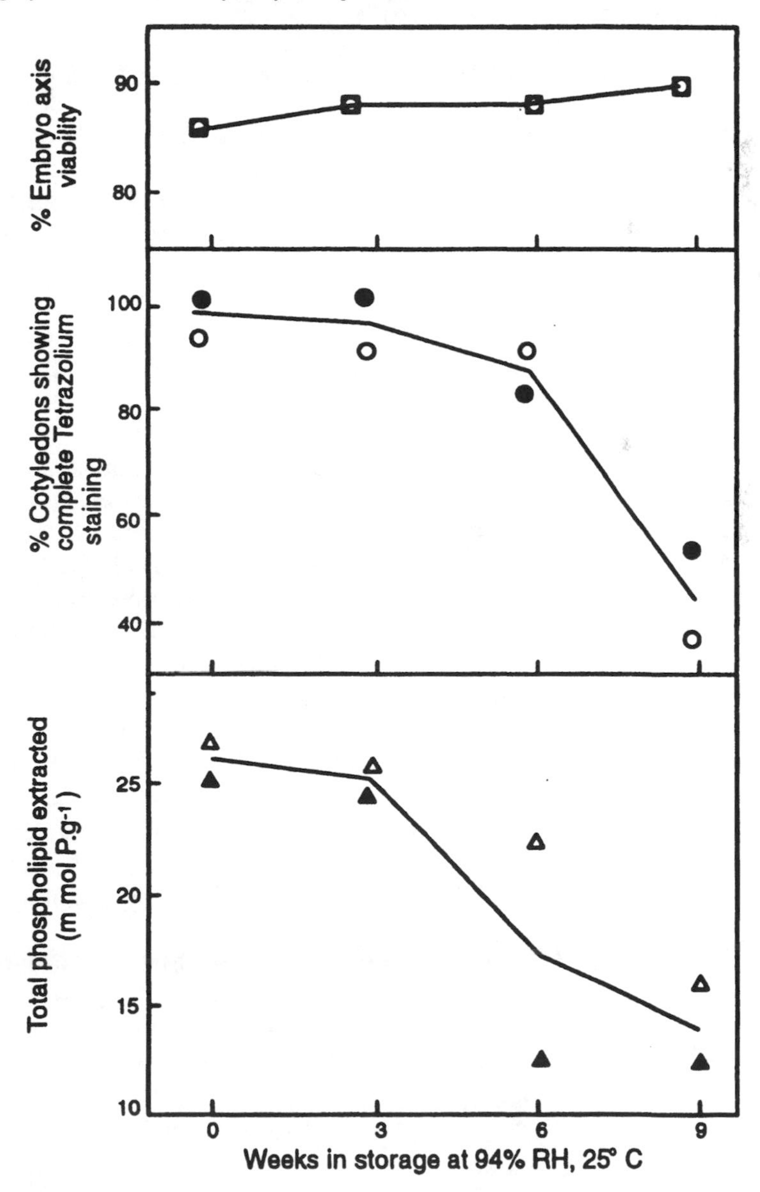

detected. It is possible that the moisture content used in this rapid-aging study was high enough to permit some measure of membrane repair and thus delay the loss of phospholipid while other events resulted in the loss of seed viability.

Changes in Fatty Acid Composition and Lipid Peroxidation

If membrane changes are a crucial event in seed deterioration on many occasions, the next question is what precipitates membrane damage itself? Is it the result of hydrolytic damage (for example the action of phospholipases) or is it due to oxidative damage of the membrane system? It is well documented, for instance, that high partial pressures of O_2 accelerate the deterioration of dry seeds (see Chapters 1 and 6). We are now in the arena for one of the most topical arguments about the mechanisms of seed deterioration.

There is substantial evidence that oxidative damage to membrane systems in seeds can occur via free-radical-driven lipid peroxidation (see reviews by Wilson and McDonald, 1986b; Benson, 1990). Free radicals are groups of atoms with an unpaired electron which renders them highly reactive and unstable. The two most important in this context are the hydroxyl ($^{\bullet}OH$) and superoxide ($O_2^{-\bullet}$) radicals. Both require the presence of oxygen for their production. Once present in a cell these free radicals can initiate highly damaging oxidative chain reactions, especially with polyunsaturated fatty acids such as linoleic acid to form lipid hydroperoxides (Figure 4). Note that as each hydroperoxide molecule is formed, an organic free radical is also produced capable of reacting with another lipid molecule. Such peroxidation reactions may be spontaneous or catalyzed by lipoxygenase enzymes, enzymes which may be active in a nonaqueous environment. Apart from initiating chain reactions within the membrane itself, free radicals can be highly damaging to DNA and other biomolecules.

While free radicals can be produced in both high and low moisture content environments, there is substantial evidence to suggest that they are trapped (and thus presumably able to do little damage) at very low seed moisture contents (Priestley et al., 1985). When the moisture content is increased (such as during the first few hours of imbibition) lipid hydroperoxides are likely to become highly reactive and may react with other macromolecules within the cell and/or form mobile secondary products which are also extremely cytotoxic. One of these, malondialdehyde, a result of the peroxidation of 18:3 linolenic acid, can do significant damage to membrane proteins by inducing cross-linking reactions. At the same time, however, the increase in seed moisture allows the seed to activate its defenses to free radicals. Seeds possess detoxification enzymes such as superoxide dismutase (which can destroy the peroxide free radical itself by converting it to hydrogen peroxide which is in turn broken down by catalase or peroxidase action), or the glutathione peroxidase/reductase system which can neutralize lipid hydroperoxides (Benson, 1990).

Stewart and Bewley (1980) showed that viable soybean seeds had the capacity to produce superoxide dismutase (SOD) within the first few hours of imbibition, but that this activity was absent in nonviable material. Similar results were obtained by Francis (1985) in tomatoes. Puntalaro and Boveris (1990) have demonstrated a correlation between reductions in SOD activity produced in the first two hours of imbibition and decreases in vigor in deteriorating seeds. On the basis that dry seeds do not appear to have any SOD activity, Priestley (1986) questioned whether the activity of this enzyme has much impact on the viability of stored seeds subject to free radical damage. Once again the key question is at what stage do free radicals do most damage to seed material, in relatively dry tissue during storage, or in the first few hours of imbibition?

A wide range of non-enzymic compounds can also act as antioxidants in seeds. For example, tocopherols, ascorbate, and glutathione all have demonstrated antioxidant properties and may be active against free radicals (Benson, 1990). Attempts to demonstrate changes in these compounds which might be associated with a protective role in seed deterioration have been inconclusive (e.g., Sharma, 1977; Pearce and Abdel Samad, 1980; Chen and Fu, 1986). Nevertheless, several workers have demonstrated the effectiveness of certain antioxidant treatments in protecting seeds from deterioration during storage (e.g., Woodstock et al., 1983; Gorecki and Harman, 1987).

All this adds up to a great deal of uncertainty both about the extent of free radical formation in deteriorating seeds under different environmental conditions and about how much damage these

FIGURE 4. An outline reaction scheme for organic free radical formation, conjugated diene production, and lipid peroxidation. This type of reaction may be autolytic or catalyzed by the enzyme lipoxygenase.

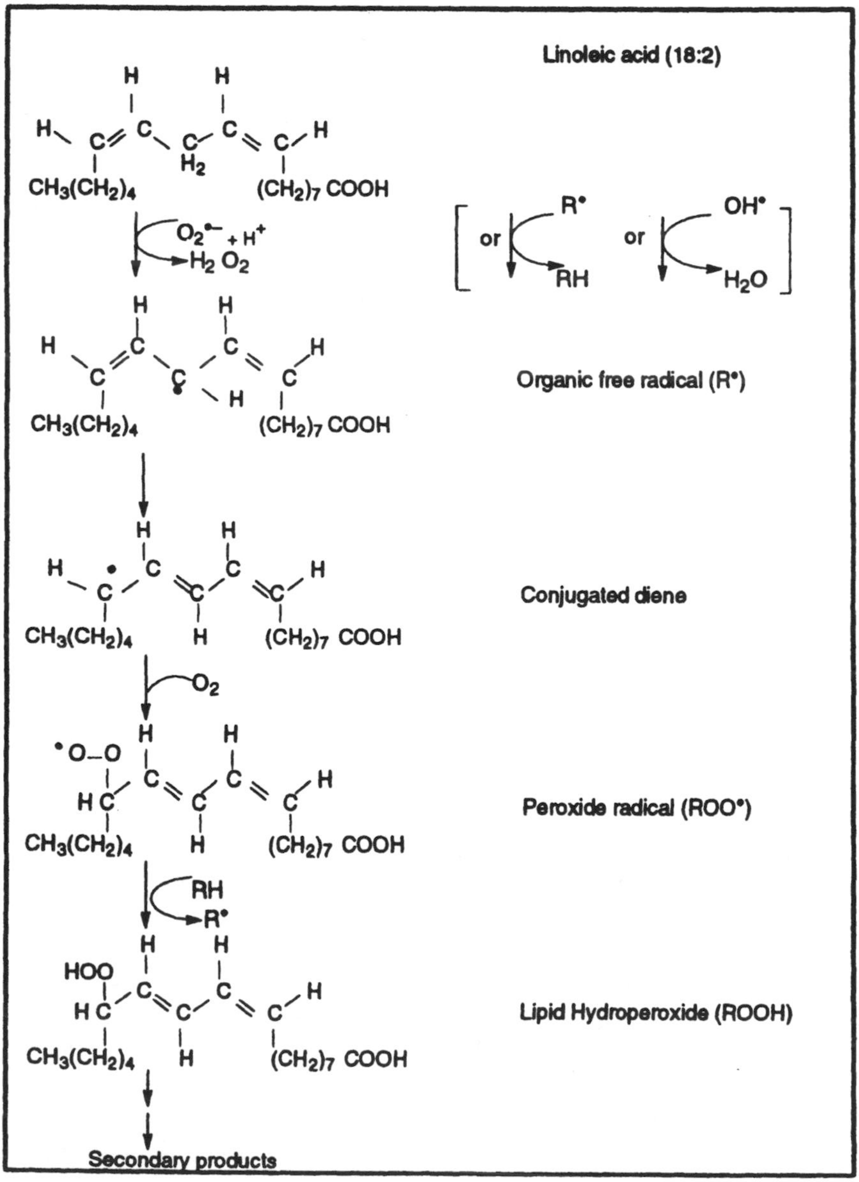

Linoleic acid (18:2)

$CH_3(CH_2)_4$ ... $(CH_2)_7COOH$

$O_2^{\bullet-}$ + H^+ → H_2O_2 [or R• → RH or OH• → H_2O]

Organic free radical (R•)

Conjugated diene

O_2

Peroxide radical (ROO•)

RH → R•

Lipid Hydroperoxide (ROOH)

Secondary products

TABLE 2. The different types of approaches used on studies in free radical driven lipid perioxidation in seeds

Direct detection of free radicals
Detection of primary products, i.e., hydroperoxides and conjugated dienes
Loss of substrate, i.e., preferential deletion of polyunsaturated fatty acids
Detection of secondary products, e.g., malondialdehyde, hexaldehyde, other volatiles
Monitoring changes in antioxidant levels in seeds
Effects of antioxidant treatments

species are capable of doing before the seeds' natural defense mechanisms destroy them. As discussed in all three recent major reviews in this area (Priestley, 1986; Wilson and McDonald, 1986b; Benson, 1990), current evidence makes it difficult to reach any kind of conclusion. Experiments to detect free radical driven oxidation in seeds fall into the six categories listed in Table 2. Each method is fraught with difficulties of interpretation and it is hardly surprising that no consensus on the evidence has been obtained to date. If only for the fact that some of the problems here epitomize the difficulties in the study of the physiology of seed deterioration in general and the validity of the evidence thereby obtained, it is worth considering some of these approaches in detail.

Direct detection of free radicals. Free radicals may be detected directly by the use of electron spin resonance (ESR) techniques to determine the presence of the unpaired electrons on the radicals themselves, or via low level chemiluminescence. Superficially this approach might be expected to provide some definitive answers, however, free radicals are highly unstable and, not unexpectedly, the small amount of experimental evidence available in this area is full of contradictions. Conger and Randolf (1968) concluded from their studies of several species held in dry long-term storage that free radicals had little to do with seed deterioration, as levels did not change significantly with seed age. Overall, the trend was a slight decline with increased storage time, although individual results with different accessions were highly variable. Three other research groups have used this technique on soybean. Both have produced data which conflicts with Conger and Randolf's conclusions, but equally there are direct contradictions between each set of data under different storage conditions and the part of the seed analyzed (Table 3). Clearly the history and moisture content of the tissue is of crucial importance to the nature of any results obtained (e.g., Priestley et al., 1985). Puntalaro and Boveris (1990) report a major increase in free radical production (measured by chemiluminescence) when seeds begin to take up water on early germination.

Even if the data were more clear-cut, it is not just a question of how many free radicals are present at any given time, but how much damage has been caused, or is likely to be caused, in this way. As each free radical is capable of participating in a chain reaction of organic free radical formation, absolute levels do not indicate how much hydroperoxide (or other products) has been formed.

Detection of primary products of free radical damage. Hydroperoxides formed during lipid peroxidation can be quantified by assaying the amount of iodide ions oxidized to iodine by extracted lipid. There are, however, questions both about the specificity of this approach and, of course, the stability of the moiety to be assayed. Once again the literature contains some conflicting results, ranging from substantial increases in sesame, cotton, and castor bean after 18 months commercial storage (Sharma, 1977) to no detectable peroxides present in peanuts stored at either low moisture and temperature for over three years or up to one month at 38°C and 90 percent relative humidity (RH) (Pearce and Abdel Samad, 1980).

Recently, in a most carefully thought out and comprehensive study, Gidrol et al. (1989) have measured the extent of conjugated diene (Figure 4) production in sunflower seeds subjected to rapid high moisture content aging for up to 8d. During this time conjugated dienes increased tenfold as seeds lost 25 percent viability. Most significantly, these changes were not associated with any increases in leachate conductivity and they could find no conjugated dienes in the microsomal fraction from either aged seeds or unaged controls. Here, then, is thought-provoking evidence that lipid peroxidation is occurring in lipid reserves, but not affecting the integrity of cell membranes.

Loss of substrate. Probably the most used approach in this type of study is to look for preferential deletion of polyunsaturated fatty

TABLE 3. Organic free radical levels as determined by electron spin resonance studies in deteriorating soybean

	Conclusions on Free Radical Levels	
	Low SMC, long-term storage	Rapid high SMC aging
Axes		
Priestley et al. (1985)	Little change	~100% increase
Buchvarov and Gantcheff (1984)	~90% increase	~70% increase
Puntalaro and Boveris (1990)	Major increase	Major increase
Cotyledons		
Priestley, McBride, and Leopold, (1980)	Little change	Little change
Buchvarov and Gantcheff (1984)	No free radicals detected	No free radicals detected

acids (particularly 18:2 and 18:3 compounds) in the lipid fractions of aged seeds. Because these compounds have the ability to partially stabilize the organic free radical as a conjugated diene (Figure 4) they are much more prone to free radical damage than other fatty acid components. In their classic study, Harman and Mattick (1976) demonstrated that during high humidity aging of pea seeds, loss of germinability was associated with loss of linoleic (18:2) and linolenic (18:3) acids with no concurrent changes in other major (saturated or monounsaturated) fatty acid components. They took this to be a clear demonstration of the occurrence of lipid peroxidation during seed deterioration. Since then other workers have looked for similar types of evidence. To say that the overall impression from subsequent reports is confusing would be an understatement. Perhaps most work of this type has been done on soybeans, and results for this species are summarized in Tables 4 and 5 for slow (low SMC) and rapid aging, respectively. Clearly the results depend on the cultivar used (and perhaps even the seedlot), the nature of the storage conditions, the extent of deterioration during storage, the tissue or cell fraction from which the lipid was assayed, and the type of lipid analyzed. As might be expected, other species show differing responses.

Genetic Damage

Apart from loss of membrane integrity, damage to the genome is most commonly cited as a likely primary cause of seed deterioration. A little damage may result in the accumulation of point mutations as discussed earlier, where small changes are nonlethal and carried through in a sufficient number of daughter cells to affect plant morphology or function at either a later stage of growth (abnormal seedling production, pollen sterility, etc.) or to be carried through as recessive genes to later generations (e.g., Dourado and Roberts, 1984, who demonstrated the inheritance of recessive photosynthetic mutations caused by aging in pea and barley). More severe genetic damage may be lethal for individual cells, the impact of this on seed germinability and survival being dependent on the location and frequency of damage within the seeds. It seems that for each genotype there is a critical proportion of cells which can be damaged before seed viability is prejudiced. Generally this level does not exceed 15 percent (Priest-

TABLE 4. Changes in proportions of fatty acids in deteriorating soybean due to slow aging at low seed moisture contents

Cultivar	Change in % viability	Tissue assayed	Lipid fraction analyzed	Result
Chippewa[1]	98→28%	Whole Seed	TL	↓ 18:3, small ↓ 18:2
			PL	↓ 18:3
		Axes	TL	Large ↓ 18:3, ↓ 18:2
Desoto[2]	No change, but ↓ vigor	Cotyledons	TL	No change
		Axes	TL	No change
			PL	No change
		Axis Mitochondria	TL	↓ 18:3
Maple Arrow[3]	93→16%	Axis Microsomes	TL	No change
Pride x 005[4]	No change	Axes	PL	No change
Union[2]	No change, but ↓ vigor	Whole Seed	TL	No change
		Axes	TL	No change
			PL	No change
		Axis Mitochondria	TL	No change in PUFAs
Wayne[1]	98→86%	Whole Seed	TL	↓ 18:3, small ↓ 18:2

TL= Total lipid; PL = Polar lipid; PUFAs = Polyunsaturated fatty acids.
[1]Priestley and Leopold (1983), [2]Ferguson, TeKrony, and Egli (1990b); [3]Senaratna, Gusse, and McKersie (1988); [4]Stewart and Bewley (1980).

TABLE 5. Changes in proportions of fatty acids in deteriorating soybean under rapid aging (usually high SMC) regimes

Cultivar	Change in % viability	Tissue assayed	Lipid fraction analyzed	Result
Amsoy[1]	89→8%	Whole seed	PL	No change
Chippewa[2]	100→5%	Whole seed	TL	No change
Davis[3]	100→85%	Axes	TL	No change
		Cotyledons	TL	No change
Pride x 005[4]	60→0%	Axes	PL	↓ 18:3, ↓ 18:2
Wayne[5]	98→38%	Whole Seed	TL	No change
Williams[6]	?→0%[7]	Cotyledons	TL	No change
			PL	No change
		Axes	TL	No change
			PL	Small ↓ 18:3 and 18:2

TL = Total lipid; PL = polar lipid
[1]Ohlrogge and Kernan (1982); [2]Priestley and Leopold (1979); [3]Sakunnarak (1992); [4]Stewart and Bewley (1980); [5]Priestley and Leopold (1983); [6]Priestley, Werner, and Leopold (1985). [7]Initial viability not stated; in this study seeds were held at 105°C for up to 6d in oxygen.

ley, 1986) and may be considerably lower, e.g., 4 percent for barley, (Abdalla and Roberts, 1968). Such damage is the result of unrepaired breaks in the genome which may be caused by a variety of agents such as free radical damage, hydrolytic enzyme activity, or mutagenic compounds which may accumulate in deteriorating seeds (Roos, 1982).

Much of the work in this field has concentrated on correlative evidence between the extent of visible genetic damage appearing during the first cycle of cell divisions during germination and loss of seed viability. Some of the resulting relationships are surprisingly strong over a range of different storage conditions (e.g., Abdalla and Roberts, 1968; Roos, 1982), giving rise to suggestions that this kind of damage may be an important event in seed deterioration. However, some workers (e.g., Abdalla and Roberts, 1968; Rao and Roberts, 1990) have noted that this kind of correlation breaks down when seeds are stored at elevated moisture contents (usually around 18 percent or higher) when fewer aberrations are found for any given loss of viability. Obviously, under these kinds of conditions other deteriorative events must be proceeding at faster rates. Also, there is evidence that such high moisture content conditions may begin to favor repair activity (Rao and Roberts, 1990; and see later discussion this section).

Distinctions have been made between different types of genetic damage in this kind of study, viz. chromosomal aberrations where both chromatids are damaged at the same point (suggesting damage before the first round of germinative DNA replication), and chromatid aberrations where only one of a pair of chromatids is damaged (a lesion likely to have occurred after the first S-phase of cell division on germination). Their relative significance in seed deterioration is unclear except that preponderance of chromosomal-type aberrations would be more indicative of damage to the genome *per se* during storage. In Rao and Roberts' (1990) study on lettuce, both types of damage were frequent in seeds stored at high temperature and low seed moisture, with chromosome-type aberrations dominating. However, at 18 percent seed moisture, chromosomal aberrations were almost completely absent.

That damage to DNA does occur in the dry seed has been demonstrated by Osborne (1980) and her coworkers using a variety of techniques to examine extracted DNA directly. There is also substantial ultrastructural evidence of damage to nuclei as seeds age, a typical feature being that nuclei lose their even granular appearance as clumps of electron-dense chromatin develop (Villiers, 1980).

Cheah and Osborne (1978) suggested that damage to DNA was due to slow hydrolytic activity (see later discussion on enzyme changes during aging). Alternatively, damage may be due to free radical activity, either direct interactions, especially with the hydroxyl free radical on thymine, or interactions with the secondary products of lipid peroxidation, such as malondialdehyde, known to have mutagenic properties (Benson, 1990). Recently, Guy, Smith, and Black (1991) used a restriction fragment length polymorphism (RFLP) technique to monitor breakdown of ribosomal DNA in deteriorating wheat seeds. Reproducible lesions were detected well before any losses in seed vigor. It is, at first sight, surprising that the pattern of breakdown of DNA in this study appeared to be so repeatable. This suggests that parts of the DNA are more vulnerable to damage than others. This may indicate the action of highly specific endoDNase enzymes. Benson (1990) notes, however, that there is evidence that some base sequences render thymine much more susceptible to free radical attack than others and that free radical damage to the genome might be much less random than might be initially supposed. Specific lesions to DNA may also arise as a function of the arrangement of the molecule on its protecting histone proteins, leaving some segments more open to damage than others. Certainly, Guy, Smith, and Black's (1991) study has opened up a whole new area for investigation.

As the presence of chromatid-type lesions would imply, mechanisms of DNA synthesis may also suffer damage during seed aging. Direct confirmation of this has been provided by several workers, for example, Dell'Aquila and Margiotta (1986) found that the deterioration of isolated wheat grains was closely paralleled by changes in tritiated thymidine incorporation into DNA in excised embryos. Changes were accompanied by delays in (and a diminution of) mitotic activity. Apart from generally slowing down the process of germination in deteriorated but still viable seeds, damage to DNA metabolism may be crucial to a seed's ability to recover from lesions incurred during storage. There is no doubt that partially deteriorated seeds do have the ability to repair damaged strands of DNA by replacing segments using the complementary strand as a template (e.g., Villiers, 1974; Osborne, 1982; Guy, Smith, and Black, 1991).

Changes in Respiratory Activity

During deterioration seeds generally suffer damage to their respiratory capacity for early germination. Often this can be demonstrated to be a consequence of damage to mitochondrial membranes and ultrastructural evidence for this has already been mentioned. Once again this topical area has been well reviewed by Priestley (1986). Many deteriorated seeds show reduced oxygen uptake and elevated respiratory quotients (the ratio of CO_2 evolved to O_2 taken up) on early imbibition. For example, Woodstock, Furman, and Solomos (1984) found loss of respiratory activity to be associated with reduced vigor in soybean. Low vigor was also accompanied by increased levels of toxic by-products such as ethanol and aldehydes, indicating the inability of damaged mitochondria to keep up with glycolytic activity, resulting in anaerobic catabolism (Woodstock and Taylorson, 1981).

Abu-Shakra and Ching (1967) noted that, in many species, oxygen uptake by deteriorated seeds did not in fact decrease much compared to higher-vigor seeds. Nevertheless, their study confirmed that loss of respiratory efficiency did occur. They compared mitochondria from four-day-old seedlings grown from both newly harvested and three-year-old soybeans. Mitochondria from the older, lower-vigor seedlot showed clear signs of ultrastructural damage and, although they took up at least as much, if not more, oxygen than mitochondria from unaged controls, they produced less than half the ATP. Clearly there was a significant uncoupling of the oxidative phosphorylation system in these organelles, itself suggesting disruption of mitochondrial membrane integrity. Recently, the hypothesis that mitochondrial membrane damage is a very early event in soybean deterioration has been confirmed in a careful time course study by Ferguson, TeKrony, and Egli (1990a, 1990b). In their study, both oxygen uptake and ATP production by isolated mitochondria were depressed after very short periods of aging. In one of the two cultivars studied, loss of mitochondrial activity was shown to be associated with deletion of polyunsaturated fatty acids from mitochondrial membranes, indicating peroxidation damage to these organelles even though there was no clear evidence of similar damage to total or polar lipids in the axis (Table 4).

Obviously, a decrease in respiratory efficiency will mean a decrease in the capacity of the seed to produce ATP for all the synthetic and repair reactions required for early germination. Several authors have suggested that direct measurement of ATP on early imbibition may be a useful index of deterioration. The problems with this approach have recently been summarized (Hampton and Coolbear, 1990). Apart from the fact that initial production of ATP may not be dependent on oxidative phosphorylation (Perl, 1987), the problem is analogous to that of measuring free radicals directly: low ATP levels in a germinating seed may be a function of either low respiratory efficiency or very rapid utilization in actively growing healthy tissue. Anderson and Gupta (1986) have carefully reviewed the effects of seed deterioration on both ATP and the other major nucleotides, emphasizing that reduced levels of any of these components may have direct consequences for specific synthetic reactions, especially (in the context of previous discussion) nucleic acid and phospholipid biosynthesis.

One other well-documented feature of aging damage in seeds is the observation that alternative respiratory pathways may be less sensitive to deterioration than normal glycolysis and oxidative phosphorylation. For instance, in rape seed, Takayanagi (1977) found that the proportion of glucose catabolized by the pentose phosphate pathway rather than the normal glycolysis route was considerably elevated in aged seeds.

Enzyme and Protein Changes

Enzyme changes during seed deterioration have been investigated by several workers. In general it would appear that enzymes involved in catabolic reactions, especially the hydrolases, are more stable than those enzymes involved in synthetic reactions (e.g., Osborne, 1980). For example, Dey and Mukherjee (1986) found that storage of mustard, maize, and soybean was associated with high levels of lipase and accumulation of free fatty acids. In contrast, dehydrogenase and peroxidase levels decreased. Their results are most interesting in that they were able to show that prestorage, hydration-dehydration treatments of mustard and maize, which greatly increased germinability after storage, also caused a large reduction in lipase activity in these kernels.

Probably the most comprehensive study has been undertaken by Perl, Luria, and Gelmond (1978), in sorghum. These authors found that seeds stored at a constant SMC of 17 percent at 30°C showed a small increase in vigor after six days storage and then a gradual decline in seed quality as seeds were aged for up to 50d (Gelmond et al., 1978). One of the most significant features of this study was that the measurements were taken as seeds were losing vigor rather than germinability (at the end of the storage period only 10 percent germination had been lost), thus the enzyme changes measured were early events in the aging process. A wide range of enzymes were assayed in the seeds aged in this way, with amylase, ribonuclease, glutamate decarboxylase, and glutamic pyruvic transaminase showing an initial increase in activity and then a marked decline as seed vigor began to fall. Acid phosphatase and general dehydrogenase activity declined from the start of aging while, in contrast, protease activity increased during the aging period. Perl, Luria, and Gelmond (1978) considered this latter enzyme to be a likely key factor in seed deterioration because of the potential for protease to destroy other enzymes in the seed.

Although this remains a most interesting study, not all other work supports these findings. While it has been generally agreed that reductions in glutamate decarboxylase activity are useful indices of seed deterioration (e.g., Delouche and Baskin, 1973; Ram and Weisner, 1988), studies on other enzymes have produced much more variable results. For example, Osborne (1980) noted the high stability of ribonuclease in deteriorating wheat grains, while in our laboratory (Nath et al., 1990; Nath, 1991), α-amylase levels have been shown to increase in wheat grains during storage at 15 percent seed moisture, 35°C, while protease levels were found to be generally constant during early deterioration and then to drop dramatically as seeds lost germinability. Both Petruzzelli and Taranto (1990) and Nath (1991) have demonstrated that seed deterioration results in decreased ability to produce amylase on subsequent germination.

In any case, care must be taken in the interpretation of data on enzyme changes of this type. Apparent changes in activity may be the result of changes in enzyme cofactors or inhibitors. For example, the increase in protease activity noted by Perl, Luria, and Gelmond (1978) is likely to be due to preferential destruction of protease inhibitors rather than the synthesis of new protease during high moisture content aging. Cheah and Osborne (1978) similarly suggested that inhibitor breakdown might give rise to DNase activity in deteriorating seeds. Alternatively, assay conditions may give misleading results in that enzyme optima, especially pH, may change with seed deterioration due to either partial denaturation of enzyme protein or a change in the isozyme profile of the particular group of enzymes being studied. For example, Livesley and Bray (1991) have shown that aleurone tissue from aged wheat seeds lost its ability to produce high pI group isozymes of α-amylase even though total protein synthetic activity of the tissue was unimpaired. In earlier work on the aging responses of crimson clover (*Trifolium incarnatum* L.) seeds, Ching (1972) demonstrated that membrane-bound phosphatase was much more resistant to deterioration than soluble acid phosphatase.

Whatever changes in extracted enzyme activity are observed during seed deterioration, these effects are largely irrelevant unless it can be demonstrated that the enzyme is active *in vivo*. This is particularly important when it is recognized that many processes within plants are ultimately substrate rather than enzyme controlled. Little work appears to have been done in this area, although twenty years ago, Berjak and Villiers (1972b) were able to demonstrate that, in aged maize embryos, damaged cells of the root cap had lost their ability to compartmentalize acid phosphatase in lysosomes and it was dispersed throughout the cytoplasm of these cells. Thus, even if enzymes like phosphatase, protease, or ribonuclease are observed to be unchanged or even declining in activity in crude extracts, each may have the capacity to cause major damage *in vivo* if it is no longer compartmentalized away from vulnerable parts of the cell.

One very promising approach that has recently been reported is the characterization of the storage behavior of mutants known to be deficient in enzymes thought to be involved in the deteriorative process. Wang et al. (1990) looked at the storage characteristics of two soybean mutants, one lacking lipoxygenases 1 and 3, the other 2 and 3, in comparison to the normal type which contains all three enzymes. If lipoxygenase-induced free radical production was a major factor in aging in this cultivar, it might be expected that the mutants would store much better than the normal line. In fact this was not the case. Not only did the levels of all three lipoxygenases fall during aging, but the germinability of the mutant deficient in lipoxygenases 2 and 3 deteriorated more rapidly than that of the other two. Not only does this argue against the involvement of enzyme-driven lipid peroxidation in this case, but it also gives rise to the novel idea that lipoxygenase 2 may in fact have a role in preventing loss of seed viability.

Once germination has started, changes in enzyme activities in aged seeds may equally be a function of the loss of ability of seeds to undertake protein synthesis. Several workers (e.g., Abdul-Baki, 1969; Ching, 1972; Perl, Luria, and Gelmond, 1978; Blowers, Stormonth, and Bray, 1980) have demonstrated that early seed deterioration is associated with reduced protein synthesis. Damage to the protein synthetic machinery can take place at both the translational and transcriptional level. Osborne (1980) and Priestley (1986) have reviewed the evidence which shows that the transcriptional components (i.e., tRNAs, associated enzymes, elongation factors, and the ribosomes themselves) are damaged during seed deterioration. Such damage may be either enzymic (especially due to the action of ribonuclease) or free radical induced. Recently Davidson, Taylor, and Bray (1991) have shown that fragmented ribosomal RNA in low-vigor leek (*Allium porrum*) seeds is replaced during germination-enhancing osmotic presowing treatments. Other work (e.g., Blowers, Stormonth, and Bray, 1980; and reviews by Osborne, 1980 and Priestley, 1986) has shown that low-vigor seeds have reduced levels of both conserved mRNA and a decreased ability for transcriptional activity on subsequent germination. While there is still considerable debate on the role of conserved mRNA in seed germination, it is evident that a diminished ability to undertake new transcription would represent a crucial lesion in deteriorated seed.

A key consequence of seed aging impairing enzyme synthesis in subsequent germination may be the loss of the ability of seeds to produce enzymes essential for detoxification and repair. Thus, as already mentioned, Stewart and Bewley (1980) and Francis (1985) were able to show that nonviable soybean and tomato seeds, respectively, were unable to develop superoxide dismutase activity on early imbibition. In an interesting contrast to these studies, Gidrol (1989) found large increases in superoxide dismutase activity as a result of accelerated aging in soybean. This was accompanied by decreases in catalase and peroxidase activities, leading to their suggestion that hydrogen peroxide may accumulate as seeds attempt to neutralize free radicals and that this itself may be toxic to the seed. These results were not, however, confirmed by Puntalaro and Boveris (1990), who reported that both low and high SMC aging induced marked reductions in superoxide dismutase activity appearing during early imbibition. Catalase and peroxidase levels produced during early imbibition increased as a result of aging but then tended to decline rapidly after 10 to 15h. Glutathione peroxidase levels increased slightly after aging and tended to remain constant. The ability of seeds to counteract high levels of free radical activity during early imbibition is crucial to their survival, and is clearly an area where further research is needed.

Changes in Endogenous Plant Growth Regulators

Another potential cause of seed deterioration, suggested to be of importance by several workers, is the disruption of the phytohormonal control regarded as essential to the germination process. While it is likely that changes in enzyme activities may have an impact on endogenous plant growth regulator (PGR) levels within the seed, it is difficult to draw any conclusions about whether this might be a major limiting factor for seed germination performance after deterioration in storage.

Several workers have shown that application of PGRs such as gibberellins (GAs), cytokinins, and ethylene are very effective in improving the vigor of aged, still viable seed (e.g., Harrington, 1973; and review by Priestley, 1986). Work by Puls and Lambeth (1974) provides an interesting example here. Application of a combination of kinetin and KNO_3 improved the germination rates of ten-year-old viable tomato seeds, although neither was effective by itself. Several aspects of metabolism, especially hydrolytic activity, were enhanced by these treatments, although some were also promoted by gibberellic acid applications which did not improve germination rate significantly. Apart from illustrating the pitfalls of simple correlative evidence, this kind of data would tend to suggest that beneficial effects of PGRs (and KNO_3) on germination are more likely to be the result of a general promotion of metabolism favoring repair activity at appropriate stages of the germination process, rather than the reinstatement of any specific control mechanism.

There have also, however, been a few reports on the effective use of PGR treatments prior to storage to protect seeds from deterioration during aging. Most notably, Petruzzelli and Taranto (1985) were able to show that treatment of dry grains with acetone solutions of gibberellic acid or ethephon before storage maintained the viability of durum wheat (*Triticum durum*) in store at 14.5 percent SMC and 30°C. In marked contrast, Bhattacharjee, Roychowdhury, and Choudhuri (1986) have shown that pretreatment of seeds with aqueous solutions of growth retardants including CCC, an inhibitor of GA biosynthesis, prolonged the storage life of jute (*Corchorus* species) seeds. Unfortunately, in this case only seeds soaked in distilled water were used as the controls and no information was given on the storage behavior of untreated seeds. It is always possible that the 6h hydration-dehydration treatment of these seeds in water was detrimental to their storability (cf. Nath, Coolbear, and Hampton, 1991).

The other aspect of this story is that seeds may lose their sensitivity to endogenous PGRs as they deteriorate. This would, in theory, be quite likely to happen if disruption of membrane integrity were to cause a reduction in the affinity of membrane-bound hormone receptors for a particular PGR, or if, for example, a change in membrane permeability reduced the efficiency of the resulting response. Aspinall and Paleg (1971) showed that as radicle emergence rates of wheat decreased with long-term dry storage, the ability of endosperm sections to produce α-amylase in response to GA fell dramatically. This work has recently been re-examined in our laboratory (Nath, 1991; Figure 5). These data indicate that aleurone tissues of aged seeds tend to lose their response capacity rather than affinity for the PGR which still produces the response with the same lag phase and at similar threshold and saturation concentrations as it does in unaged tissue. Although stored grain in this study showed loss of vigor (measured as rate of radicle emergence), they still retained similar viability levels to unaged controls.

FIGURE 5. (Main graph) α-amylase production by endosperm halves cut from wheat grains aged for 40d at 15 percent SMC, 35°C (□) in response to changes in GA_3 concentration; O: unaged controls. Half grains were incubated for 42h at 25°C before assay. (Inset) Time course of response of endosperm halves incubated in $2x10^{-6}$ M GA_3 at 25°C. Data from Nath (1991).

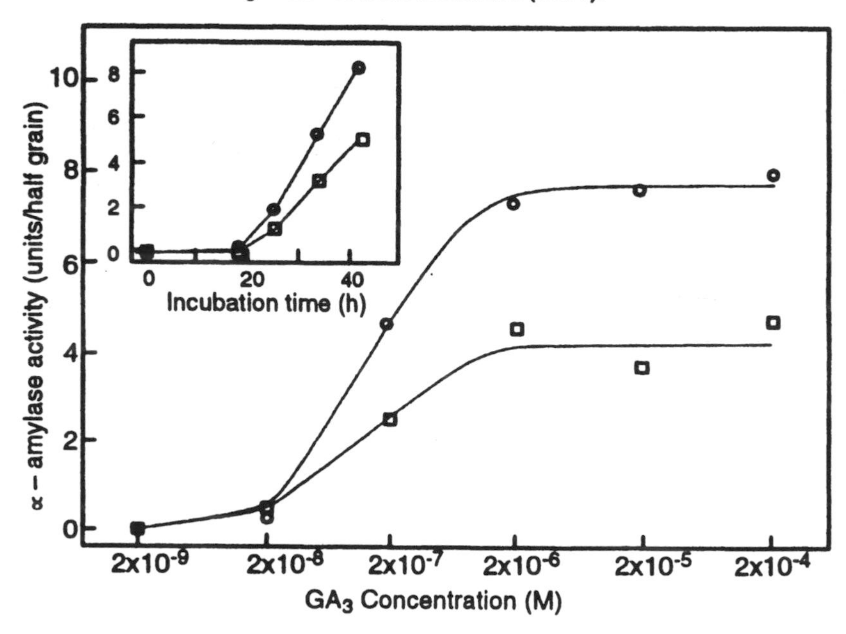

Accumulation of Toxic Metabolites

Already mentioned in the foregoing discussion is the possible involvement of accumulated toxic metabolites in the deterioration process. Such compounds may include ethanol (as a result of anaerobic respiration), aldehydes (anaerobic respiration or lipid peroxidation), short chain fatty acids (lipid breakdown), and phenolics (possible secondary products of lipid peroxidation). All are inhibitory to seed germination.

Woodstock and Taylorson (1981) demonstrated that the levels of accumulated ethanol and acetaldehyde were inversely correlated with subsequent seedling growth of soybean as seeds deteriorated in both accelerated and slow aging storage systems. As noted previously, they suggested that high levels of these two compounds were the result of the inability of mitochondria to initiate efficient aerobic respiration causing accumulated pyruvate produced by glycolysis to be metabolized into these compounds. Application of an inhibitor of cytochrome c oxidase, sodium azide, mimicked this condition in high-vigor seed, while allowing low-vigor seed to imbibe water more slowly by using an inert osmoticum as a substrate for germination, and allowed further time for recovery and a reduction of ethanol and aldehyde levels (Woodstock and Tao, 1981).

Similarly, Wilson and McDonald (1986a) demonstrated that evolution of aldehydes during the first 48h of germination of soybean was inversely correlated with field performance and other measures of seed vigor. In contrast to Woodstock's group, however, these authors argued that the aldehyde produced is largely a by-product of lipid peroxidation. Another volatile of similar origin which is produced in higher quantities in low-vigor seeds is hexaldehyde (hexanal) (Castro and Sediyama, 1990). One of the most researched compounds in this context is malondialdehyde, which, as previously mentioned, is a relatively stable secondary product of free radical damage (e.g., Stewart and Bewley, 1980; Figure 6; Dey and Mukherjee, 1986). Amongst other effects listed previously, this compound can react with DNA and thus has mutagenic properties (Benson, 1990). In many cases, healthy seeds have inbuilt mechanisms to neutralize the effects of such toxic metabolites. For example, in Stewart and Bewley's (1980) studies on malondialdehyde production in soybean, unimbibed unaged seeds had high levels of this compound. The key difference due to aging was the seed's ability to reduce the concentration of the metabolite during the first few hours of imbibition.

Once again, the general conclusion must be that results of this type must be interpreted with caution. The detection of toxic metabolites *per se*, does not necessarily provide evidence that they have caused damage within the seed, they may simply be symptoms of damage rather than its cause. There is a need for accurate information on both their location within the seed and whether that particular tissue is capable of processing the metabolite safely. Even then, it is open to question whether these compounds are more important in the deterioration process than the damage which produced them (or allowed them to accumulate) in the first place.

The Impact of Microorganisms on Seed Deterioration

The final facet of the deterioration story which needs to be considered is often a neglected one, the role of microorganisms. Associations of microorganisms with seeds are discussed in detail in Chapter 5. Here, for completeness, I will simply highlight the key issues which pertain to the physiology of seed deterioration. These have recently been summarized in an excellent short review by Halloin (1986). Ultimately his article becomes a cry from the heart for more cooperation between seed physiologists and pathologists so that each discipline does not neglect the other in evaluating causes of seed damage in storage.

Of the microorganisms which can initiate or accelerate the deterioration process, saprophytic fungi are particularly important. These essentially fall into two groups, the field fungi which require free water for growth, and the storage fungi which can grow at reduced relative humidities (some species as low as 65 percent RH) and are thus active on seeds at low moisture contents. Field fungi such as *Fusarium*, *Cladosporium*, *Nigrospora*, *Curvularia*, and *Alternaria* species are often associated with seeds at harvest and require moisture contents in starchy seeds of at least 22 percent for growth. Much deterioration of seeds in the field, indicated, for example, by discoloration, shriveling or damage to the embryo, may be caused by the invasion of these types of fungi, which thus must be regarded as a major component of the weathering process (Christensen and Kaufmann, 1974). Under storage at low moisture contents, these genera cannot grow and may even die out depending on conditions, while the storage fungi, especially *Aspergillus* and *Penicillium* species, begin to dominate. Significant growth of *Aspergillus restrictus*, the most xerophytic of the major storage fungi, will occur at around 75 percent RH. Cereal grains in equilibrium with this humidity will have moisture contents of around 14 percent, while oily seeds such as sunflower and groundnut may be as low as 7 percent SMC. *Penicillium* species will begin to become a problem when the RH is around 85 percent (Halloin, 1986). Christensen (1967) noted how careful one needs to be to obtain stored seeds completely free

FIGURE 6. Changes in the proportion of linolenic acid (18:3) in the polar lipid fraction (o—o), and malondialdehyde (MDA) levels (•- - -•) in embryo axes of deteriorating soybeans held at approximately 100 percent RH, 45°C for up to 4d. Loss of geminability (here measured as the ability to produce radicles 48h after the start of imbibition) is shown by the closed histograms. MDA levels given here are those at 1h after the start of imbibition. Initial levels for dry seeds were all high, irrespective of aging. Data replotted from Stewart and Bewley (1980).

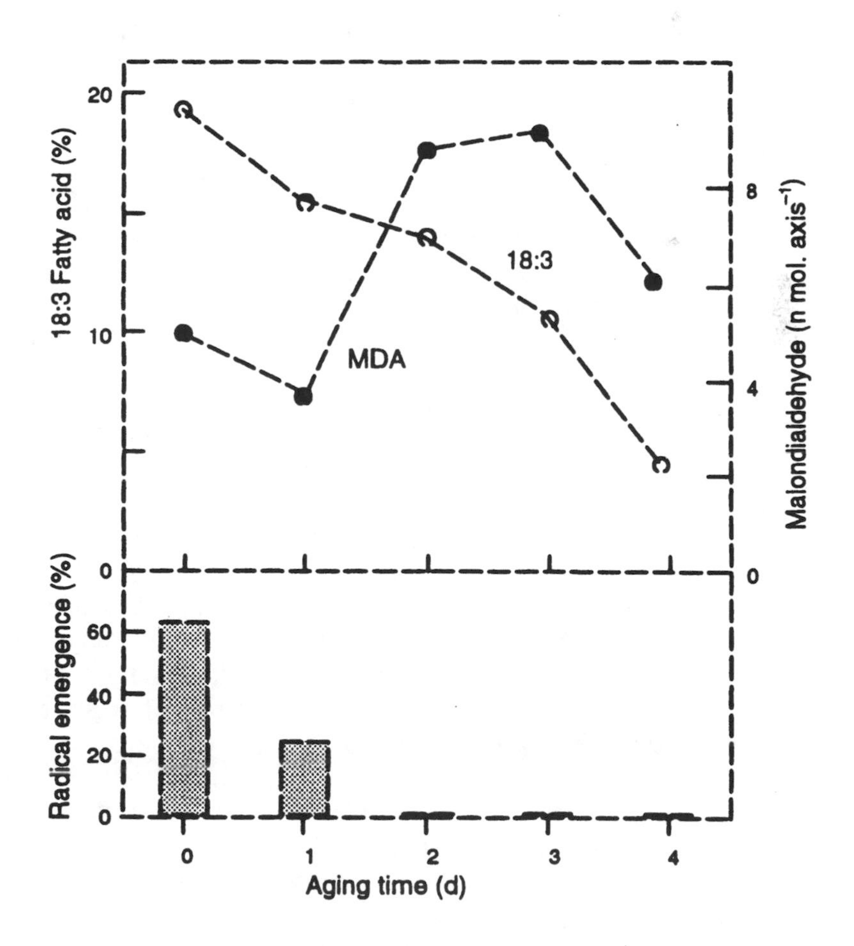

of infections of *Aspergillus* or *Penicillium* species under laboratory conditions: spores of these genera are, to all intents and purposes, ubiquitous. Clearly, then, any account of seed deterioration in storage conditions above these minimal RHs or moisture contents, especially artificial aging systems, should take the effects of storage fungi into account–it is surprising how few do.

Experiments on the survival of seeds inoculated with fungi, for example Fields and King (1962) and Harman and Granett (1972) both working with peas, showed that various species of *Aspergillus*, especially *A. candidus* and *A. flavus*, could rapidly initiate deterioration in otherwise healthy seed. Other species, such as *A. restrictus* and *A. glaucus*, are important for their "lead in" effects, in that they can grow actively at lower seed moisture contents than *A. flavus* and *A. candidus*, and their respiratory activity can rapidly induce localized areas of raised RH in poorly ventilated conditions and thus allow the latter two species to take hold (Christensen, 1972). The extent of infection of any seedlot will depend on the level of mechanical damage present, as the seed coat is a very effective barrier to fungal infection in many species. Similarly, internal tissues of many crops may have some degree of chemical resistance to fungal invasion (Halloin, 1986).

It is now well established that both storage and field fungi can damage seed by two major mechanisms, either by the production of exocellular hydrolytic enzymes or by the production of toxins. Measurements of increased leakage in seeds may thus sometimes be the result of cellulase and pectinase activity secreted by storage fungi, while increases in short chain fatty acids may similarly be due to exocellular lipase activity (Halloin, 1986). *Aspergillus* species have also been shown to produce significant amounts of protease. Fungal infection can result in ultrastructural changes similar to some of the physiological ones already discussed. Thus pea seeds inoculated with *Aspergillus ruber* (a member of the *A. glaucus* group) were found to show separation of the plasmalemma from cell walls, coalescence of lipid bodies and, in particular, damage to mitochondria (Harman and Granett, 1972).

In the previously discussed classic enzyme study by Perl, Luria, and Gelmond (1978), no precautions against fungal infection were recorded as having been taken despite aging conditions of 17 percent SMC and 30°C, which would provide excellent conditions for storage growth of storage fungi. Accordingly, a contribution of fungal activity to the enzyme changes noted (particularly the observed increase in protease activity) cannot be ruled out. The difficulty is that surface sterilization treatments (such as rinsing seeds with hypochlorite) do themselves add an extra variable to the studies involved. Such treatments may in some circumstances cause soaking injury (e.g., to soybean; Sakunnarak, 1992) or chemically interfere with seed metabolism (Abdul-Baki, 1974).

Most, if not all, species of storage and field fungi produce toxic compounds, collectively referred to as mycotoxins (Christensen and Kaufmann, 1974). These have been mostly studied in terms of their effects on animals and humans who might eat spoiled grain, but it is clear that many of these compounds have effects on seed quality too (Halloin, 1986). Mycotoxins constitute a wide diversity of different compounds, ranging from simple substituted organic acids, such as ß-nitroproprionic acid, to extremely complex multicyclic systems (Cole and Cox, 1981). Many of these compounds are known to inhibit nucleic acid and protein synthesis, but some may also have antirespiratory or mutagenic properties (Moulé, 1984). Other mycotoxins have antifungal or insecticidal effects (Betina, 1984). Documented effects of mycotoxins on plants include inhibition of germination, impaired photosynthetic activity, inhibition of cell extension, and membrane damage (Betina, 1984; Halloin, 1986).

UNRAVELING THE EXPERIMENTAL EVIDENCE

The most striking conclusion about all the research on mechanisms summarized on the previous pages must be its inconclusiveness. In no instance, whether it is loss of phospholipids from membranes, changes in activity of different enzymes, or some other aspect of seed function, can any general conclusions be drawn about the importance of that particular process in seed deterioration. Data are frequently highly variable and direct contradictions are not uncommon. It is clear that there is a need for new and carefully thought out research strategies to escape from this apparent impasse. Assuming the research techniques are appropriate for any chosen investigation, there are three areas which need to be addressed in this context. These are summarized in Table 6 and will be dealt with in turn.

Sources of Variation in Studies on the Physiology of Deterioration

One apparent cause of variable results in deterioration studies is differences between cultivars and within cultivars between lots. There are well-documented instances of genotypic variation in storability (see review by Priestley, 1986), while the importance of the prestorage history of the seed is also crucial. It follows, then, that more attention should be paid to cultivar (and seedlot) comparisons in studies of this type.

Another key source of variation between different studies has been the storage conditions employed. *A priori*, there is no basis for the assumption that seeds will deteriorate via the same mechanisms under high humidities and temperatures as they will under less stressful, longer-term storage conditions. In many ways it is surprising that the behavior of seeds under rapid aging controlled deterioration conditions can be used to predict the behavior of seeds in long-term storage, but the pioneering work of the Reading University, U.K. group have shown that this is so (Chapter 6). Nevertheless, mathematical descriptions of loss of germination performance, however accurate, cannot be extrapolated to explain the physiology of deterioration.

In recent years, two major advances in approach have been made in this area. Firstly, an increasing number of studies are undertaking comparative studies between rapid and slower aging approaches (e.g., Priestley and Leopold, 1979, 1983; Tables 4 and 5). In all too few of these studies, however, are careful comparisons made between seeds at the same stage of germination performance reached via different durations of storage under different environmental conditions. This should be a critical consideration in this type of approach (e.g., Petruzzelli and Taranto, 1984; Nath, Coolbear, and Hampton, 1991). Secondly, some workers are beginning to look more critically at the role of water in metabolism (e.g., Roberts and Ellis, 1989; Vertucci and Roos, 1990). Discussion of these two papers is beyond the scope of this chapter, but this is clearly an area where more work needs to be focused. Certainly this particular

TABLE 6. Areas for consideration in developing studies on the physiology of seed deterioration

Sources of variation	cultivar differences
	seedlot differences (the prestorage history of the seed)
	clear definitions of germination performance
	differences in metabolic changes between different parts of the seed
	unidentified contributions of microflora to the deterioration process
The timing of deterioration	although lesions of various kinds obviously accumulate in seeds during storage, when do these lesions cause major damage: during storage or during subsequent early germination? When do badly deteriorated seeds actually die?
Weaknesses of correlative evidence	inability to demonstrate cause and effect

subject is not without its controversy, especially concerning the long-term storage of seeds at extremely low relative humidities (Ellis, Hong, and Roberts, 1991; Vertucci and Roos, 1991).

A much more easily resolved area of confusion in the literature, but nevertheless one which can cause major difficulties in comparing different pieces of work, is the lack of uniformity (and sometimes clarity) in identifying what aspect of germinative activity is being measured in various studies. Most physiologists tend to use the term germination in *senso strico* (i.e., to mean the combination of processes leading to visible growth), and thus score seeds as germinable on the basis of radicle emergence. Other workers, especially those with a seed testing background, tend to use germination to mean both this process and the subsequent development of a "normal" seedling (i.e., the production of a seedling having all the structures and attributes required for successful establishment in the field). To complicate this picture still further, some papers cite germination percentages scored after an arbitrary number of days which, if germination is incomplete, becomes a crude measure of germination rate rather than a measure of final germination capacity. Yet other confusions arise between the terms viability and germinability. In the absence of uniform definitions it is imperative that all published papers are absolutely clear and precise about the details of the methods used to evaluate seed performance.

The two other areas which need to be developed in this kind of research have already been discussed elsewhere in this chapter. Over the past ten years there has been an increased general awareness that different parts of the seed, even tissues of the same genetic origin, can age at different rates. Somewhat more reluctantly (no doubt because of the technical problems involved), more seed physiologists are beginning to appreciate that they also need to take a fuller account of the role of microflora in physiological studies on seed deterioration.

The Timing of Deterioration and the Possibility of Its Repair

A crucial question raised in the introduction to this chapter and which needs to be fully addressed in future research concerns the interaction of deteriorative events during storage with subsequent germinative metabolism. How extensive are the repair mechanisms involved in the recovery of a partly deteriorated (low-vigor) seed as it begins to germinate? What factors affect the successful outcome of the processes? In order to fully understand what the key events in the deteriorative matrix are (Figure 2) we must also take into account what happens when a deteriorated seed is given the opportunity to germinate. The recently published studies by Puntalaro and Boveris (1990) on free radical and free radical defense enzyme activity during the early hours of soybean germination (discussed earlier) exemplify a new level of sophistication in this kind of approach.

Correlative Evidence: Its Weaknesses

The majority of the work reviewed in this chapter is simply correlative evidence indicating aspects of what is occurring in particular seeds during a process of loss of germination performance under a particular set of storage conditions. While valuable, this kind of data is limited in that it provides little information about cause and effect. Without some insight into this, more general and definitive conclusions about the processes of seed deterioration cannot be developed. Simply because phospholipid is lost from seed at a rate closely related to loss of seed germination, or chromosomal aberrations occur at a comparable frequency, does not mean that either type of damage is a primary cause of deterioration. The tomato data from my own work, contrasting the time course of phospholipid changes in two varieties (Francis and Coolbear, 1984, 1987), is a clear illustration of this kind of problem.

It is inevitable that, because of limitations in resources and techniques, much research will still be of the correlative kind, but if this is the case there is a need for much more stringent time course studies, preferably under storage regimes representative of commercial practice. The crucial issue is to identify events in seed deterioration at very early stages of the process, i.e., just when (or even before) seeds begin to lose germination performance (however this is chosen to be measured). In this context, the studies of Ferguson, TeKrony, and Egli (1990a), discussed previously, are an excellent example of the type of careful approach required. In the course of a long-term storage experiment, they took samples each month and were able to identify changes in respiratory activity and embryo axis leakage which clearly preceded changes in seed vigor. This study also has two other strengths in that it (1) focused on changes on a specific part of the seed, i.e., the embryo axis, and (2) made comparisons between two cultivars.

Just as the study of orthodox seed survival in storage has gone beyond simply placing seeds under different storage conditions in order to monitor changes in germinability and has moved into predictive modeling (Chapter 6), we need to rethink research approaches so that we can overcome the inherent weaknesses of correlative studies and begin to make statements about causality. The following section offers some suggestions about how this might be achieved.

Some Suggestions for Future Research Strategies

Given that there appears to be considerable variation between different cultivars and seedlots in their storage behavior, one avenue of research might be to revisit some of these investigations with a view to undertaking carefully designed comparative studies. This may provide a way forward in resolving some of the arguments about techniques which have been touched on in this review (e.g., appropriate strategies for the quantification of free radical damage). It may also allow us to identify which events are more important than others in the deterioration process, especially if projects are designed to provide information on a range of deteriorative events in the same tissues simultaneously.

A relatively novel dimension to this kind of approach, which has been utilized only infrequently, is the exploitation of mutations in deterioration studies. There would seem to be great potential for comparative studies on the storability of different mutant lines deficient in known aspects of metabolism to determine their role in the deterioration process. The study by Wang et al. (1990) on lipoxygenase deficient mutants of soybean, discussed previously, is one of the few documented pieces of research taking this line and would suggest that the search for other appropriate mutants (e.g., showing deficiencies in proteases) would be most worthwhile, especially if the studies were to be extended to a range of storage conditions. It is worth remembering that this kind of approach provided a way out of similar limitations to correlative evidence in studies on endogenous plant growth regulators.

Failing the availability of appropriate mutants, another avenue is to attempt to manipulate the deterioration process by employing pre-sowing seed treatments (such as antioxidant or hydration-dehydration methods which may either protect seeds during storage or allow repair afterwards, e.g., Coolbear, Francis, and Grierson, 1984; Woodstock et al., 1983; Gorecki and Harman, 1987; Nath, Coolbear, and Hampton, 1991) as investigative tools. If, for instance, a range of metabolic changes, A, B, and C, are demonstrated to occur during storage as germination performance is lost, but B and C continue to occur when a protective treatment is successfully applied to stored seeds, this strengthens the case for metabolism involving A being a primary cause of seed deterioration. Similarly, if a post-aging repair treatment more effectively reversed change A than B or C, this would enhance the case still further.

The key to this kind of study is to first find a reliable and reproducible treatment system for a particular species. Most recently my colleagues and I have been concentrating our efforts on what is probably the most studied species of all, soybean. We have been looking at the effectiveness of both antioxidant and hydration-dehydration treatments for enhancing the storability of this species, and as yet have met with little success (Sakunnarak, 1992): treatments reported as successful elsewhere in the literature have proved neither reliable nor reproducible in our hands. In particular, in this and other species, we have been beset by seedlot variation to the treatments used. Nevertheless it still seems to be a line worth pursuing and our current work with wheat (Nath, Coolbear, and Hampton, 1991) is more promising in this respect. The plant growth regulator treatments used by Petruzzelli and Taranto (1985) which increase the storability of wheat grains are also most interesting in this context.

CONCLUSION

As this review has shown, researchers know quite a lot about the processes which can occur in individual cases during seed deterioration, but very little in terms of their general importance as primary causes of loss of germination performance as opposed to secondary lesions. There is little doubt that deterioration is a matrix of interrelated events rather than a simple sequence. Superimposed on this matrix is, remember, the seed's capacity both to resist damage and to repair lesions as they occur. Depending on the storage conditions and the time elapsed, some of these processes will occur during the storage period, while others will be initiated during the first few hours of imbibition.

It is clear too from this review that achieving unequivocal, definitive research in this area is no easy task. There are a huge number of variables involved, and it is probably unrealistic to ask that any one piece of research should take them all into account. Nevertheless, future studies should pay greater attention to intraspecific differences in the process, not only between different cultivars, but also between seedlots and even tissue within the same seed. They should also take into account their interactions between different storage conditions and the seed's prestorage history. Nor can the possible involvement of microflora in these deterioration processes be ignored, especially under the high humidity aging regimes exploited by many workers as a convenient entry into this problem.

While the study of the physiology of seed deterioration is a fascinating one in its own right, the practical implications are immense. One of the initial motivations for studies on seed deterioration was the search for biochemical tests for seed vigor. Our present knowledge would indicate that no one aspect of seed metabolism is likely to provide a clear universal index of seed deterioration, for there are too many different events going on at the same time. Nevertheless, a better understanding of what is occurring as seeds lose quality in different situations would give us clearer ideas of how to assess the planting value of seeds prior to sowing. It might also open up ways of improving the performance of valuable seed stock which has already begun to lose vigor. For genetic conservation, a detailed understanding of all the factors which might impinge on genetic damage is essential to the successful maintenance of germplasm collections. In many agricultural systems, particularly in the tropics, safe conventional storage is beyond the reach of many farmers and seed merchants and good quality seed is at a premium. The ultimate goal of this research discipline is to provide an information base for the development of new lines of commercially important crops such as soybean with much improved storage potential, or, failing that, alternative seed treatments which might delay the deterioration process.

REFERENCES

Abdalla, F. H., and E. H. Roberts. Effect of temperature, moisture and oxygen on the induction of chromosome damage in seeds of barley, broad beans and peas during storage. *Ann. Bot.* 32 (1968): 119-136.

Abdul-Baki, A. A. Relationship of glucose metabolism to germinability and vigour in barley and wheat seeds. *Crop Sci.* 9 (1969): 732-737.

Abdul-Baki, A. A. Pitfalls in using sodium hypochlorite as a seed disinfectant in ^{14}C incorporation studies. *Plant Physiol.* 53 (1974): 768-771.

Abdul-Baki, A. A., and J. D. Anderson. Physiological and biochemical deterioration of seeds. In *Seed Biology, Volume II*, ed. T. T. Koslowski (New York: Academic Press, 1972), pp. 283-315.

Abu-Shakra, S. S., and T. M. Ching. Mitochondrial activity in germinating new and old soybean seeds. *Crop Sci.* 7 (1967): 115-118.

Aldana, A. B., R. C. Fites, and H. E. Pattee. Changes in nucleic acids, protein and ribonuclease activity during maturation of peanut seeds. *Plant & Cell Physiol.* 13 (1972): 515-521.

Anderson, J. D., and K. Gupta. Nucleotide alterations during seed deterioration. In *Physiology of Seed Deterioration*, eds. M. B. McDonald, Jr. and C. J. Nelson (Madison, Wisconsin: Crop Science Society of America, Special Publication No. 11, 1986), pp. 47-63.

Aspinall, D., and L. G. Paleg. The deterioration of wheat embryo and endosperm function with age. *J. Exp. Bot.* 22 (1971): 925-935.

Benson, E. E. Free Radical Damage in Stored Plant Germplasm (Rome: International Board for Plant Genetic Resources, 1990).

Berjak, P., and T. A. Villiers. Ageing in plant embryos. II. Age-induced damage and its repair during early germination. *New Phytol.* 71 (1972a): 135-144.

Berjak, P., and T. A. Villiers. Ageing in plant embryos. III. Acceleration of senescence following artificial ageing treatment. *New Phytol.* 71 (1972b): 513-518.

Betina, V. Biological effects of mycotoxins. In *Mycotoxins: Production, Isolation, Separation and Purification*, ed. V. Betina (Amsterdam: Elsevier, 1984), pp. 25-36.

Bewley, J. D. Membrane changes in seeds as related to germination and the perturbations resulting from deterioration in storage. In *Physiology of Seed Deterioration*, eds. M. B. McDonald, Jr. and C. J. Nelson (Madison, Wisconsin: Crop Science Society of America Special Publication No. 11, 1986), pp. 27-46.

Bhattacharjee, A., S. Roychowdhury, and M. A. Choudhuri. Effects of CCC and Na-dikegulac on longevity and viability of seeds of two jute cultivars. *Seed Sci. & Technol.* 14 (1986): 127-139.

Biddle, A. J. Production factors affecting vining pea-seed quality. In *Seed Production*, ed. P. D. Hebblethwaite (London: Butterworths, 1980), pp. 527-534.

Blowers, L. E., D. A. Stormonth, and L. M. Bray. Nucleic acid and protein synthesis and loss of vigour in germinating wheat embryos. *Planta* 150 (1980): 19-25.

Buchvarov, P. Z., and T. Gantcheff. Influence of accelerated and natural aging on free radical levels in soybean seeds. *Physiol. Plant.* 60 (1984): 53-56.

Castro, C. A. S., and C. S. Sediyama. Liberacion del aldehido hexanal como indice para estimar el vigor de semilla de soya. *Semillas* 15 (1990): 7-10.

Cheah, K. S. E., and D. J. Osborne. DNA lesions occur with loss of viability in embryos of ageing rye seed. *Nature* 272 (1978): 593-599.

Chen, G. Y., and J. R. Fu. (Deterioration of groundnut seeds and perosidation). *Acta Scient. Nat. Univ. Sunyat Seni* 3 (1986): 69-75.

Ching, T. M. Ageing stresses on physiological and biochemical activities of crimson clover (*Trifolium incarnatum* L. var. Dixie) seeds. *Crop Sci.* 12 (1972): 415-418.

Christensen, C. M. Germinability of seeds free of and invaded by storage fungi. *Proc. AOSA* 57 (1967): 141-143.

Christensen, C. M. Microflora and seed deterioration. In *The Viability of Seeds*, ed. E. H. Roberts (London: Chapman and Hall, 1972), pp. 59-93.

Christensen, C. M. Loss of viability in storage: Microflora. *Seed Sci. & Technol.* 1 (1973): 547-562.

Christensen, C. M., and H. H. Kaufmann. Microflora. In *Storage of Cereal Grains and Their Products*, ed. C. M. Christensen (St. Paul, Minnesota: American Association of Cereal Chemists, 1974), pp. 158-192.

Cole, R. J., and R. H. Cox. *Handbook of Toxic Fungal Metabolites* (New York: Academic Press, 1981).

Conger, A. D., and M. L. Randolf. Is age-dependent genetic damage in seeds caused by free radicals? *Rad. Bot.* 8 (1968): 193-196.

Coolbear, P., A. Francis, and D. Grierson. The effect of low temperature pre-sowing treatment on the germination performance and membrane integrity of artificially aged tomato seeds. *J. Exp. Bot.* 35 (1984): 1609-1617.

Copeland, L. O., and M. B. McDonald, Jr. *Principles of Seed Science and Technology*, second edition (Minneapolis: Burgess Publishing, 1985).

Davidson, P. A., R. M. Taylor, and C. M. Bray. Changes in ribosomal RNA integrity in leak (*Allium porrum* L.) seeds during osmopriming and drying-back treatments. *Seed Sci. Res.* 1 (1991): 37-44.

Dell'Aquila, A., and B. Margiotta. DNA synthesis and mitotic activity in germinating wheat seeds aged under various conditions. *Env. & Exp. Bot.* 26 (1986): 175-184.

Delouche, J. C. Environmental effects on seed development and seed quality. *HortScience* 15 (1980): 775-780.

Delouche, J. C., and C. C. Baskin. Accelerated ageing techniques for predicting the relative storability of seed lots. *Seed Sci. & Technol.* 1 (1973): 427-452.

Dey, G., and R. K. Mukherjee. Deteriorative change in seeds during storage and its control by hydration-dehydration pre-treatments. *Seed. Res.* 14 (1986): 49-59.
Dourado, A. M., and E. H. Roberts. Phenotypic mutations induced during storage in barley and pea seeds. *Ann. Bot.* 54 (1984): 781-790.
Ellis, R. H., T. D. Hong, and E. H. Roberts. Seed moisture content, storage, viability and vigour. *Seed Sci. Res.* 1 (1991): 275-277.
Ferguson, J. M., D. M. TeKrony, and D. B. Egli. Changes during early soybean seed and axes deterioration: 1. Seed quality and mitochondrial respiration. *Crop Sci.* 30 (1990a): 175-179.
Ferguson, J. M., D. M. TeKrony, and D. B. Egli. Changes during early soybean seed axes deterioration: II. Lipids *Crop Sci.* 30 (1990b): 179-182.
Fields, R. W., and T. H. King. Influence of storage fungi on deterioration of stored pea seed. *Phytopathology* 52 (1962): 336-339.
Filho, C. P., and R. H. Ellis. The development of seed quality in spring barley in four environments II. Field emergence and seedling size. *Seed Sci. Res.* 1 (1991): 179-185.
Francis, A. The effect of ageing and low temperature pre-sowing treatment on the membrane status and germination performance of tomato seeds. PhD Thesis, Luton College of Higher Education, UK, 1985.
Francis, A., and P. Coolbear. Changes in membrane phospholipid composition of tomato seed accompanying loss of germination capacity caused by controlled deterioration. *J. Exp. Bot.* 35 (1984): 1764-1770.
Francis, A., and P. Coolbear. A comparison of changes in the germination responses and phospholipid composition of naturally and artificially aged tomato seeds. *Ann. Bot.* 59 (1987): 167-172.
Gelmond, H., I. Luria, L. W. Woodstock, and M. Perl. The effect of accelerated aging of sorghum seeds on seedling vigour. *J. Exp. Bot.* 29 (1978): 489-495.
Gidrol, X. Accumulation de peroxyde d' hydrogene et vieillissement accelere des semences de soja. *C. R. Acad. Sci. Paris* 308 (1989): 223-228.
Gidrol, X., H. Serghini, A. Noubhani, B. Mocquot, and P. Mazliak. Biochemical changes induced by accelerated ageing in sunflower seeds. I. Lipid peroxidation and membrane damage. *Physiol. Plant.* 76 (1989): 591-597.
Gordon, I. L., D. R. Smith, and V. Sereepraseri. Grain development and sprouting damage in wheat. Massey University Faculty of Agricultural and Horticultural Sciences Research Report (1990): 16.
Gorecki, R. J., and G. E. Harman. Effects of antioxidants on viability and vigour of ageing pea seeds. *Seed Sci. & Technol.* 15 (1987): 109-117.
Guy, P., S. Smith, and M. Black. Changes in DNA associated with vigour and viability in wheat as revealed by RFLP analysis: A preliminary study. *J. Exp. Bot.* 42 (1991). Supplement, p. 27.
Hallam, N. D. Fine structure of viable and non-viable rye and other embryos. In *Seed Ecology*, ed. W. Heydecker (London: Butterworths, 1973), pp. 115-145.
Halloin, J. M. Microorganisms and seed deterioration. In *Physiology of Seed Deterioration*, eds. M. B. McDonald, Jr. and C. J. Nelson (Madison, Wiscon-

sin: Crop Science Society of America, Special Publication, No. 11, 1986), pp. 89-99.

Hampton, J. G., and P. Coolbear. Potential versus actual seed performance–Can vigour testing provide an answer? *Seed Sci. & Technol.* 18 (1990): 215-228.

Harman, G. E., and A. L. Granett. Deterioration of stored pea seed: Changes in germination, membrane permeability and ultrastructure resulting from infection by *Aspergillus ruber* and from ageing. *Physiol. Plant Pathol.* 2 (1972): 271-278.

Harman, G. E., and L. R. Mattick. Association of lipid oxidation with seed ageing and death. *Nature* 260 (1976): 323-324.

Harrington, J. F. Seed storage and longevity. In *Seed Biology Volume III*, ed. T. T. Kozlowski (New York: Academic Press, 1972), pp. 145-245.

Harrington, J. F. Biochemical basis of seed longevity. *Seed Sci. & Technol.* 1 (1973): 453-461.

Heydecker, W. Vigour. In *Viability of Seeds*, ed. E. H. Roberts (Syracuse, N.Y.: Syracuse University Press, 1972), pp. 209-252.

Humphrey-Taylor, V. J., and N. G. Larsen. Sixty years of sprout damage in New Zealand wheats. *New Zealand J. Crop. & Hort. Sci.* 18 (1990): 105-113.

Justice, O. L., and L. N. Bass. *Principles and Practices of Seed Storage* (Washington: U.S. Department of Agriculture, 1978).

Kermode, A. R., J. D. Bewley, J. Dasgupta, and S. Misra. The transition from seed development to germination: A key role for desiccation? *HortScience* 21 (1986): 1113-1118.

Livesley, M. A., and C. M. Bray. The effects of ageing upon α-amylase production and protein synthesis by wheat aleurone layers. *Ann. Bot.* 68 (1991): 69-73.

Matthews, S., A. A. Powell, and N. E. Rogerson. Physiological aspects of the development and storage of pea seeds and their significance to seed production. In *Seed Production*, ed. P. D. Hebblethwaite (London: Butterworths, 1980), pp. 513-525.

McDonald, M. B., Jr., and C. J. Nelson, eds. *Physiology of Seed Deterioration* (Madison, Wisconsin: Crop Science Society of America, Special Publication No. 11, 1986).

McDonald, M. B., Jr., and D. O. Wilson. ASA-610 ability to detect changes in soybean seed quality. *J. Seed Technol.* 5 (1980): 56-66.

Mitchell, B., C. Armstrong, M. Black, and J. Chapman. Physiological aspects of sprouting and spoilage in developing *Triticum aestivum* L. (Wheat) grains. In *Seed Production*, ed. P. D. Hebblethwaite (London: Butterworths, 1980), pp. 339-356.

Moulé, Y. Biochemical effects of mycotoxins. In *Mycotoxins: Production, Isolation, Separation and Purification*. ed. V. Betina (Amsterdam: Elsevier, 1984), pp. 37-44.

Nath, S. Changes in germination performance and hydrolytic enzyme activity in wheat seeds (*Triticum aestivum* L.) caused by ageing and pre-sowing treatments. PhD Thesis, Seed Technology Centre, Massey University, New Zealand, 1991.

Nath, S., P. Coolbear, and J. G. Hampton. Hydration-dehydration treatments to protect or repair stored "Karamu" wheat seeds. *Crop Sci.* 31 (1991): 822-826.

Nath, S., P. Coolbear, J. G. Hampton, and C. A. Cornford. The efficiency of hydration-dehydration seed treatments for improving the storage of wheat seeds. *Proc. Agron. Soc. New Zealand* 20 (1990): 51-57.

Ohlrogge, J. B., and T. P. Kernan. Oxygen dependent ageing of seeds. *Plant Physiol.* 70 (1982): 791-794.

Osborne, D. J. Senescence in seeds. In *Senescence in Plants*, ed. K. V. Thimann (Boca Raton, Florida: CRC Press, 1980), pp. 13-37.

Osborne, D. J. Deoxyribonucleic acid integrity and repair in seed germination: The importance in viability and survival. In *The Physiology and Biochemistry of Seed Development, Dormancy and Germination*, ed. A. A. Khan (New York: Elsevier Biomedical Press, 1982), pp. 435-463.

Pearce, R. S., and I. M. Abdel Samad. Changes in fatty acid content of polar lipids during ageing of seeds of peanuts (*Arachis hypogea* L.). *J. Exp. Bot.* 31 (1980): 1283-1290.

Perl, M. Biochemical aspects of the maturation and germination of seeds. *Advances in Research and Technology of Seeds* 10 (1987): 1-27.

Perl, M., I. Luria, and H. Gelmond. Biochemical changes in sorghum seeds affected by accelerated aging. *J. Exp. Bot.* 29 (1978): 497-509.

Petruzzelli, L., and G. Taranto. Phospholipid changes in wheat embryos aged under different storage conditions. *J. Exp. Bot.* 35 (1984): 517-520.

Petruzzelli, L., and G. Taranto. Effects of permeation with plant growth regulators in acetone on seed viability during accelerated aging. *Seed Sci. & Technol.* 13 (1985): 183-191.

Petruzzelli, L., and G. Taranto. Amylase activity and loss of viability in wheat. *Ann. Bot.* 66 (1990): 375-378.

Powell, A. A., and S. Matthews. Deteriorative changes in pea seeds (*Pisum sativum* L.) stored in humid or dry conditions. *J. Exp. Bot.* 28 (1977): 225-234.

Powell, A. A., and S. Matthews. Association of phospholipid changes with early stages of seed ageing. *Ann. Bot.* 47 (1981): 709-712.

Priestly, D. A. *Seed Ageing: Implications for Seed Storage and Persistence in the Soil* (Ithaca, N.Y.: Cornell University Press, 1986).

Priestly, D. A., and A. C. Leopold. Absence of lipid oxidation during accelerated aging of soybean seeds. *Plant Physiol.* 53 (1979): 726-729.

Priestley, D. A., and A. C. Leopold. Lipid changes during natural ageing of soybean seeds. *Physiol. Plant.* 59 (1983): 467-470.

Priestley, D. A., M. B. McBride, and A. C. Leopold. Tocopherol and organic free radical levels in soybean seeds during natural and accelerated ageing. *Plant Physiol.* 66 (1980): 715-719.

Priestley, D. A., B. G. Werner, and A. C. Leopold. The susceptibility of soybean seed lipids to artificially-enhanced atmosphere oxidation. *J. Exp. Bot.* 36 (1985): 1653-1659.

Priestley, D. A., B. G. Werner, A. C. Leopold, and M. B. McBride. Organic free

radical levels in seeds and pollen: The effects of hydration and ageing. *Physiol. Plant.* 64 (1985): 88-94.
Puls, E. E., and V. N. Lambeth. Chemical stimulation of germination rate in aged tomato seeds. *J. Amer. Soc. Hort. Sci.* 99 (1974): 9-12.
Puntalaro, S., and A. Boveris. Effect of natural and accelerated ageing on the hydroxide metabolism of soybean embryonic axes. *Plant Sci.* 68 (1990): 27-32.
Ram, C., and L. E. Weisner. Glutamic acid decarboxylase activity (GADA) as an indicator of field performance of wheat. *Seed Sci. & Technol.* 16 (1988): 11-18.
Rao, N. K., and E. H. Roberts. The effect of oxygen in seed survival and accumulation of chromosome damage in lettuce (*Lactuca sativa* L.). *Seed Sci. & Technol.* 18 (1990): 229-238.
Roberts, E. H., and R. H. Ellis. Water and seed survival. *Ann. Bot.* 63 (1989): 39-52.
Roos, E. E. Induced genetic changes in seed germplasm during storage. In *The Physiology and Biochemistry of Seed Development, Dormancy and Germination*, ed. A. A. Khan (Amsterdam: Elsevier Biomedical Press, 1982), pp. 409-434.
Saha, R., and R. N. Basu. Invigoration of soybean seed for the alleviation of soaking injury and ageing damage on germinability. *Seed Sci. & Technol.* 12 (1984): 613-622.
Sakunnarak, N. An evaluation of antioxidant and hydration treatments for the improvement of the storability of soybean (Glycine max (L.) Merr.) Seeds. PhD Thesis, Seed Technology Centre, Massey University, New Zealand, 1992.
Senaratna, T., J. F. Gusse, and B. D. McKersie. Age-induced changes in cellular membranes of imbibed soybean seed axes. *Physiol. Plant.* 73 (1988): 85-91.
Sharma, K. D. Biochemical changes in stored oil seeds, *Indian J. Agric. Res.* 11 (1977): 137-141.
Stewart, R. R. C., and J. D. Bewley. Lipid peroxidation associated with accelerated ageing of soybean axes. *Plant Physiol.* 65 (1980): 245-248.
Takayanagi, K. (An examination of seed vitality, with special reference to a method using seed exudate). *Nogyo Gijutsu Kenkyusko Hokoku D. Seiri Iden Sakumotsu Ippan* 28 (1977): 1-87 (cited by Priestley, 1986).
TeKrony, D. M., D. B. Egli, and J. Balles. The effect of the field production environment on soybean seed quality. In *Seed Production*, ed. P. D. Hebblethwaite (London: Butterworths, 1980), pp. 403-425.
Vertucci, C. W., and E. E. Roos. Theoretical basis of protocols for seed storage. *Plant Physiol.* 94 (1990): 1019-1023.
Vertucci, C. W., and E. E. Roos. Seed moisture content, storage, viability and vigour: Response. *Seed Sci. Res.* 1 (1991): 277-279.
Villiers, T. A. Ageing and the longevity of seeds in field conditions. In *Seed Ecology*, ed. W. Heydecker (London: Butterworths, 1973), pp. 265-288.
Villiers, T. A. Seed ageing: Chromosome stability and extended viability of seeds stored fully imbibed. *Plant Physiol.* 53 (1974): 875-878.
Villiers, T. A. Ultrastructural changes in seed dormancy and senescence. In *Plant Senescence*, ed. K. V. Thimann (Boca Raton, Florida: CRC Press, 1980), pp. 39-66.
Vishnyakova, J. A., N. P. Krasnook, R. I. Povaronva, and E. A. Morgunova. Ultra-

structure of cells of the embryos of viable and unviable rice seeds in the course of swelling. *Soviet Plant Physiol.* 23 (1976): 307-311.

Walker-Simmons, M. ABA levels and sensitivity in developing wheat embryos of sprouting resistant and susceptible cultivars. *Plant Physiol.* 84 (1987), pp. 61-66.

Wang, J. Y., K. Fujimoto, T. Miyazawa, Y. Endo, and K. Kitamera. Sensitivities of lipoxygenase-lacking soybean seeds to accelerated ageing and their chemiluminescence levels. *Phytochemistry* 29 (1990): 3739-3742.

Wang, Y. R. A study of the relationship between seed vigour and seed performance in *Trifolium pratense* L. cv. Grasslands Pawera. MAgrSc Thesis, Seed Technology Centre, Massey University, New Zealand, 1989, pp. 79-83.

Wilson, D. O., and M. B. McDonald, Jr. A convenient volatile aldehyde assay for measuring soybean seed vigour. *Seed Sci. & Technol.* 14 (1986a): 259-268.

Wilson, D. O., and M. B. McDonald, Jr. The lipid peroxidation model of seed ageing. *Seed Sci. & Technol.* 14 (1986b): 269-300.

Woodstock, L. W., and K-L. J. Tao. Prevention of imbibitional injury in low vigour soybean embryonic axes by osmotic control of water uptake. *Physiol. Plant.* 51 (1981): 133-139.

Woodstock, L. W., and R. B. Taylorson. Ethanol and acetaldehyde in imbibing soybean seeds in relation to deterioration. *Plant Physiol.* 67 (1981): 424-428.

Woodstock, L. W., K. Furman, and H. R. Leffler. Relationship between weathering deterioration and germination, respiratory metabolism, and mineral leaching from cotton seeds. *Crop Sci.* 25 (1985): 459-466.

Woodstock, L. W., K. Furman, and T. Solomos. Changes in respiratory metabolism during ageing in seeds and isolated axes of soybean. *Plant & Cell Physiol.* 25 (1984): 15-26.

Woodstock, L. W., S. Maxon, K. Faul, and L. Bass. Use of freeze-drying and acetone impregnation with natural and synthetic antioxidants to improve storability of onion, pepper and parsley seeds. *J. Amer. Soc. Hort. Sci.* 108 (1983): 692-696.

Chapter 9

Variety Identification: Modern Techniques and Applications

Robert J. Cooke

What does seed quality mean? To many seed technologists, the answer to this question would be quite straightforward–seed quality is defined primarily by purity, germination, and freedom from disease. However, there are good reasons for believing that the scope of the definition will need to be extended over the next few years to include, among other things, varietal (cultivar) identity and purity. The increasing involvement of private companies in plant breeding and the necessity for these companies to be rewarded financially for their investment, mean that there will be a growing need for better and more precise varietal description and protection, through various plant breeders rights and related schemes. In addition, seed certification, which forms a link between variety registration and seed production, involves an assessment of both varietal identity and purity to assure the quality of seed marketed to the farmer or grower. Finally, the ultimate consumers of the harvested seed also often need to be certain that they are purchasing the correct variety, particularly if the grain is to be used for mechanized processing (bread making, for instance). It is thus clearly important from many points of view to be able to distinguish between and identify crop varieties.

Traditionally, variety identification has been carried out by means of what might be termed a classical taxonomic approach. This in-

It is a pleasure to acknowledge the assistance of the many friends and colleagues at NIAB and elsewhere who have contributed to the work and ideas described in this chapter. Andy Tiley is thanked for the photography and the NIAB typists for all their efforts.

volves a detailed study of seeds and growing plants and the observation and recording of a number of morphological characters or descriptors. In practice, such an approach is extremely successful and largely forms the basis, for instance, of distinctness testing procedures prior to variety registration. However, this can be a time-consuming and expensive process, requiring large areas of land and highly skilled personnel making what are often subjective decisions. Also, many of the morphological descriptors used are multi-genic, quantitative, or continuous characters, the expression of which can be altered by environmental factors. Again, in some species the number of descriptors is limited or is no longer sufficient for identification of all varieties. There are thus compelling reasons to find more rapid and cost-effective procedures which could augment this morphologically based approach and which are directly applicable to seeds. The observed morphology (phenotype) of a seed or plant arises from an interaction between its genetic make-up, or genotype, and the environment. In general, it is an increasingly desirable objective to reduce or eliminate the environmental influence so that the genotype of a variety can be observed more directly.

This chapter is concerned with modern techniques which address the question of the environmental factor and which are currently applied to variety identification. They are of two main types:

1. the use of computerized systems to capture and process morphological information (machine vision).
2. the use of biochemical methods to analyze various components of seeds (chemotaxonomy).

THE USE OF MACHINE VISION

Morphology-based taxonomy utilizes characters which are often multi-genic and continuously expressed (quantitative), requiring replicated measurements in order to apply statistical means of establishing identity or distinguishing between individuals. The environmental interaction with the characters again necessitates replication of sampling and analysis. This requires time-consuming and labor-intensive manual methods of measuring, recording, and processing the information. An alternative to this is to make use of machine vision systems. Simply, machine vision refers to the acquisition of data (shape, size, etc.) via a video camera or similar system and the subsequent computer analysis of these data following suitable processing. The term "image analysis" has also been used in this context, but it more strictly refers to the extraction of numerical data from an acquired image. Apart from providing an automated means of obtaining measurements, a great advantage of machine vision is that it is possible, given the appropriate computer software, to make detailed comparisons of sets of data, i.e., to operate pattern recognition systems. This makes the technology very attractive for both variety description and identification.

According to Draper and Keefe (1988), the basic elements of a machine vision system are:

1. image capture, via a video camera or other electronic system (e.g., a charge-coupled device).
2. analogue to digital conversion of the image data (not necessary if a charge-coupled device is used).
3. computerized manipulation of the image data, to obtain a version suitable for analysis (e.g., removal, via the software, of extraneous and anomalous objects; "grey-level partition" to achieve a binary image or silhouette). This is termed image processing.
4. image analysis–the extraction of information from the processed image.
5. pattern recognition, either statistical or syntactic, to sort and compare objects (e.g., varieties). In statistical systems, the results of the image analysis are analyzed statistically, while in syntactic pattern recognition, the image itself is searched for particular features, e.g., angles in close proximity to one another.
6. computerized decision making and presentation of results, including assessments of statistical significance.
7. an automated or robotic mechanism for movement of the sample (or the camera), so that operator intervention is minimized (desirable for a useful practical system).

Machine Vision and Variety Identification

The prospect of using instrumental methods to carry out the direct observations made by trained laboratory staff in assessing seed quality prompted a major examination of machine vision by workers at the National Institute of Agricultural Botany (NIAB) (Cambridge). Using a low-cost image capture and processing system, Draper and Travis (1984) demonstrated that it was possible to analyze the shape of seeds and various vegetative structures. The measurements were relatively simple, including length, width, area, and perimeter, which enabled two further variates to be derived, namely aspect ratio (width/length) and shape factor (or thinness ratio) (4π. area/perimeter^2). Using seeds of 49 different crop and weed species, they further showed that most species could be differentiated on the basis of the shape factor in combination with seed length (Travis and Draper, 1985). This clearly indicated the taxonomic potential of machine vision and encouraged the extension of the work to variety identification.

A preliminary study examined five wheat (*Triticum aestivum* L.) varieties and illustrated how such identification might be achieved (Keefe and Draper, 1986). This work required a more sophisticated system, consisting of a Cambridge Instruments Quantimet 10 (Q10) image analyzer and custom-written computer software. Individual wheat seeds were placed crease-side down on a horizontal surface, with their longitudinal axes perpendicular to the camera lens and their embryos to the left. The seeds were viewed in side elevation using transmitted light, enabling a binary image of the seed and its support to be recorded. The support silhouette was removed using the computer software, leaving the image of the seed.

Figure 1 illustrates some of the parameters measured from the image. By combining these measurements to produce shape descriptors which were largely independent of the size of the seed, and so minimizing the effect of environmental and other factors, it was possible to distinguish the five varieties from one another. This work has now been extended to include a greater range of varieties and a wider array of shape descriptors. The 69 descriptors used are sufficient to recover virtually all of the shape information present in the outline silhouette image of a wheat seed (Draper and Keefe, 1989). Work is in progress to improve the statistical approach used to distinguish be-

FIGURE 1. A diagram of a wheat seed, illustrating some of the parameters used by Keefe and Draper (1986) for shape description.

Germ height (GH): the vertical distance (A-H) of the lower edge of the scutellum above the stage.

Germ length (GL): the length of the scutellar region (A-B).

Germ angle (GA): the angle subtended by a line drawn through A-B to the horizontal.

Length: the total horizontal length of the seed including the embryo.

High point: the horizontal distance from A-C divided by the total length of the seed.

Dorsal angle (DA): as germ angle, but through points D-E.

Height: the total vertical height of the seed.

Brush height (BH): the vertical distance (F-G) of the brush above the stage.

Horizontal axis (HA): germ height/brush height.

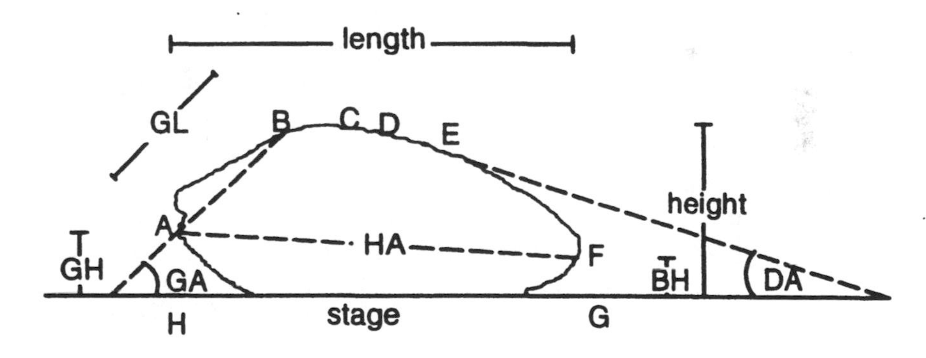

tween varieties, with a view to being able to determine not only varietal identity but also purity. In conjunction with a seed company, this technology is being transferred from the research machine to a dedicated commercial analyzer. The prototype of this uses a CCD camera mounted on a threaded rod which moves linearly, enabling the analyzer to make 33 measurements on 50 separate seeds and process the data, all within three to four minutes. The potential for machine vision demonstrated by this research has been confirmed in other laboratories.

Much of this work has been directed toward discrimination between different classes of wheat, rather than variety identification. Thus in the U.S.A., Zayas and co-workers (Zayas, Lai, and Pomeranz, 1986; Zayas, Pomeranz, and Lai, 1989) have reported the discrimination between wheat and nonwheat components in grain samples and also between hard red winter, soft red winter, and hard red spring varieties, using a Quantimet 720 system. A statistical approach somewhat different from that used at Cambridge utilized two class models to assign an individual seed to its appropriate class. Different morphological parameters (both measured and derived) were also used. Correct assignment was achieved in 78 to 85 percent of cases for the eight varieties tested. Bushuk's group at the University of Manitoba have tried to overcome the constraints imposed by the necessity for manual orientation of seeds prior to imaging and have also examined more complex species admixtures (Sapirstein et al., 1987). Initially, they were able to successfully separate mixtures of wheat, oats, barley, and rye, with incorrect classification of only 1 percent of over 1,100 seeds. The samples were randomly orientated before measurement. Noncereal admixtures (e.g., rapeseed) and weeds (wild oats) were also correctly classified. Later work considered class and varietal identification (Neuman et al., 1987). Although categorization according to class was generally effective, discrimination of varieties within a class was inconclusive, with correct classification occurring in 96 percent of cases for some varieties and only 15 percent of cases for others. This is partly explicable by the parameters chosen for measurement and discrimination, which included some based on absolute size. Environmental and other factors influencing seed size inevitably confound the discrimination (Keefe and Draper, 1986). This is also, of course, a problem which occurs when subjective human examination of seed morphology is used to assess varietal identity. It is clear that careful selection of measured and derived parameters, as well as effective methods of statistical analysis, are required in order to extract the maximum benefit from machine vision technology.

Using five Australian wheat varieties, Myers and Edsal (1989) have confirmed that analysis of silhouette images allows discrimination between varieties. The crease-down side view alone (as used by Keefe and Draper, 1986) did not give particularly good results, with two of the five varieties being correctly identified in only 50 percent or less of cases. By combining these data with parameters measured from the side-down view, however, and by using statistical classifiers, the identification rate was greatly improved.

More recently, the use of three-dimensional image analysis has been reported for the classification and discrimination of two wheat varieties of similar two-dimensional profile (Thomson and Pomeranz, 1991). Although only preliminary, such an approach offers the prospect that researchers will be able to study surface features more closely and hence increase the available character set.

To date, the vast majority of the research effort in this area has been directed toward the identification of wheat varieties or classes. However, preliminary work at Cambridge with barley (*Hordeum vulgare* L.), using the same kind of approach as that taken with wheat, was promising in that it was possible to discriminate between a small group of varieties (Keefe and Purchase, unpublished). Because of the presence of many useful surface characteristics in barley seeds, the analysis of silhouette images may not ultimately be the best approach for this species. There seems to be little doubt that other cereals such as triticale, rye, and oats could be successfully analyzed. Also, maize (*Zea mays* L.) and particularly rice (*Oryza sativa* L.) varieties have been shown to possess different and observable seed characteristics. There is thus a need for work on these and other important crops.

The use of machine vision technology clearly offers considerable potential for variety identification. The main advantages are the speed of analysis and ease of operation of the equipment. However, this method does rely on the necessary research having been conducted to establish appropriate useful character sets for identification and suitable statistical methods of analysis, and also on the compilation of databases of information relating to existing varieties.

BIOCHEMICAL METHODS

Chemotaxonomists have recognized two groups of compounds that are generally useful for the classification of organisms:

1. episemantic, or secondary compounds (pigments, fatty acids, etc.)
2. semantides, or "sense-carrying" molecules (proteins, nucleic acids)

Although the semantides have proved to be far more useful for variety identification, particularly from the seed, there are several instances of the successful use of secondary compounds.

Analysis of Secondary Compounds

A range of different tests is available for the analysis of secondary compounds in seeds and vegetative parts of plants. The tests range from simple color tests to complex chromatographic separations of anthocyanins, flavonoids, and other compounds.

Probably the best known example of a widely used color test is the phenol test, used to distinguish between varieties of wheat by the differential oxidation of phenol (and hence coloration) of the seeds. This test has also been used for varietal identification in rice and Kentucky Bluegrass (*Poa pratensis*). The achievable discrimination can be improved by considering, for instance, the color reaction of the outer glumes of wheat, in addition to the seeds (Singhal and Prakash, 1988). Another simple color test is the use of acidified vanillin reagent to detect the presence of tannins in the testa of field bean (*Vicia faba*) seeds (Cooke et al., 1985).

Two principal types of chromatography have been used, depending on the analyte of interest. Thus gas-liquid chromatography (GLC) has been used for the separation of fatty acids from seeds of oilseed rape (*Brassica napus*), and White and Law (1991) demonstrated how fatty acid composition could be used to distinguish between rape varieties. GLC has also been used for glucosinolate analysis in *Brassica* and related species (Morgan, 1989), although in only a few cases have differences in seed glucosinolate composition been found to be useful for variety discrimination. Heaney and Fenwick (1980), for example, reported that it was possible to distinguish between 22 varieties of Brussels sprout from the glucosinolate profiles of the seeds.

The other major type of chromatography, high performance (or pressure) liquid chromatography (HPLC), has also been used for glucosinolate analysis, but has an important chemotaxonomic role in the identification of varieties of horticultural species by the separation of anthocyanin and flavonoid pigments from the flowers (see Morgan, 1989, for references).

Secondary compounds are thus very useful for variety identification. However, there can be no question that the analysis of semantides, and in particular the dating of proteins and enzymes, has been far more successful.

Proteins and Variety Identification

The successful exploitation of proteins for variety identification purposes is based on the fact that proteins are the direct products of gene transcription and translation. Proteins can thus be regarded as markers for the structural genes that encode them. The proximity of the process of protein synthesis to the primary genetic information (DNA) also greatly reduces or even eliminates any environmental interaction in protein composition. Hence an analysis of protein composition becomes, albeit one step removed, an analysis of gene expression, and methods for comparing protein composition provide a measure of the genetic variation between individuals and populations.

For variety identification, it is necessary to utilize proteins that exist in multiple molecular forms (i.e., are polymorphic), and also preferably that are present in relatively large amounts and are easy to extract. For these reasons, seed proteins of all types are extremely useful for identification purposes and have been widely used. This includes albumins (water-soluble proteins, mainly enzymes), globulins (the typical salt-soluble storage proteins of legume seeds), prolamins (the typical alcohol-soluble storage proteins of cereal seeds), and glutelins (detergent-soluble structural or enzymic proteins). However, a wide range of vegetative enzymes has also been found to be useful (see reviews by Cooke, 1984, 1988).

The Use of HPLC

The chromatographic methods that have been employed for the analysis of secondary compounds (above) have also been successfully applied to proteins and subsequently used to identify varieties. Following the first report of Bietz (1983), it has been demonstrated unequivocally that HPLC will separate seed proteins of wheat, barley, oats, rice, maize, and other cereals, and that the resultant protein profiles can be used to distinguish between varieties (see

Morgan, 1989, for references). Generally, the analyses have been of the alcohol-soluble prolamins (storage proteins) of cereals–gliadins (wheat), hordeins (barley), zeins (maize), avenins (oats), etc.–although there are methods involving albumins and glutelins (e.g., glutenins in wheat). Several different reversed-phase (RP) HPLC systems and methods have been developed and the varietal protein profiles produced by a typical RP-HPLC analysis of seed proteins are illustrated in Figure 2. Varieties can be distinguished from one another by the qualitative absence or presence of particular protein peaks detectable at specific points (elution or retention times) on the profiles. In addition, information regarding the amount of protein in a particular peak (peak area or peak height) can readily be obtained with confidence, particularly if computerized integration of peak data is available. This means that quantitative differences between varieties can also be used as an aid to identification.

There can be no doubt that the HPLC-based comparison of seed protein profiles allows varieties to be distinguished from each other and to be identified. The techniques can be extremely discriminating and several authors have published catalogues of profiles, primarily for cereals (see, for instance, Bietz, 1983; Bietz et al., 1984; Marchylo and Kruger, 1984; Allison and Bain, 1986; Smith, 1986; Lookhart et al., 1987), but also for other crops such as soybeans (Buehler et al., 1989).

The attractions of this approach, in addition to the potential removal of environmental effects and greater discrimination possibilities, include speed and automation. An HPLC separation is generally completed within an hour, and can be much quicker, particularly if ion-exchange rather than reversed-phase chromatography is used (Wingad et al., 1986). This means that results are rapidly available. However, it must be remembered that this is only the time for the analysis of one sample, which could be an individual seed. Any kind of varietal purity estimation would clearly require the analysis of a number of such individual seeds. This could make the process rather lengthy, although there are ways to minimize this. In addition, the possibility exists for substantial automation of HPLC separation procedures, by using programmable sample injection systems for instance, and the data produced cannot only be readily quanti-

FIGURE 2. The seed storage protein profiles of different varieties of (A) wheat and (B) barley, following analysis by typical RP-HPLC procedures. Note the polymorphic nature of the proteins (multiple peaks) and the existence of both qualitative (absence/presence of peaks eluting at a particular time) and quantitative (peak heights/areas) differences between varieties. The varieties are (A) 1-Brimstone, 2-Galahad, 3-Norman, 4-Rapier, 5-Avalon; (B) 1-Maris Otter, 2-Keg, 3-Patty, 4-Kym, 5-Egmont.

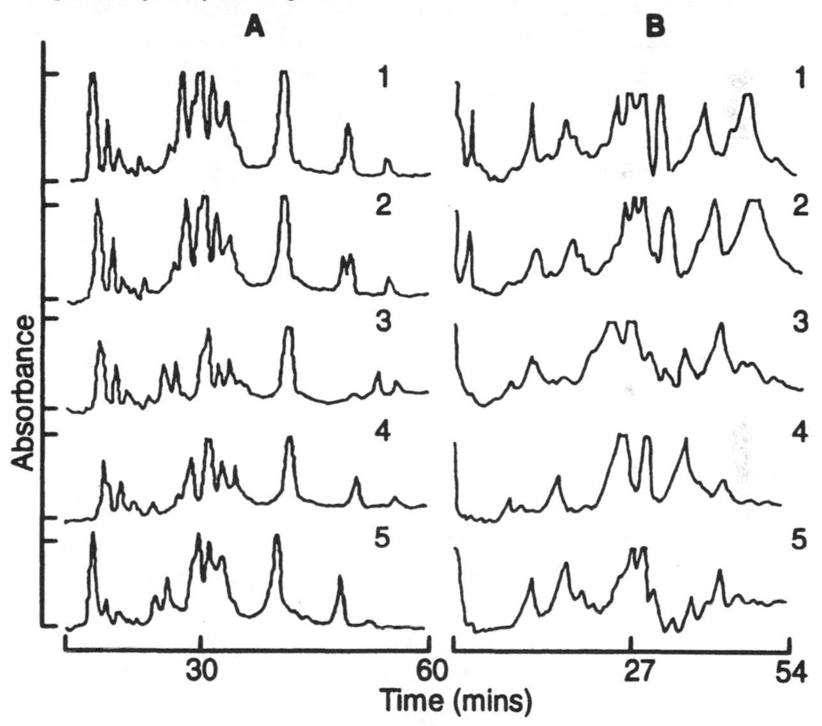

fied, which enhances discrimination possibilities, but can also be stored and processed by computer.

However, HPLC does inevitably have some disadvantages, the two primary ones being (1) the relatively high capital and operating costs, and (2) the long-term reproducibility of the analyses. An HPLC system suitable for protein separation could cost anything up to 40,000 dollars (at 1991 prices), especially if equipped for automation. HPLC equipment represents highly sophisticated technology, and care must be taken to ensure that it isadequately maintained. For example, the performance of the pump used to force the mobile phase through the column must be monitored, since small changes in flow rate can result in significant changes in varietal protein profiles. Again, the analytical columns used have a limited life and their performance declines with time. This could be especially serious if computerized capture and comparison of profiles were being utilized. Although there are ways of protecting the columns, for instance by the use of "guard" or precolumns (which act effectively as sieves, removing particulate material from extracts), these accordingly lengthen the separation time and increase the running costs. There is also some evidence that columns of apparently the same type from different manufacturers can produce different sample resolutions, which again could seriously hamper the automated comparison of profiles. Hence, although a computerized approach to wheat varietal identification has been suggested, involving automatic data capture, pattern matching of profiles, and searching of databases, the long-term variations in retention times and gradual deterioration of column performance might be expected to hinder the general applicability of such systems (Scanlon, Sapirstein, and Bushuk, 1989a, 1989b).

The tendency toward the simplification of chromatography equipment via a modular approach and the increasing use of plug-in cartridge column systems has reduced the technical sophistication of HPLC systems and will continue to lead to improvements. In addition, the computerized systems for data manipulation which are available for HPLC do raise some intriguing possibilities. Perhaps the most interesting of these is the analysis of a bulked sample, or flour, of a particular seedlot, in place of the examination of individual seeds. Then only one or two analyses might be required of a given sample to provide information on both varietal identity and purity. Such a system depends on the availability of computer software to evaluate mixtures of varietal patterns. Software does exist which enables protein profiles to be compared and differences between them to be highlighted. The actual processing is still rather slow and requires considerable operator interaction. However, this is largely a function of computing power and advances in this area are likely. The potential of such an approach to varietal purity testing has been examined in relation to hybrid purity assessment. Thus McCarthy and coworkers (1990) reported that it was possible to estimate the purity of F1 hybrid wheat samples by duplicate RP-HPLC analysis of the gliadin profiles of samples of bulked and milled seeds. The results compared favorably with those obtained by the sequential analysis of individual seeds, although clearly much depends on the level of purity required and its statistical significance.

Although HPLC has been found to be extremely useful for the identification of varieties from the seed, it is without doubt the various techniques of protein electrophoresis which have had the largest impact in this area.

Variety Identification by Electrophoresis

The uses of electrophoresis for variety identification have been comprehensively reviewed and summarized in recent years (see, for example, Cooke, 1988; Smith, and Smith, 1992; Wrigley, 1992). The precise way in which the techniques are utilized varies according to the species in question. However, two main approaches have been recognized (Cooke, 1989):

1. the direct (multi-locus) approach, in which proteins that are polymorphic and genetically encoded at multiple loci are analyzed. Cereal seed storage proteins provide a good example. They are encoded by multigenic loci and the products of a single locus can comprise several electrophoretically separable bands. The criterion for distinctness between varieties is taken as the presence or absence of a particular protein band (or set of bands) occurring at a defined position or positions on the gel.
2. the indirect (single locus) approach, involving the examination of proteins which, although polymorphic, are derived from a single locus (isozymes or allozymes). Varietal distinctness is demonstrated either as the occurrence of different isozyme phenotypes (banding patterns) in self-pollinated and vegetatively propagated species or as differences in the frequency of occurrence of isozyme phenotypes in cross-pollinated species (this is explored further below).

It is generally the case that the optimal way in which electrophoresis can be used for identification purposes is governed by the mode of reproduction and hence the genetic structure of the variety in question. It is convenient to consider two groups of species in this context:

1. self-pollinating species (including vegetatively propagated and asexually reproducing crops).
2. cross-pollinating species.

Self-Pollinating Species

Self-pollinating or autogamous crop species have been particularly thoroughly researched with respect to the application of electrophoresis for variety identification (see Cooke, 1988, for references). Generally, the direct (multi-locus) approach has been taken. As a result of the importance of this group of species in modern world agriculture, there is a plethora of methodology available. For example, more than 27 methods for wheat variety identification by analysis of seed proteins (gliadins) have been published (summarized in Cooke, 1988). New methods for wheat and barley continue to be produced (Weiss, Postel, and Gorg, 1991; Wrigley, Gore, and Manusu, 1991). This proliferation of techniques has caused problems, for instance in comparing results from different laboratories. However, thanks to the work of organizations such as the International Seed Testing Association (ISTA), this situation is slowly being rationalized. The ISTA has tested and defined a standard reference method for the electrophoretic identification of wheat and barley varieties from the seed (Cooper, 1987) and is currently evaluating a method for pea (*Pisum sativum* L.) varieties. The method for wheat and barley involves the analysis of the alcohol-soluble prolamins (gliadins or hordeins) by polyacrylamide gel electrophoresis (PAGE) at pH 3.2 and can also be applied to other cereal such as oats (*Avena sativa* L.) or triticale (x *Triticosecale* Wittmack). Typical results are illustrated in Figure 3.

Other organizations have also tried to standardize electrophoresis methodology in this area. The International Standards Organization (ISO) currently has a draft acid PAGE method for wheat variety identification under consideration by its members. The method is similar to the International Association for Cereal Science and Technology (ICC) Standard Method No. 143. Others, such as the Royal Australian Chemical Institute, recommend the use of com-

FIGURE 3. The use of the ISTA acid PAGE procedure to analyze the seed storage proteins of varieties of (A) wheat, (B) barley, and (C) oats. Each track represents the protein pattern of a single seed of a different variety. Note the highly polymorphic nature of the seed storage proteins and the variety-specific banding patterns. The origin and anode are at the top. The varieties are (left to right) Brimstone, Apollo, Avocet, Copain, Digger, Goldspear, Cameo, Maris Osprey, Maris Oberon, Margam, and Rhiannon.

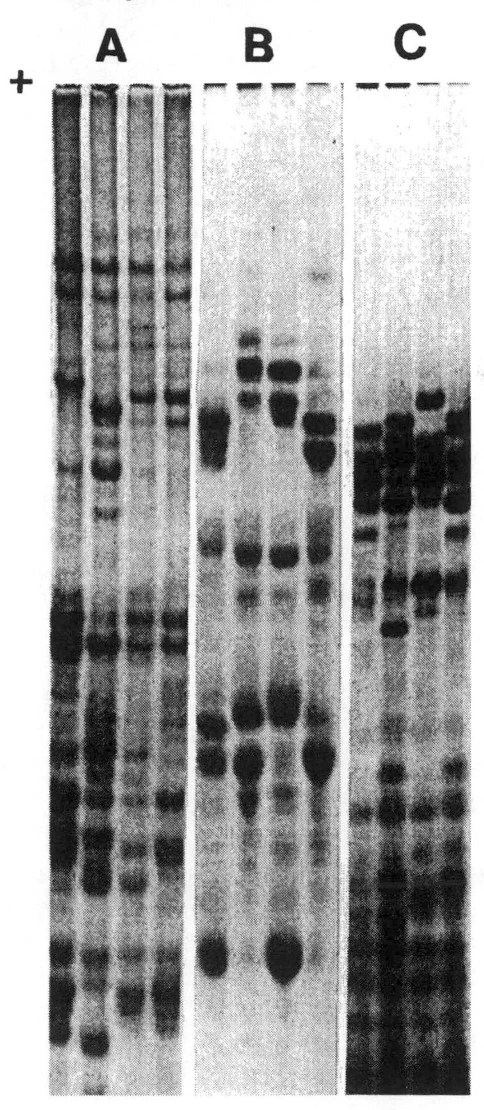

mercially available gradient gels run at pH 3.1 for cereal variety identification.

These methods, along with systems for the nomenclature of protein bands and subsequent classification of varieties, have been well described and discussed (Cooke, 1984, 1988; Wrigley, Autran, and Bushuk, 1982; Wrigley, 1992). Such relatively straightforward acid PAGE techniques have considerable resolving power, with about 40 individual gliadin bands and some 30 hordein bands being evident on the gels. This extensive protein polymorphism allows impressive levels of discrimination between varieties to be achieved. For instance, work at NIAB (White and Cooke, unpublished) has shown that it is possible to identify uniquely about 90 percent of the more than 200 wheat varieties examined, while with barley a collection of 400 varieties has been classified into 70 groups based on their hordein composition.

Authors from several countries have published catalogues of the electrophoretic patterns and formulae of wheat and barley varieties in their national collections. A practical problem is the lack of a standardized or widely recognized system for the nomenclature of gliadin and hordein bands or patterns (Cooke, 1988, 1989). The existence of such a generally accepted system would be of considerable international benefit.

In addition to acid PAGE, other kinds of electrophoresis have proved extremely useful for identification of cereals and other self-pollinating species. PAGE in the presence of sodium dodecyl sulphate (SDS-PAGE) is particularly suitable for legume species (Figure 4), but has been widely applied to barley hordeins and wheat glutenin proteins. Again catalogues of protein patterns have been produced by various authors to facilitate identification (see Cooke, 1984, 1988; also Wrigley, Autran, and Bushuk, 1982; and Wrigley, 1992, for references). Isoelectric focusing (IEF) of seed proteins and enzymes has also been used to distinguish between and identify varieties (Cooke, 1985, 1988; Cooke and Draper, 1983).

In many cases, the use of different kinds of electrophoresis can improve the resolution achieved between a given collection of varieties. For example, White and Cooke (unpublished) have shown that IEF of seed esterases can distinguish between some barley varieties with identical hordein compositions. Electrophoresis methods can also be combined in a two-dimensional (2D) approach. In this, a sample is first analyzed by one technique (e.g., IEF) and the resultant gel is then rotated for a second separation by a different technique (e.g., SDS-PAGE), carried out at right angles to the first. The "maps" produced by 2D analysis can contain over 100 protein spots, providing considerable potential for discrimination between even closely related genotypes. Such methods have been widely applied to self-pollinating crops (see Cooke, 1984, 1988, for references), but the technical complexity of both the analysis and the interpretation of the data has limited the widespread use of 2D electrophoresis for routine variety identification purposes.

Because of their method of breeding and selection, most varieties of self-pollinating crops consist of a single, homozygous line, and hence display a high degree of both phenotypic and genotypic uniformity. However, in a proportion of varieties of all species, there are found two or more electrophoretic lines or biotypes, a consequence of the lack of selection for protein homogeneity (Cooke, 1988, 1989). The existence of biotypes does not detract from the overall power of electrophoresis for identification purposes, but does highlight the necessity of analyzing a sufficient number of individual seeds of varieties in order to detect and record any non-uniformity.

The direct, multi-locus approach has also been very successful when applied to vegetatively propagated or asexually reproducing species. Varieties of such crops are essentially clones of phenotypically and genotypically identical individuals, each containing the same fixed complement of genes. Several crops of this type have been investigated electrophoretically, including bluegrass (*Poa pratensis* L.), bananas (*Musa* spp), roses (*Rosa* spp.), strawberries (*Fragaria* spp.), pears (*Pyrus* spp.) and other fruits (Cooke, 1988; Gilliland, 1989), but without doubt the most thoroughly researched is the potato (*Solanum tuberosum* L.). Due primarily to the work of Stegemann and his group in Braunschweig, over 15,000 potato varieties, species, subspecies, and wild types have been examined by electrophoresis (Huaman and Stegemann, 1989). The most useful and generally applied method is PAGE at pH 7.9 or 8.9 of the soluble tuber proteins or esterases expressed from potato sap, al-

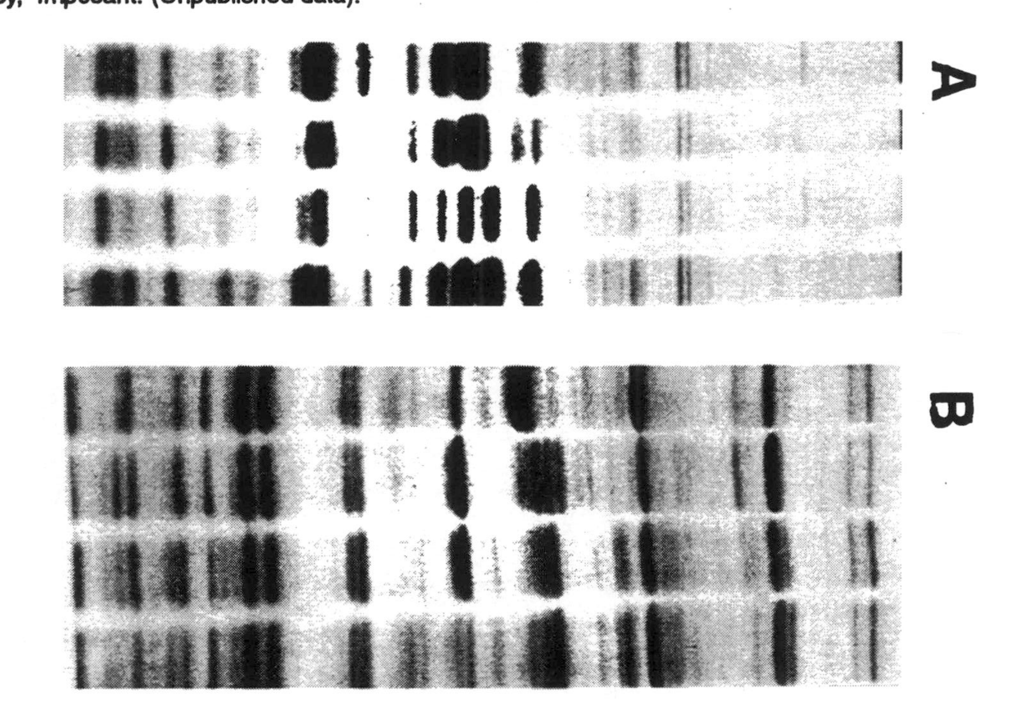

FIGURE 4. The use of SDS-PAGE to analyze the seed storage proteins of varieties of (A) *Phaseolus* beans and (B) peas. Each track represents the storage protein composition of a single seed of a different variety. Note the polymorphism of the storage proteins and the variability in pattern between varieties. The varieties are (left to right) Brilliant, Solare, Fantastico, Borres, Nadja, Tercober, Filby, Imposant. (Unpublished data).

though IEF and various 2D approaches have also been widely used with considerable success (see Cooke, 1988).

Cross-Pollinating Species

Varieties of cross-pollinating (allogamous) species are populations of individuals, expressing a range of phenotypic characters. These individuals can also be genetically distinct, containing different combinations of homozygous and heterozygous genes, including those encoding seed storage proteins or isozymes. The variation is maintained in equilibrium over generations of seed production. Because of this genetic structure, such varieties present special problems from the electrophoretic point of view. There are two ways of approaching identification.

1. Analyze a bulk extract of seeds to obtain an overall varietal protein profile. Effectively, this means that varieties can then be treated as self-pollinating crops and the direct approach, examining seed storage proteins, can be used.
2. Utilize the indirect approach to examine individual plants of a variety, determine the degree of variability within varieties, and compare them statistically. Because of the difficulties involved in accurately reading gels containing complex storage protein profiles, it is more usual with this approach to analyze single locus isozymes which produce relatively simple electrophoretic patterns.

Both approaches have been successfully used in cross-pollinating species. A good example is provided by the work which has been conducted with ryegrass (*Lolium* spp.; see Cooke, 1988; Gilliland, 1989). SDS-PAGE has been used to analyze the seed storage protein composition of ryegrass varieties. Although individual seeds within a variety have a range of storage protein patterns, bulk samples (approximately 2g) of the same variety display a consistent pattern and the SDS-PAGE profiles of bulk samples can thus be used to discriminate between varieties (Figure 5). The ISTA has evaluated this technique and has proposed a standardized version of SDS-PAGE for inclusion in the International Rules, as a technique for distinguishing between and identifying commercial seedlots of ryegrass varieties and species.

FIGURE 5. SDS-PAGE analysis of the seed proteins of the Italian ryegrass variety Matador. The first ten tracks are extracts of single seeds–note the variability of protein pattern within the variety. The remaining eight tracks are bulk extracts of about 2g of seed, from seed lots grown in different years at different locations. The overall varietal profile is constant, and can be used to distinguish between varieties (Unpublished data).

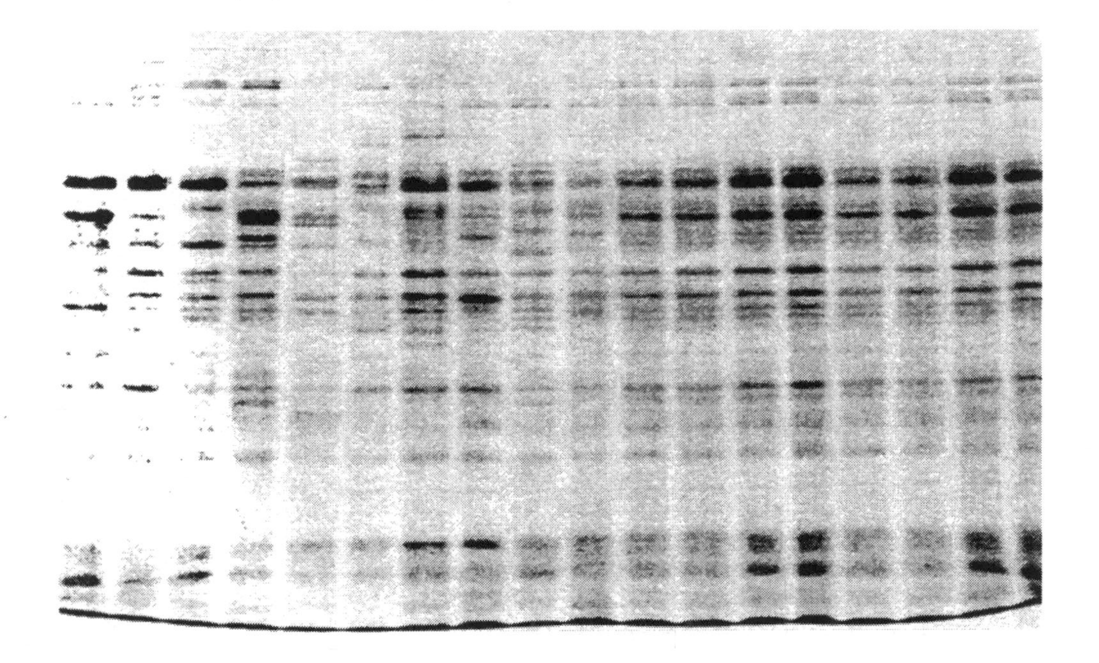

Ryegrass varieties have also been thoroughly investigated by the indirect approach. Most of these studies have used starch gel electrophoresis to examine a range of leaf enzymes such as phosphoglucoisomerase (PGI) and glutamate-oxaloacetate transaminase (GOT), with PGI being particularly well researched (see Greneche, Lallemand, and Migaud, 1991). There are four common alleles of PGI in ryegrass, which give rise to ten possible electrophoretic phenotypes (Figure 6). Homozygous plants display a single-banded phenotype, whereas heterozygotes have three bands, one corresponding to each allele and one of intermediate mobility (a heterodimer), arising from the dimeric structure of PGI. For variety identification, extracts of leaves from a number (50 to 100) of individual plants of each variety are taken, electrophoresed, and the starch gels stained for PGI activity. The frequency of occurrence of each phenotype (or genotype) in a variety is determined, and then these frequencies are compared statistically, using a chi-squared analysis. Varieties are thus discriminated from each other on the basis of differences in the frequency of occurrence of phenotypes or alleles. This approach can be very powerful. It has been shown, for instance, that some 70 percent of a collection of 149 perennial and

FIGURE 6. A diagram of the ten possible electrophoretic phenotypes of a dimeric enzyme with four alleles (a-d), such as PGI in ryegrass.

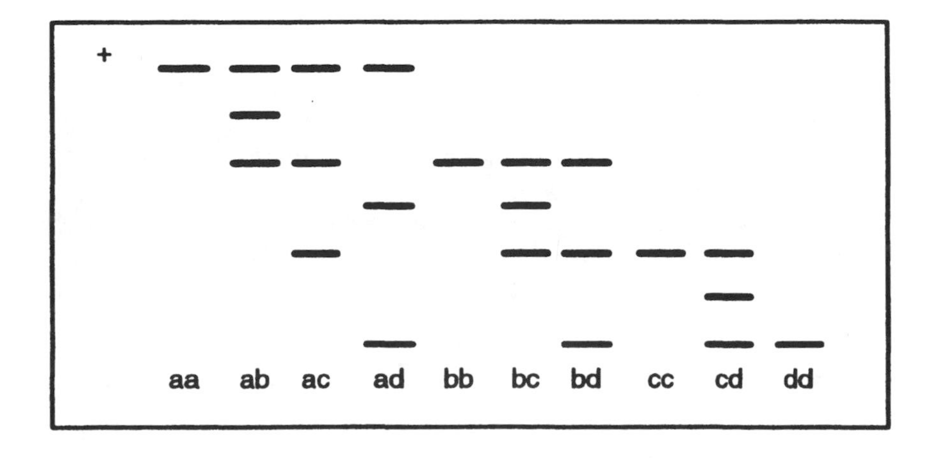

Italian ryegrass varieties could be distinguished (P < 0.05) solely on the basis of PGI allele frequencies (Gilliland, 1989).

Various other cross-pollinating species have been investigated using both of the above approaches, and there are reports of the successful use of electrophoresis for variety identification in many cases, including rye (*Secale cereale* L.), Faba beans (*Vicia faba* L.), oilseed rape (*Brassica napus* L.), sugar beet (*Beta vulgaris*), alfalfa (*Medicago sativa* L.), pasture species including fescue (*Festuca* spp.) and timothy (*Phleum* spp.), and horticultural species such as onions (*Allium cepa* L.) and others (see Cooke, 1985, 1988; Gardiner and Forde, 1988; Arulsekar and Parfitt, 1986; Mundges, Kohler, and Friedt, 1990).

Thus there can be no doubt that the various techniques of protein electrophoresis have had an enormous impact in the area of crop variety identification and that the use of electrophoresis is both widespread and growing. For example, in the reviews by Cooke (1988) and Smith and Smith (1992), over 50 species are considered for which electrophoresis has been reported to be successful for identification purposes, and this coverage, although comprehensive, is by no means complete. However, there are good reasons for progressing one step beyond proteins and considering the potential of nucleic acid analysis for variety identification.

The Use of Nucleic Acid Analysis for Variety Identification

Electrophoresis and HPLC are methods which can reveal genetic variability in protein composition. If such variability exists, then it follows that there must be variation in the underlying genetic material (DNA). Indeed, it is virtually certain that there will be far more DNA variability, since not all differences will occur in regions of the DNA which are expressed phenotypically. Thus an extremely powerful tool for variety identification would be available if DNA polymorphism could be detected and assessed. Advances in molecular biology now allow such detection, and it is becoming increasingly possible to identify variations between individuals at the DNA level. There are several approaches available for examining this variation.

Restriction Fragment Length Polymorphisms (RFLPs)

One way of detecting variability in the DNA of different individuals would be to compare the sequence of nucleotide bases. However, DNA molecules have extremely high molecular weights (of the order of 10^9), and this would represent an unrealistic task. In order to assist in this undertaking, it is necessary to cut the DNA into smaller fragments. This became possible in the 1970s when restriction enzymes (endonucleases) were purified and made available. Such enzymes recognize short specific DNA sequences, usually four to eight nucleotides in length, and cut (restrict) the DNA within or adjacent to these sites. For example, the enzyme *EcoRl* recognizes the six-base sequence GAATTC in DNA and cuts between the G and A nucleotides. Variations in DNA sequences can be detected as changes in the number or more usually the length of the DNA fragments resulting from such enzymic restriction. These variations have been termed restriction fragment length polymorphisms (RFLPs).

Hence at the simplest level, RFLPs can be seen as variations in the size of restriction enzyme-generated fragments of DNA, which can be detected by separation on an agarose electrophoresis gel (see Figure 7). However, genomic DNA is generally cleaved by restriction enzymes into an extremely large number of fragments. For example, an enzyme with a six-base recognition sequence would cut an average DNA molecule once every 4,000 nucleotide bases and so in wheat would produce about 4 million fragments. Clearly, the separation of this number of fragments could not be seen on a stained electrophoresis gel, and hence a restriction digest of genomic DNA generally appears as a continuous smear down the gel, although frequently repeated fragments may appear as faint bands. Consequently, a method has to be found to highlight fragments of interest and this is achieved by utilizing DNA probes. A probe is a small piece of cloned DNA which is homologous to part or all of the fragment of interest and hence will hybridize with it. In practice, the fragments from the agarose electrophoresis gel are transferred to a nylon or nitrocellulose membrane filter, by a process known as Southern blotting. The filter (or blot) is then exposed to a radioactively labeled DNA probe under conditions which promote hybrid-

FIGURE 7. A diagrammatic representation of RFLPs. (A) A restriction map of part of the DNA from four individuals (varieties), which differ in their sites specific for the restriction enzyme R. (B) Gel patterns for the DNA fragments produced from the restriction of the DNA of the individuals shown in (A), following electrophoresis.

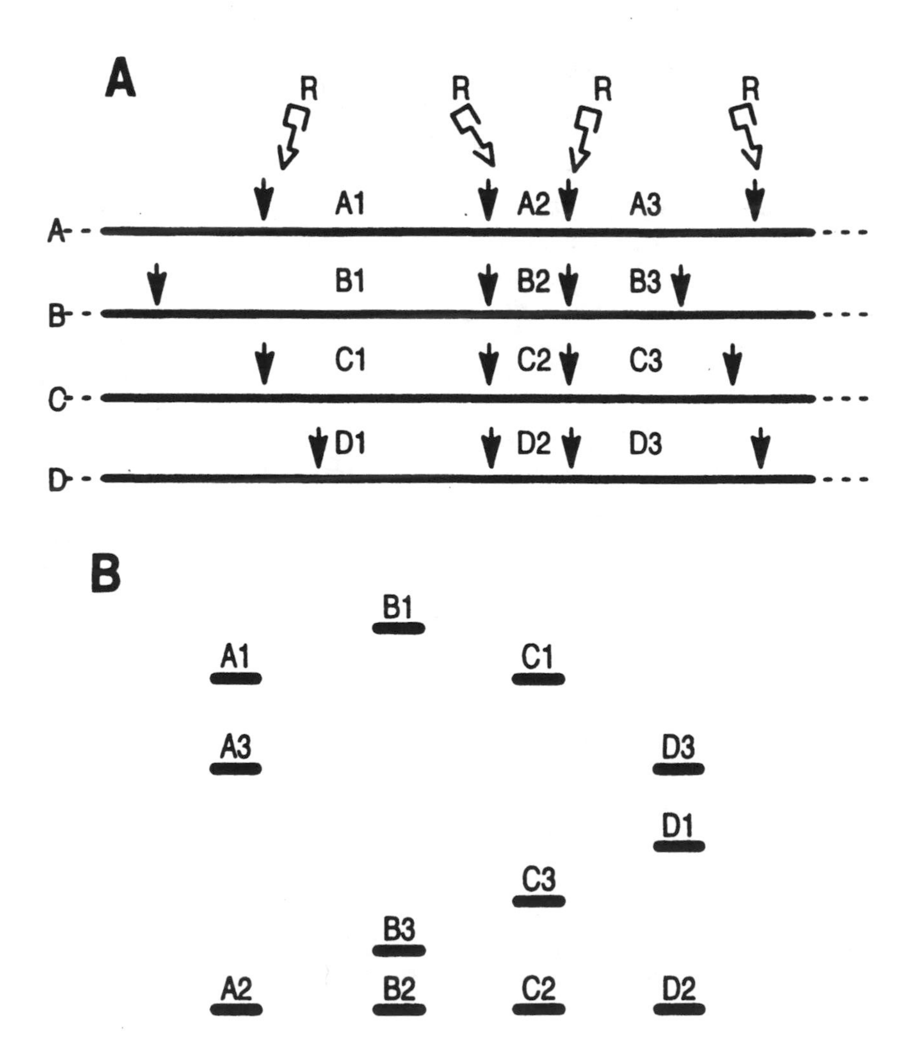

ization and, after a suitable time, unhybridized probe DNA is removed by washing. The filter is dried and exposed to X-ray film, and the DNA fragments that have hybridized to the labeled probe will subsequently appear as discrete bands on the resultant autoradiograph. Thus any differences in the relative positions of restriction enzyme sites between individuals, arising for example from point mutations, deletions, insertions, or DNA rearrangements, will show up as a polymorphism (an RFLP) on an autoradiograph when a filter is probed with a DNA probe spanning part or all of the appropriate region. For instance, in Figure 7, a probe homologous to the A_1 region would highlight a polymorphism (hybridizing to fragments A_1,B_1,C_1, and D_1). However, a probe homologous to the A_2 region would display no polymorphism, hybridizing to fragments A_2,B_2,C_2, and D_2, which are all of the same size and hence migrate to the same place on the electrophoresis gel.

RFLP technology could offer some considerable advantages over other identification techniques (see Ainsworth and Sharp, 1989). The major advantage lies in the availability of an almost unlimited number of combinations of restriction enzyme and probe. There are several hundred commercially available restriction enzymes, and various laboratories throughout the world are beginning to put together collections of cloned DNA fragments which could be used as probes for different crop species. Given a suitable set of probes, a much larger portion of the genome can be sampled for RFLPs than is the case for morphological or protein descriptors. Again, the degree of variation exhibited by RFLPs is likely to be greater than that shown by gene products (proteins). Minor changes in the DNA sequence may not alter protein electrophoretic mobility, but could be detected by RFLP analysis, which will also identify silent (not expressed) variation in introns and flanking sequences. A further advantage is that the same RFLP will be detected in DNA isolated from any tissue or organ of a given plant, irrespective of age and developmental stage.

By far the major use of RFLP analysis to date has been by plant breeders, as a source of new genetic markers and to construct genetic maps (see, for example, Beckmann and Soller, 1986). Variability at RFLP loci has been identified in several species, including wheat, barley, maize, peas, soybeans, rice, and potatoes, and a limited number of reports has now appeared of the direct use of this variability for distinguishing between and identifying varieties (e.g., Bunce et al., 1986, in barley; Wang and Tanksley, 1989 in rice; Ruiz and Hemleben, 1991, in cucurbits; Smith, Smith, and Wall, 1991, in maize). However, there are disadvantages to the use of RFLPs for identification purposes, these primarily being time, expense, technical difficulty, and the restricted availability of probes. The use of ^{32}P to label DNA probes could also clearly cause some problems if employed on a routine basis, in seed testing laboratories for instance. Such difficulties might be alleviated by, for example, the use of restriction enzymes with rare cutting sites and direct visualization of total DNA, which obviates the need for probes. There are also other approaches to DNA analysis now coming to the forefront which address some of these disadvantages.

DNA Fingerprinting

It is now well established that a substantial portion of the genome of plants and animals consists of repetitive DNA sequences. Such sequences contain a series of tandem repeats of a core consensus sequence made up from a varying number of base pairs. Short sequences (between two and five base pairs) tend to be known as "simple repetitive sequences" or "micro-satellites," whereas longer ones have been termed "mini-satellites." These are often dispersed throughout the genome and thus can represent many loci. At a given locus, numerous alleles may occur, which usually differ in the number of core repeats (giving rise to the term "variable number of tandem repeats," VNTRs), and which have been used to produce genetic fingerprints when hybridized with a suitable probe. The use of DNA fingerprinting in humans is based on the polymorphism of such mini-satellites (Jeffreys, Wilson, and Thein, 1985).

An interesting and perhaps slightly surprising finding of the past three years is that several of the probes used to produce DNA fingerprints in man and other animals can also reveal polymorphisms in plant species (Dallas, 1988; Rogstad, Patton, and Schaal, 1988). The DNA fingerprints thus produced can be used for variety identification purposes. The most widely utilized probe has been the M13 repeat probe, which has been used, particularly by Nybom and colleagues, to distinguish between and identify varieties of several horticultural species such as apples, blackberries, and raspberries (see Nybom, 1990). Such fingerprinting is essentially the same technically as an RFLP analysis, using a restriction enzyme which cuts at the ends of the multiple repeat unit and a VNTR probe. The pattern of DNA fragments produced by this process will be more or less complex, depending on the number of loci revealed and, with an appropriate choice of enzyme and probe, can be variety specific.

In a variation of this approach, Weising and coworkers have used repetitive oligonucleotide sequences as probes.In a recent review they clearly demonstrated how such probes, and in particular the $(GATA)_4$ probe, would elicit varietal differences in the DNA fingerprints of a wide range of crop species, including chickpeas, rapeseed, lentils, sugarbeet, and tobacco (1991). An interesting feature of this work was the demonstration that nonradioactively labeled probes could be used. Such biotinylated or digoxygenated probes show considerable promise for variety identification purposes and, as with RFLPs, the existence of a wide range of restriction enzymes and probes enables extremely powerful resolving techniques to be developed.

Polymerase Chain Reaction

The polymerase chain reaction (PCR) is a relatively new technique which produces the selective amplification of a specific DNA sequence by as much as 10^6, thereby greatly facilitating subsequent manipulations (Saiki et al., 1988). Normally, PCR uses two oligonucleotide primers that flank the DNA sequence of interest and involves repeated cycles of DNA heat denaturation, annealing of the primers to their complementary sequences, and extension of the primers with an enzyme, DNA polymerase. The action of this enzyme effectively doubles the amount of target DNA present and, by using several cycles of denaturation, annealing, and extension, the amount of DNA can be increased exponentially. The use of a heat-stable enzyme from *Thermus aquaticus,* known as *Taq* polymerase, has enabled the PCR process to be largely automated and it has now become an invaluable tool for many aspects of molecular biology research.

For variety identification, the attraction of PCR-based techniques is that they offer an alternative to RFLP analysis which is quicker, simpler, and avoids the use of radioactive probes. Two approaches can be envisaged. The first requires prior knowledge of the precise sequence of the portions of DNA of interest and then PCR amplification of these sequences and electrophoresis of the subsequent fragments. For example, a polymorphism of a gliadin-encoding gene which might be of use for distinguishing between wheat genotypes and varieties has been recently reported (D'Ovidio, Tanzarella, and Porceddu, 1990). The second approach may ultimately prove to be the best choice for variety identification. It involves a new PCR-based assay which has been termed RAPDs (Random Amplified Polymorphic DNA, and pronounced "rapids") (Rafalski, Tingey, and Williams, 1991). The use of RAPDs is based on the observation that a short oligonucleotide of arbitrarily chosen DNA sequence, when used in a PCR reaction, will prime the amplification of several fragments of genomic DNA. The fragments are separated by electrophoresis on agarose gels and directly stained with ethidium bromide. No knowledge of the sequence of the DNA being assayed is required. The nature of the amplified fragments depends upon the arbitrary primer sequence and on the DNA being analyzed. Different primers give rise to different amplified bands and polymorphisms at the priming sites result in the disappearance of an amplified band (Figure 8). Thus PCR with random primers is a method for detecting polymorphisms distributed throughout the genome, with a primer usually amplifying several bands, each of which will probably originate from a different locus. The preliminary work which has so far been reported has shown clear differences between accessions and varieties of soybeans and rice (Welsh and McClelland, 1990, Williams et al., 1990). Initial results at NIAB have demonstrated differences between varieties of barley (Reeves, unpublished), indicating the considerable potential of this approach for discrimination and identification. A variation of this procedure, using primers as short as five nucleotides in length of arbitrary sequence, less stringent PCR conditions, and PAGE plus silver staining for visualization of fragments, has also been reported to demonstrate differences between soybean varieties (Caetano-Anolles, Bassam, and Gresshoff, 1991).

Evidently it is fairly early for an evaluation of nucleic acid analysis for plant variety identification. However, enough is now known

FIGURE 8. A simple example of RAPDs analysis in barley varieties. DNA was extracted from different varieties, subjected to amplification using an arbitrary primer and the products separated by agarose gel electrophoresis. Note the polymorphism between some of the varieties, detected as the absence/presence of bands at a specific point. Different primers would reveal different polymorphisms. The varieties are (left to right): Pastoral, Blenheim, Triumph, Puffin, Marinka and Halcyon. (Unpublished data from Dr. J. C. Reeves, NIAB, Cambridge.)

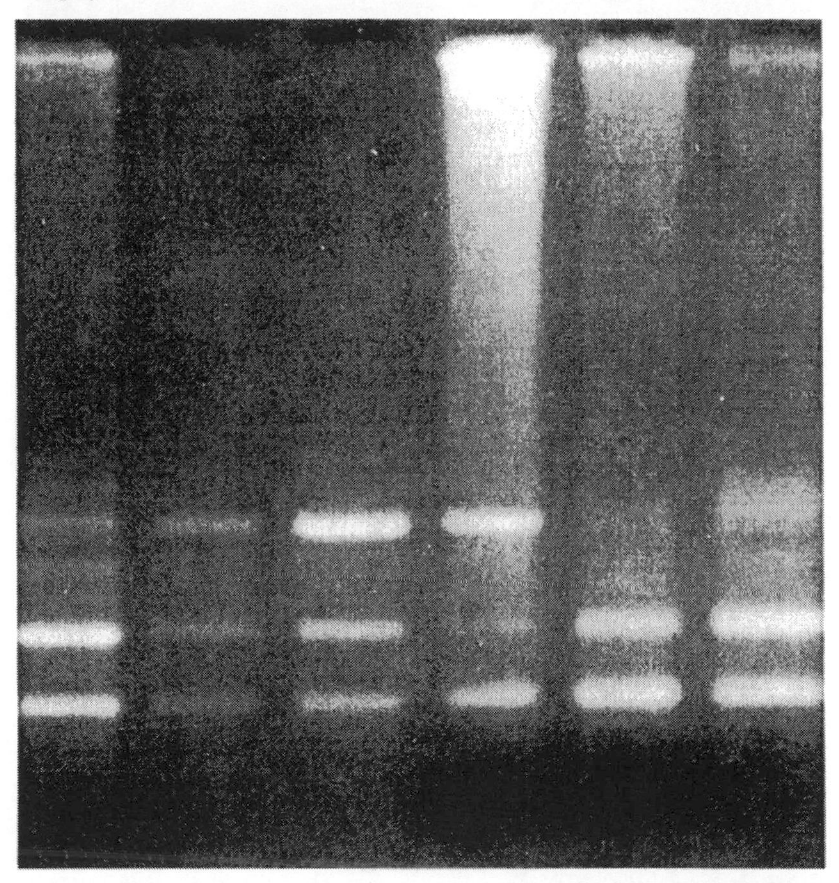

of the various possible approaches to indicate the exciting potential available. The next steps are to undertake a systematic and comparative study of the various approaches, to investigate the degree of discrimination possible between collections of varieties of different species, and to begin the development of simple analytical protocols. Only then will the maximum benefit begin to be derived from this powerful technology.

PRACTICAL APPLICATIONS

The ability to distinguish between and identify varieties of crop species by laboratory-based methods has engendered many practical applications within the seeds and allied industries. The use by plant breeders of isozymes, proteins, and RFLPs as genetic markers and for the construction of genetic maps is beyond the scope of this chapter (see Ainsworth and Gale, 1987; Cooke, 1988; Smith and Smith, 1992; Tanksley et al., 1989, for reviews). Some of the other important applications are considered in this section.

Hybrid Purity Testing

Hybrid varieties of many crops are now commonplace and the improvements in yield, uniformity, and disease resistance in such varieties are well known. The production of F1 hybrids involves crossing highly inbred parental lines and depends on the induction of male sterility in the female parent, by genetic or other mechanisms. In seed production terms, it is most important to be able to determine the success of the crossing procedure, i.e., to measure hybrid (or genetic) purity. This can be done by field growing-out procedures, but these are expensive, time-consuming, and liable to environmental influence. Fortunately, protein analysis by electrophoresis provides an alternative approach. The great attraction of electrophoresis for this purpose lies in the fact that isozyme or protein bands are inherited in a Mendelian fashion and expressed codominantly. This means in simple terms that when two inbred lines are crossed for F1 hybrid production, the resultant progeny will contain and express all of the protein bands from both parents.

Impurities arising from self-pollination of the female parent or pollination by an unintended male parent can be readily identified. Electrophoresis is used to assess hybrid purity in crops such as brassica vegetables, cotton, and tomatoes, but it is undoubtedly in maize that the most extensive use is made. Commonly, starch gel electrophoresis is used (Stuber et al., 1988) to examine 11 or 12 isozyme systems (equivalent to 21 or 22 loci). Such technology has been reported to be a much more accurate and cost-effective means of hybrid purity testing than field-based morphological observation (see Smith and Wych, 1986).

The ISTA is currently evaluating a method for hybrid purity assessment in maize using IEF to analyze the seed storage proteins (zeins). This has the advantage that seed material can be used, rather than seedlings as in the isozyme technique, and is already utilized by some seed testing laboratories in Europe.

The possible use of HPLC for hybrid purity assessment has been mentioned previously (see McCarthy et al., 1990) and may prove to be attractive in the future especially if bulked samples rather than individual seeds can be analyzed. RFLP analysis may also become more commonplace, particularly where isozyme or protein polymorphisms are lacking. One such instance, in pepper, has already been reported (Livneh et al., 1990).

Quality Control

An increasingly important and widespread use of modern variety identification methods is in quality control checking of grain in trade, primarily milling wheat and malting barley. Since the operation of automated machinery for processing cereals is dependent on the quality of the grain used, varietal identity and purity are critical elements in assessing the grain entering factories. Protein electrophoresis methods are now routinely used throughout North America, Europe, and Australia to check grain for processing. The economic and technological advantages of using electrophoresis are considerable. For instance, it has been estimated that for one British company alone, the use of electrophoresis to monitor wheat grain cargoes for milling results in a saving of about 10 million pounds per year.

HPLC can likewise be used to check varietal identity and purity as a quality control check, but its application to date has been far less than that of electrophoresis. The use of DNA analysis in this context is for the future and its success will depend on the ready availability of reasonably rapid and simple techniques. Perhaps the most exciting development in this area is the application of machine vision technology. As mentioned previously, a commercial wheat grain analyzer is under development at Cambridge. Much of the work in North America and Australia into machine vision is aimed at categorization of wheat on a commercial scale. This would provide an extremely rapid method of analysis, suitable for factory intake purposes, and further developments are eagerly awaited.

Distinctness Testing and Variety Registration

The power of the methods considered in this chapter for distinguishing between varieties implies that they might be particularly useful for distinctness testing of new varieties prior to registration. Indeed, it has been shown on several occasions at the NIAB and elsewhere that electrophoresis of proteins and/or enzymes would enable morphologically identical pairs or groups of varieties of a range of species to be discriminated. At the present time, the International Union for the Protection of New Varieties of Plants (UPOV) does not recognize that differences between varieties demonstrated by electrophoresis are sufficient in themselves to allow varieties to be regarded as distinct. However, this situation is being reviewed at present and, for self-pollinating cereal species at least, may well change in the near future.

Regardless of this, many countries analyze new varieties by electrophoresis and/or HPLC of proteins and use the data to supplement traditional distinctness testing methods. There are documented instances of varieties granted distinctness on the basis of electrophoresis tests. The use of protein analysis methods for distinctness testing will help to establish the ground rules for when more powerfully resolving DNA techniques become commonplace (see Smith and Smith, 1992). Machine vision perhaps presents fewer problems in this context than the other methods, especially when it is used as a measuring device to replace less accurate or more subjective tools. In the UK, for example, onion bulb shape characters measured by a machine vision technique have been successfully used for distinctness testing purposes. Overall there may well be substantial cost benefits to be gained from the considered and correct application of laboratory-based methods of analysis to variety registration.

Seed Testing and Certification

As the assessment of both species and varietal identity and purity are important elements in seed testing and certification, it is not too surprising that modern identification techniques can play a significant role in the operation of the procedures. This can range from routine checking of the identity of a particular seed sample to investigating the origin of off-types in specific seed crops (see Cooke, 1988, for examples). Both electrophoresis and machine vision can provide a means of distinguishing between closely related species and for categorizing species whose presence in seedlots is controlled by specific standards, such as wild oats in British seed crops. These techniques can also be used to replace special tests such as disease tests, which occasionally can be the only means of distinguishing between pairs or groups of otherwise very similar varieties.

Documentation of Genetic Resources

Since HPLC and/or electrophoresis of proteins and DNA analysis can provide a wealth of information about genetic variation, and machine vision not only allows existing characters to be measured more accurately but also provides a supply of potential new descriptors, it can be readily appreciated that these methods should be of considerable use for cataloguing genetic resources in gene banks and similar collections. Until now, extensive use has only really been made of electrophoresis, and there are reports of the electrophoretic characterization of collections of potatoes, wheat, barley, peas, beans, maize, and cocoa (see Smith, 1986; Cooke, 1988; Huaman and Stegemann, 1989). The Vavilov Institute in Leningrad (St. Petersburg), one of the largest germplasm resources in the world, uses protein electrophoresis to screen and catalogue its material. It can be stated with a reasonable degree of confidence that the various methods of DNA analysis will be increasingly widely used in this area, allowing as they do a much more substantial part of the genome to be analyzed (see Bernatzky and Tanksley, 1989). Such fingerprinting will surely prove to be one of the best methods of assessing genetic variation, detecting duplication of accessions, and cataloging germplasm collections. It also offers the prospect of being able to reduce the number of accessions that need to be stored in order to preserve the genetic integrity of a species.

FUTURE TRENDS

What does the future hold for this area of seed science research? It has already been mentioned that the various molecular biological methods of DNA analysis will begin to increase in importance and scope of application (see also Smith and Smith, 1992). This depends on sufficient resources being made available to enable the development and evaluation of suitable techniques. There are already reports of rapid methods for the extraction of plant genomic DNA prior to PCR analysis, enabling the processing of hundreds of individual samples in a working day (Edwards, Johnstone, and Thompson, 1991). This would begin to make DNA analysis feasible in a more routine, nonresearch laboratory, and also begin to match the sample throughput possible with protein electrophoresis. The trend in this latter area is toward the use of copies or blots from polyacrylamide gels, to facilitate multiple enzyme staining, and the use of smaller, thinner, commercially available gels which offer consistent gel quality and allow separations to take place in ten to 20 minutes (e.g., McDonald 1991; Wrigley, Gore, and Manusu, 1991). HPLC separations are already possible in this kind of time scale and the future here probably lies in the development of new and consistent column packing materials, the wider availability of cheaper and less sophisticated equipment, and standardization of the methodologies employed for seed and variety work.

A new electrophoresis process known as capillary zone electrophoresis (CZE), which is conducted in thin columns, is currently much in favor among separation scientists and can be used for protein separations. It can be anticipated that the application of CZE to variety identification will soon be common. Methods of protein detection based on immunological reactions and in particular the use of monoclonal antibodies have been suggested as being potentially useful for variety identification (e.g., Skerritt, Wrigley, and Hill, 1988). However, the relative lack of specificity of storage protein antibodies probably limits their use to distinguishing quality types or classes of grain, rather than for the identification of specific varieties. Nonetheless, this could still be valuable commercially and further developments here can be expected. The wider availability of a machine-vision-based variety identification system draws ever nearer, as does the application of this technology to the investigation of a greater range of species.

Hence overall, we can see that there is no doubt that biochemical techniques and computer-aided machine vision systems provide a rapid, reliable, and highly discriminating means for the identification of crop varieties. Some of these methods have already found widespread acceptance within certain sectors of the seeds and allied industries. The use of modern manipulation techniques in plant breeding and the imminent availability of transgenic (genetically modified) varieties, together with the need for overall improved seed quality control, will ensure that more use is made of these methods in all areas of the industry in the future.

REFERENCES

Ainsworth, C. C., and M. D. Gale. Enzyme structural genes and their exploitation in wheat genetics and breeding. In *Enzymes and their Role in Cereal Technology*. J. E. Kruger, D. Lineback, and C. E. Stauffer, eds. (St. Paul, Minnesota: American Association of Cereal Chemists Inc., 1987), pp. 53-82.

Ainsworth, C. C., and P. J. Sharp. The potential role of DNA probes in plant variety identification. *Plant Varieties & Seeds* 2 (1989): 27-34.

Allison, M. J., and H. Bain. The use of reversed-phased high performance liquid chromatography as an aid to the identification of European barley cultivars. *Euphytica* 35 (1986): 345-351.

Arulsekar, S., and D. E. Parfitt. Isozyme analysis procedures for stone fruits, almond, grape, walnut, pistachio and fig. *HortScience* 21 (1986): 928-933.

Beckmann, J. S., and M. Soller. Restriction fragment length polymorphisms and genetic improvement of agricultural species. *Euphytica* 35 (1986). 111-124.

Bernatzky, R., and S. D. Tanksley. Restriction fragments as molecular markers for germplasm evaluation and utilisation. In *The Use of Plant Genetic Resources*, A. D. Brown, D. R. Marshall, O. H. Frankel, and J. T. Williams, eds. (Cambridge: Cambridge University Press, 1989), pp. 353-362.

Bietz, J. A. Separation of cereal proteins by reversed-phase high performance liquid chromatography. *J. Chromat.* 255 (1983): 219-238.

Bietz, J. A., T. Burnouf, L. A. Cobb, and J. S. Wall. Wheat varietal identification and genetic analysis by reversed-phase high-performance liquid chromatography. *Cereal Chem.* 61 (1984). 129-135.

Buehler, R. E., M. B. McDonald, T. T. Van Toai, and S. K. St. Martin. Soybean cultivar identification using high performance liquid chromatography of seed proteins. *Crop Sci.* 29 (1989): 32-37

Bunce, N. A. C., B. G. Forde, M. Kreis, and P. R. Shewry. DNA restriction fragment length polymorphism at hordein loci: Application to identifying and fingerprinting barley cultivars. *Seed Sci. & Technol.* 14 (1986): 419-429.

Caetano-Anolles, G., B. J. Bassam, and P. M. Gresshoff. DNA amplification fingerprinting using very short arbitrary oligonucleotide primers. *Bio Technology* 9 (1991): 553-557.

Cooke, R. J. The characterisation and identification of crop cultivars by electrophoresis. *Electrophoresis* 5 (1984): 59-72.

Cooke, R. J. Characterisation of cultivars of *Allium cepa* L. (onions) by ultrathin-layer isoelectric focussing of seed esterases. *Electrophoresis* 6 (1985): 572-573.

Cooke, R. J. Electrophoresis in plant testing and breeding. *Advances in Electrophoresis* 2 (1988): 171-261.

Cooke, R. J. The use of electrophoresis for the distinctness testing of varieties of autogamous species. *Plant Varieties & Seeds* 2 (1989): 3-13.

Cooke, R. J. Unpublished data, NIAB, Cambridge.

Cooke, R. J., and S. R. Draper. Potential applications of ultrathin-layer isoelectric focusing for the characterization of cultivars of crop species. *Nat. Inst. Agric. Bot.* 16 (1983): 173-181.

Cooke, R. J., J. Higgins, A. G. Morgan, and L. J. Evans. The use of a vanillin test for the detection of tannins in cultivars of *Vicia faba* L. *Nat. Inst. Agric. Bot.* 17 (1985): 139-143.

Cooper, S. R. Report of the Rules Committee. *Seed Sci. & Technol.* 15 (1987): 555-575.

Dallas, J. F. Detection of DNA "fingerprints" of cultivated rice by hybridisation with a human minisatellite DNA probe. *Proc. Natl. Acad. Sci. U.S.A.* 85 (1988): 6831-6835.

D'Ovidio, R., O. A. Tanzarella, and E. Porceddu. Rapid and efficient detection of genetic polymorphism in wheat through amplification by polymerase chain reaction. *Plant Mol. Biol.* 15 (1990): 169-171.

Draper, S. R., and P. D. Keefe. Electrophoresis of seed storage proteins and whole-seed morphometry using machine vision: Alternative or complementary methods for cultivar identification? *Proceedings ISTA Symposium*, Leningrad 1988, pp. 27-35.

Draper, S. R., and P. D. Keefe. Machine vision for the characterization and identification of cultivars. *Plant Varieties & Seeds* 2 (1989): 53-62.

Draper, S. R., and A. J. Travis. Preliminary observations with a computer-based system for analysis of the shape of seeds and vegetative structures. *Nat. Inst. Agric. Bot.* 16 (1984): 387-395.

Edwards, K., C. Johnstone, and C. Thompson. A simple and rapid method for the preparation of plant genomic DNA for PCR analysis. *Nucleic Acids Res.* 19 (1991): 1349.

Gardiner, S. E., and M. B. Forde. Identification of cultivars and species of pasture legumes by sodium dodecyl sulphate polyacrylamide gel electrophoresis of seed proteins. *Plant Varieties & Seeds* 1 (1988): 13-26.

Gilliland, T. J. Electrophoresis of sexually and vegetatively propagated cultivars of allogamous species. *Plant Varieties & Seeds* 2 (1989): 15-26.

Greneche, M., J. Lallemand, and O. Migaud. Comparison of different enzyme loci as a means of distinguishing ryegrass varieties by electrophoresis. *Seed Sci. & Technol.* 19 (1991): 147-158.

Heaney, R. K., and G. R. Fenwick. Glucosinolates in brassica vegetables. Analysis of 22 varieties of brussels sprout (*Brassica oleracea* var. gemmifera). *J. Sci. Food Agric.* 31 (1980): 785-793.

Huaman, Z., and H. Stegemann. Use of electrophoretic analyses to verify morphologically identical clones in a potato collection. *Plant Varieties & Seeds* 2 (1989): 155-161.

Jeffreys, A. J., V. Wilson, and S. L. Thein. Individual-specific "fingerprints" of human DNA. *Nature* 316 (1985): 76-79.

Keefe, P. D., and S. R. Draper. The measurement of new characters for cultivar identification in wheat using machine vision. *Seed Sci. & Technol.* 14 (1986): 715-724.

Keefe, P. D., and L. V. Purchase. Unpublished data, NIAB, Cambridge.

Livneh, O., Y. Nagler, Y. Tal, S. B., Harush, Y. Gafni, J. S. Beckmann, and I. Sela. RFLP analysis of a hybrid cultivar of pepper (*Capsicum annuum*) and its use in distinguishing between parental lines and in hybrid identification. *Seed Sci. & Technol.* 18 (1990): 209-214.

Lookhart, G. L., L. D. Albers, Y. Pomeranz, and B. D. Webb. Identification of U.S. rice cultivars by high-performance liquid chromatography. *Cereal Chem.* 64 (1987): 199-206.

Marchylo, B. A., and J. E. Kruger. Identification of Canadian barley cultivars by reversed-phase high-performance liquid chromatography. *Cereal Chem.* 61 (1984): 295-301.

McCarthy, P. K., R. J. Cooke, I. D. Lumley, B. F. Scanlon, and M. Griffin. Applications of reversed-phase high-performance liquid chromatography for the estimation of purity in hybrid wheat. *Seed Sci. & Technol.* 18 (1990): 609-620.

McDonald, M. B. Blotting of seed proteins from isoelectrically focused gels for cultivar identification. *Seed Sci & Technol.* 19 (1991): 33-40.

Morgan, A. G. Chromatographic applications in cultivar identification. *Plant Varieties & Seeds* 2 (1989): 35-44.

Mundges, H., W. Kohler, and W. Friedt. Identification of rape seed cultivars (*Brassica napus*) by starch gel electrophoresis of enzymes. *Euphytica* 45 (1990): 179-187.

Myers, D. G., and K. J. Edsal. The application of image processing techniques to the identification of Australian wheat varieties. *Plant Varieties & Seeds* 2 (1989): 109-116.

Neuman, M., H. D. Sapirstein, E. Shwedyk, and W. Bushuk. Discrimination of wheat class and variety by digital image analysis of whole grain samples. *J. Cereal Sci.* 6 (1987): 125-132.

Nybom, H. Genetic variation in ornamental apple trees and their seedlings (*Malus*, Rosaceae) revealed by DNA "fingerprinting" with the M13 repeat probe. *Hereditas* 113 (1990): 17-28.

Rafalski, J. A., S. V. Tingey, and J. G. K. Williams. RAPD makers–A new technology for genetic mapping and plant breeding. *AgBiotech News & Infor.* 3 (1991): 645-648.

Reeves, J. C. Unpublished data; NIAB, Cambridge.

Rogstad, S. H., J. C. Patton, and B. A. Schaal. M13 repeat probe detects DNA minisatellite–Like sequences in gymnosperms and angiosperms. *Proc. Natl. Acad. Sci. U.S.A.* 85 (1988): 9176-9178.

Ruiz, R. A. T., and V. Hemleben. Use of ribosomal DNA spacer probes to distinguish cultivars of *Cucurbita pepo* L. and other Cucurbitaceae. *Euphytica* 53 (1991): 11-17.

Saiki, R. K., D. H. Gelfand, S. Stoffel, S. J. Scharf, R. Higuchi, G. T. Horn, K. B. Mullis, and H. A. Erlich. Primer-directed enzymatic amplification of DNA with a thermostable DNA polymerase. *Science* 239 (1988): 487-491.

Sapirstein, H. D., M. Neuman, E. H. Wright, E. Shwedyk, and W. Bushuk. An instrumental system for cereal grain classification using digital image analysis. *J. Cereal Sci.* 6 (1987): 3-14.

Scanlon, M. G., H. D. Sapirstein, and W. Bushuk. Computerized wheat varietal identification by high-performance liquid chromatography. *Cereal Chem.* 66 (1989a): 439-443.

Scanlon, M. G., H. D. Sapirstein, and W. Bushuk. Evaluation of the precision of high-performance liquid chromatography for wheat cultivar identification. *Cereal Chem*, 66 (1989b): 112-116.

Singhal, N. C., and S. Prakash. Use of phenol colour reaction on seeds and outer glumes for identification of bread wheat cultivars. *Plant Varieties & Seeds* 1 (1988): 153-157.

Skerritt, J. H., C. W. Wrigley, and A. S. Hill. Prospects for the use of monclonal antibodies in the identification of cereal species, varieties and quality types. *Proceedings ISTA Symposium*, Leningrad 1988, pp. 110-123.

Smith, J. S. C. Biochemical fingerprints of cultivars using reversed-phase high performance liquid chromatography and isozyme electrophoresis: A review. *Seed Sci. & Technol.* 14 (1986): 753-768.

Smith, J. S. C., and O. S. Smith. Fingerprinting crop varieties. *Adv. Agron.* 47 (1992): 85-140.

Smith, J. S. C., and R. D. Wych. The identification of female selfs in hybrid maize: A comparison using electrophoresis and morphology. *Seed Sci. & Technol.* 14 (1986): 1-8.

Smith, J. S. C., O. S. Smith, and S. J. Wall. Associations among widely used French and U.S. maize hybrids as revealed by restriction fragment length polymorphisms. *Euphytica* 54 (1991): 263-273.

Stuber, C. W., J. F. Wendell, M. M. Goodman, and J. S. C. Smith. Techniques and scoring procedures for starch gel electrophoresis of enzymes from maize (*Zea mays* L.). *Technical Bulletin 286*, North Carolina Agricultural Research Service, 1988.

Tanksley, S. D., N. D. Young, A. H. Paterson, and M. W. Bonierbale. RFLP mapping in plant breeding: New tools for an old science. *BioTechnology* 7 (1989): 257-264.

Thomson, W. H., and Y. Pomeranz. Classification of wheat kernels using three-dimensional image analysis. *Cereal Chem.* 68 (1991): 357-361.

Travis, A. J., and S. R. Draper. A computer-based system for the recognition of seed shape. *Seed Sci. & Technol.* 13 (1985): 813-820.

Wang, Z. Y., and S. D. Tanksley. Restriction fragment length polymorphism in *Oryza sativa* L. *Genome* 32 (1989): 1113-1118.

Weising, K., B. Beyermann, J. Ramser, and G. Kahl. Plant DNA fingerprinting with radioactive and digoxigenated oligonucleotide probes complementary to simple repetitive DNA sequences. *Electrophoresis* 12 (1991): 159-169.

Weiss, W., W. Postel, and A. Gorg. Barley cultivar discrimination: I. Sodium dodecyl sulphate-polyacrylamide gel electrophoresis and glycoprotein blotting. *Electrophoresis* 12 (1991): 323-330.

Welsh, J., and M. McClelland. Fingerprinting genomes using PCR with arbitrary primers. *Nucleic Acids Res.* 18 (1990): 7213-7218.

White, J., and R. J. Cooke. Unpublished data, NIAB, Cambridge.

White, J., and J. R. Law. Differentiation between varieties of oilseed rape (*Brassica napus*) on the basis of the fatty acid composition of the oil. *Plant Varieties & Seeds* 4 (1991): 125-132.

Williams, J. G. K., A. R. Kubelik, K. J. Livak, J. A. Rafalski, and S. V. Tingey. DNA polymorphisms amplified by arbitrary primers are useful as genetic makers. *Nucleic Acids Res.* 18 (1990): 6531-6535.

Wingad, C. E., M. Iqbal, M. Griffin, and F. J. Smith. Separation of hordein proteins from European barley by high-performance liquid chromatography: Its application to the identification of barley cultivars. *Chromatographia* 21 (1986): 49-54.

Wrigley, C. W. Identification of cereal varieties by gel electrophoresis of the grain proteins. In *Modern Methods of Plant Analysis (New Series), Vol. 14, "Seed Analysis,"* 1992.

Wrigley, C. W., J. C. Autran, and W. Bushuk. Identification of cereal varieties by gel electrophoresis of the grain proteins. *Adv. Cereal Sci. & Technol.* 5 (1982): 211-259.

Wrigley, C. W., P. J. Gore, and H. P. Manusu. A rapid (< 10 minute) electrophoresis method for identification of wheat varieties. *Electrophoresis* 12 (1991): 384-385.

Zayas, I., F. S. Lai, and Y. Pomeranz. Discrimination between wheat class and varieties by image analysis. *Cereal Chem.* 63 (1986): 52-56.

Zayas, I., Y. Pomeranz, and F. S. Lai. Discrimination of wheat and non-wheat components in grain samples by image analysis. *Cereal Chem.* 66 (1989): 233-237.

Chapter 10

Low Water Potential and Presowing Germination Treatments to Improve Seed Quality

Wallace G. Pill

Modern crop production systems require a high degree of precision in crop establishment (Salter, 1985). The need for high plant population densities and uniform plant stands for machine harvest has led to a growing interest in direct field sowing. At the same time, the widespread use of various module systems for vegetable and ornamental plant raising requires that each cell of the module contain a plant. Consequently, seed of high quality that will consistently produce rapid and uniform seedling emergence from each seed sown is required. Increasing use of expensive hybrid seed has placed additional emphasis on the performance of each seed planted.

For certain crops such as carrots or onions grown at close, within-row spacings, a seedlot of lower viability can be used provided allowance is made for the proportion of dead seed in calculating the numbers of seeds to be sown to reach a target plant density (Gray, 1989). However, seedlots of low viability usually germinate over a longer period. This wide spread of germination and the subsequent emergence in the field was shown to be the most important source of plant-to-plant variation in size or weight in a crop (Benjamin, 1984, 1990). Any initial variation in plant size will increase as the crop matures, the larger plants in the population will continue to secure proportionally more of the resources available. Finch-Savage (1986) noted in cauliflower, leek, and onion seeds that seedling length decreased and the coefficient of variation of seedling lengths increased with increasing number of days of imbibition required for germination. Slow germinating seeds produced fewer normal healthy seedlings than faster germinating seeds. In further work, germination rate within a carrot seedlot was positively related to vigor, percentage emergence, and seedling weight, and was negatively related to the spread of emergence times and the coefficient of variation of seedling weights (Finch-Savage and McQuistan, 1988a).

TeKrony and Egli (1991) reviewed the relationship between seed vigor and crop yield. They noted that seed vigor affected vegetative growth and frequently was related to the yield of crops that were harvested vegetatively or during early reproductive growth. There was usually no such relationship in crops harvested at full reproductive maturity, because seed yield at full reproductive maturity usually was not associated closely with vegetative growth. They concluded that planting high-vigor seeds can be justified for all crops to ensure adequate plant population densities across the wide range of field conditions that occur during emergence.

A generalized sequence of physiological changes has been proposed to take place during the period of aging (deterioration) before loss of viability (Delouche and Baskin, 1973). The first change observed was the reduced ability of the seed to retain solutes, resulting in increased leakage of amino acids, sugars, and electrolytes from the seeds into water. That electrical conductivities of the leachate were inversely correlated with field establishment and thus with seed vigor (Matthews and Bradnock, 1973; Oliviera, Matthews, and Powell, 1984) supported the hypothesis that the earliest stage in aging is the deterioration of cellular membranes (Delouche and Baskin, 1973; Roberts, 1973). Imbibition damage, found so far only in large legumes, resulted from rapid water uptake into cotyledons, causing cell death on the cotyledon surface, leading to high solute leakage (Powell and Matthews, 1978, 1979; Oliviera, Matthews, and Powell, 1984). Obviously, a myriad of physiological and biochemical events follow initial membrane deterioration leading to loss of seed viability.

Seed quality is influenced throughout the life of the seed, from time of fertilization on the mother plant to the moment of sowing. Brocklehurst (1985) listed the principal factors, in chronological order, as: seed genotype, environmental conditions during seed development, seed position on the parent plant, harvest timing and techniques, storage conditions, and presowing treatments. Only presowing treatments, specifically low water potential seed treatments (osmotic and matric priming) and presowing germination, will be the subject of this chapter.

It should be realized, however, that there is some controversy regarding the merits of such seed treatments with respect to existing seed vigor. Matthews and Powell (1986) state that "presowing treatments of seeds such as pelleting, priming and pregermination can be of little benefit to low-vigour seed." They further suggested that the major benefits of many seed treatments may lie in the direct influence they have in ensuring that only highly germinable, vigorous seedlots were given expensive treatments. Others have found, however, that low-quality seedlots derived greater benefit from presowing treatments. For example, preplant priming improved the performance of good- and poor-quality carrot seeds, but the improvements were greater in the poor seeds than in the more vigorous seeds (Szafirowska, Khan, and Peck, 1981; Khan, Abawi, and Maguire, 1992). Finch-Savage and McKee (1990) noted that a presowing germination (pregermination) treatment effectively eliminated differences in seedlot quality.

PREPLANT LOW WATER POTENTIAL SEED TREATMENTS

Preplant low water potential seed treatments permit partial seed hydration (usually within 10 to 20 percent of full imbibition) so that pregerminative metabolic activities proceed but germination is prevented. Such treatments influence the speed, synchrony, and percentage of seed germination. Several seed hydration procedures have been developed. One technique to shorten the time between sowing and emergence involves presowing germination (pregermination, chitting) of seeds under optimal hydration and temperature conditions. Two other approaches involve priming seeds by exposing them to low water potentials: osmotic priming involves hydration of seeds in solutions containing organic or inorganic solutes; matric priming involves hydration of seeds in solids containing limited amounts of water or aqueous solution.

Osmotic Priming

Descriptions of the osmotic priming process (also termed osmotic conditioning, osmoconditioning, priming, or osmopriming) and reviews of the effects of priming on germination are available (Heydecker and Coolbear, 1977; Bradford, 1986; Khan, 1992).

Seed vigor has been enhanced by osmotic priming so that rate, synchrony, and percentage of seedling emergence from such seeds were greater than from control seeds, especially under such adverse seedbed conditions as low temperature (Szafirowska, Khan, and Peck, 1981; Pill and Finch-Savage, 1988), high temperature (Heydecker, 1973/74; Khan, 1980/1981; Globerson and Feder, 1987; Carpenter and Boucher, 1991), reduced water availability (Akers, Berkowitz, and Rabin, 1987; Frett and Pill, 1989), or salinity (Weibe and Muhyaddin, 1987; Cano et al., 1991; Pill, Frett, and Morneau, 1991). Osmotic priming has reduced lettuce seed thermodormancy (Guedes and Cantliffe, 1980; Cantliffe, Fischer, and Nell, 1984; Perkins-Veazie and Cantliffe, 1984; Khan, 1992), and has substituted for stratification in certain *Aquilegia* species (Finnerty, Zajicek, and Hussey, 1992).

Osmotic priming has improved seedling emergence or yield in asparagus (Evans and Pill, 1989; Pill, Frett, and Morneau, 1991), beet root (Khan et al., 1983), Brussels sprouts (Khan, 1980/81), cantaloupe (Bradford, 1985), carrot (Szafirowska, Khan, and Peck, 1981; Brocklehurst and Dearman, 1983a, 1983b; Haigh et al., 1986; Haigh and Barlow, 1987a), cucumber (Passam et al., 1989), leek (Brocklehurst, Dearman, and Drew, 1984; Dearman, Brocklehurst, and Drew, 1987), lettuce (Cantliffe, Schuler, and Guedes, 1981; Valdes, Bradford, and Mayberry, 1985), melon (Passam et al., 1989), muskmelon (Bradford et al., 1988), onion (Brocklehurst and Dearman, 1983a, 1983b, 1984; Haigh et al., 1986; Haigh and Barlow, 1987a), parsley (Heydecker and Coolbear, 1977; Pill, 1986; Rabin, Berkowitz, and Akers, 1988), parsnip (Gray et al., 1984), pepper (Rivas, Sundstrom, and Edwards, 1984; Passam et al., 1989; Bradford, Steiner, and Trawatha, 1990), perennial ryegrass (Danneberger et al., 1992), sugar beet (Durrant, Payne, and Maclaren, 1983), tomato (Wolfe and Sims, 1982; Alvarado, Bradford and Hewitt, 1987; Barlow and Haigh, 1987; Haigh et al., 1986; Haigh and Barlow,

1987a; Argerich, Bradford, and Ashton, 1990; Pill, Frett, and Morneau, 1991), and watermelon (Sachs, 1977). While an inverse relationship of days to 50 percent seedling emergence from primed seeds and subsequent growth has been reported (Brocklehurst and Dearman, 1983b; Pill, 1986; Evans and Pill, 1989), the greater plant growth from primed than untreated seeds, at least for leek, resulted from earlier emergence and not from increased relative growth rates (Brocklehurst, Dearman, and Drew, 1984).

Seed osmotic priming has been less successful in improving the performance of large seeds. For instance, Knypl and Khan (1981) noted that osmotic priming of soybean seeds led to greater germination and emergence rates at suboptimal temperatures in laboratory and growth chamber studies but had little effect on seedling emergence or yield in several early spring field plantings under cold, wet conditions (Khan, 1980/81). Helsel, Helsel, and Minor (1986) found that osmotically primed soybean seeds had faster and more uniform seedling emergence than untreated seeds in early plantings, but that economic yield was little affected. Bennett and Waters (1987a, 1987b) found that osmotic priming of sweet corn reduced seedling emergence percentage in cool soils compared to untreated seeds.

Priming Conditions

Since the response to a given priming treatment can vary between seedlots of the same cultivar (Brocklehurst and Dearman, 1983a, 1983b; Brocklehurst, Dearman, and Drew, 1984), the optimal priming treatment for a seedlot is determined by trial and error (Bradford, 1986). Variable factors include the nature of the priming agent (Brocklehurst and Dearman, 1984; Bradford, 1986; Haigh and Barlow, 1987a; Frett, Pill, and Morneau, 1991; Smith and Cobb, 1991a), the duration of seed exposure to the priming agent (Pill and Finch-Savage, 1988; Evans and Pill, 1989; Frett and Pill, 1989), and the temperature and osmotic potential (Ψ_s) of the priming agent (Haigh and Barlow, 1987a; Evans and Pill, 1989; Frett and Pill, 1989; Smith and Cobb, 1991a). Typically, the priming duration is two to 21 days, Ψ_s is –0.8 to –1.6 MPa, and the priming temperature is 15 to 25°C. For species such as lettuce or celery in which seed germination is under phytochrome control, irradiation was required during prolonged priming (Khan et al., 1978; Khan, 1980/81).

A commonly used seed priming agent is polyethylene glycol (PEG) of approximately 8000 molecular weight. While PEG-priming is termed "osmotic," molecules of this size are colloidal so that the reduced water potential is derived matrically. The relationship between molarity and osmotic potential of PEG solutions has been described with high precision by second order polynomials (Michel and Kaufmann, 1973; Money, 1989). The interactive effects of solutes added to PEG 8000 on water potential have been reported (Michel, 1983). While PEG is inert and nonphytotoxic, it is costly and sufficiently viscous to hinder aeration (Mexal et al., 1975). Bujalski, Nienow, and Gray (1989) reported that humidified, enriched air (75 percent oxygen, 25 percent nitrogen) diffused through sintered-plate gas diffusers into PEG 8000 at 0.6 liters per minute resulted in greater rate, synchrony, and percentage germination of onion seeds than normal aeration. Enriched air thus is viewed as a way of aerating PEG 8000 for large-scale priming.

A number of other osmotica, including KNO_3, K_3PO_4, KH_2PO_4, $MgSO_4$, NaCl, glycerol, and mannitol have been used. Frett, Pill, and Morneau (1991) compared factorial combinations of Ca^{2+}, Na^+, or K^+ with NO_3^-, Cl^-, SO_4^{-2} or $H_2PO_4^-$ at –0.8 MPa and 20°C as priming agents for asparagus and tomato seeds. While salt solutions were more effective than PEG in speeding tomato seed germination, salt solutions provided no such benefit for asparagus seeds. Of the salt solutions, NO_3^- salts were most effective in reducing germination time of tomato and asparagus seeds. Synthetic seawater (Instant Ocean™, Aquarium Systems, Mentor, OH), at about one-third full strength, was as effective as PEG as a priming agent for asparagus seeds and was superior to PEG as a priming agent for tomato seeds. Haigh and Barlow (1987a) found that tomato seeds primed in solutions that contained KNO_3 germinated more rapidly and synchronously than those primed in solutions without KNO_3. The NO_3^- salts may be absorbed preferentially to lower the internal osmotic potential and thereby encourage water influx, an effect that would explain the alleviation of "underpriming" by inclusion of KNO_3 in the priming solution observed by Haigh and Barlow (1987a). Nutritional effects of nitrate uptake during priming in providing additional substrate for amino acid and protein synthesis cannot be discounted. Dissociated ions from such salts can penetrate seed tissues, whereas PEG molecules do not. Variable uptake of different ions from priming salts would not only influence the amount of water absorbed by the seeds along the osmotic gradient but may also exert disruptive specific ion effects on enzymes and membranes. The amounts of water taken up by beet seeds at 15°C in water, –1.2 MPa PEG, –1.2 MPa KNO_3, and –1.2 MPa NaCl solutions were 94, 57, 84 and 87 percent (dry weight basis), respectively, after a 24-hour soak corresponding values after 96 hours being 119, 61, 88, and 102 percent (Khan, 1992). The term "overpriming" was used to describe the greater water content of seeds primed in salt (KNO_3) than in PEG (Alvarado and Bradford, 1988).

Calculation of solution osmotic potential (Ψ_s) is based on the van't Hoff equation (Salisbury and Ross, 1985): $\Psi_s = -imRT$, where i is an ionization factor, m = molality (im = osmolality), R = the gas constant, and T = absolute temperature. Smith and Cobb (1991a), after comparing many salt solutions as priming agents for sweet pepper seeds, determined that the priming effect was dependent on solution Ψ_s and the duration of soak, and not on the specific salts. Suzuki et al. (1990) reported that in tertiary phosphate (K_3PO_4) priming solutions of the same Ψ_s, pH had a major effect in preventing germination, with pH 12 preventing germination and pH 6 promoting germination.

Different temperatures (15, 20, or 25°C) during priming of onion seeds had little effect on subsequent emergence characteristics (Haigh et al., 1986). These workers found, however, that increasing the duration of salt priming from one to three weeks increased subsequent emergence rate, but at the expense of decreased final percentage and synchrony of emergence. The combination of the highest water potential and temperature (–0.6 MPa and 20°C) determined to be the optimum for priming asparagus seeds (Evans and Pill, 1989) agreed with the assumptions upon which Akers and Holley (1986) based a screening procedure to determine priming conditions. Their assumptions were that priming temperature and duration should be optimum for the germination of unprimed seeds but that the water potential of the priming solution should be as high as possible without permitting germination.

Priming effects have been enhanced by prepriming water soaks (Akers, Berkowitz, and Rabin, 1987; Pill and Finch-Savage, 1988), presumably by removing germination inhibitors. Priming effects likewise have been enhanced by incorporating plant growth regulators and pesticides during either the prepriming soak or the osmotic priming.

Additives to the Priming Solution

Gibberellins (GA) added to the PEG priming solution replaced the light effect in speeding germination and in preventing dormancy induction in lettuce and celery (Khan, 1980/81). The promotive effect of GA during osmotic priming of lettuce seeds was not influenced by the addition of abscisic acid (ABA) in the PEG solution (Khan and Samimy, 1982). This lent credence to Khan's hypothesis (Khan, 1971) that hormones have at least two sites of action in the control of germination, a primary site (GA site) and a preventive-permissive site (secondary site) at which cytokinin-ABA interactions might occur. The presence or absence of any one of three classes of hormones (gibberellin, cytokinin, or inhibitor) at physiologically active concentrations, might dictate whether the seed will remain dormant or germinate. On the basis of this scheme, the seed is dormant in the absence of gibberellin, whether or not cytokinin or inhibitor is present, or in GA presence when the inhibitor also is present but cytokinin is absent. Germination occurs in the presence of GA and absence of inhibitor whether or not cytokinin is present, or in the presence of the inhibitor with cytokinin to oppose its effect. This scheme clearly gave gibberellins the primary role in the control of germination. The roles of inhibitors and cytokinins were secondary and essentially preventive and permissive, respectively.

Further work showed that aminoethoxyvinyl glycine (AVG), a specific inhibitor of the conversion of S-adenosyl methionine to 1-aminocyclopropane-1-carboxylic acid (ACC) in the ethylene biosynthetic pathway (Yang and Hoffman, 1984), failed to inhibit the germination advancement effect of osmotic priming in the light or in the presence of GA in the dark. Ethylene biosynthesis thus was not involved in the germination advancement effect of seed osmotic priming. While ethylene biosynthesis does not appear to be effective during long-term osmotic priming, it may be effective in circumventing thermoinhibition during short-term prepriming soaks. The addition of 1 mM AVG during the presoak period (for four, six, or nine hours at 15 to 20°C) prevented the relief of thermoinhibition; the effect was reversed by adding 1 mM ACC or 0.1 mM ethephon. Ethylene production from ACC by lettuce seeds is reduced under such adverse conditions as high temperature, salinity, or low water potentials (Khan and Huang, 1988; Khan and Prusinski, 1989). Khan (1992) hypothesized that the effectiveness of a short duration presoak treatment in preventing thermodormancy in unaged but not in aged lettuce seeds (Perkins-Veazie and Cantliffe, 1984) was related to the reduced ability of the aged seeds to produce ethylene. Khan (1992) pointed out that the effectiveness of short duration hydration of seeds in water or priming solutions in alleviating high temperature effects on germination (Wurr and Fellows, 1984 [lettuce]; Carpenter, 1989 [*Salvia splendens*]; Carpenter, 1990 [dusty miller]) was consistent with the suggestion that the requirement of ethylene biosynthesis for germination is induced only under stressful conditions such as high temperature. Cantliffe (1991) reported that while priming lettuce seeds in 1 percent K_3PO_4 for 20 hours in the dark reduced thermodormancy, the addition of 100 mg of cytokinin (6-benzyladenine) per liter of priming solution further increased germination percentage. This response may reflect the permissive role of cytokinins if inhibitors were present at the high germination temperatures.

Inclusion of 200 mg per liter N-substituted phthalimide (a synthetic functional gibberellin analogue) during a 14-day soak in −1.0 MPa PEG at 15°C further enhanced the priming effect in increasing the rate, synchrony, and percentage of carrot seed germination (Pill and Finch-Savage, 1988). Inclusion of GA_{4+7} (Regulex™) in the osmoticum had no effect on germination synchrony and delayed carrot seed germination (Pill and Finch-Savage, 1988), contrary to results obtained with pepper seeds using GA_3 (Sosa-Coronel and Motes, 1982) or GA_{4+7} (Watkins et al., 1985).

Combination priming treatments of plant growth regulator/fungicide and PEG 8000 treatments synergistically reduced the mean germination time of all species (*Primula*, *Impatiens*, *Verbena* and *Petunia*) and increased the germination percentage of *Primula* and *Impatiens* in the dark compared with untreated controls (Finch-Savage, 1991). The growth regulators were more effective when added to the priming solution than when applied as a prepriming soak. Oxygen enrichment (75 percent O_2, 25 percent N_2) of the priming solution generally reduced seed performance unless growth regulators were present in the solution. The most effective treatment combinations were: *Primula*, oxygen enriched priming with 10^{-5}M $GA_{4/7}$; *Impatiens*, priming with 10^{-4} M $GA_{4/7}$ plus 10^{-6} M benzyladenine; *Verbena*, priming with 10^{-4} M $GA_{4/7}$; and *Petunia*, priming alone. Iprodione was applied as a three-hour prepriming soak because this fungicide generally exerted a phytotoxic effect when added to the priming solutions. Osburn and Schroth (1989) noted that the addition of the fungicide metalaxyl (Apron) during osmotic priming of sugar beet in NaCl or PEG solutions resulted in greater control of preemergence damping-off by *Pythium ultimum*, than with fungicide or osmotic priming alone.

Biopriming, a combination of biological seed treatment and preplant hydration, is a new strategy for improving reliability of biological treatments, particularly for crops such as sweet corn that are susceptible to imbibitional chilling injury (Callan, Mathre, and Miller, 1990, 1991). Callan, Mathre, and Miller (1991) noted that sweet corn seeds coated with a 1.5 percent methylcellulose suspension of *Pseudomonas fluorescens* AB254 and allowed to hydrate for 20 hours at 23°C were protected from preemergence damping-off caused by *Pythium ultimum* Trow to the same level achieved with metalaxyl. The efficacy of adding biological treatments to osmotic priming solution has not been determined.

Storage/Drying Conditions

In many studies of seed priming, seeds are transferred directly from the priming solutions to the germination media, generally resulting in very rapid germination compared to seeds imbibed at the same time. In the absence of seed drying, this response represents the release of seeds from osmotic inhibition. The seeds have completed phases I (hydration) and II (lag phase) of germination, and only require a favorable gradient for water uptake in order to begin radicle growth. This situation is therefore more comparable to that of pregerminated seeds (Bradford, 1986) in that the hydrated, primed seeds enter immediately into phase III of imbibition (growth), while dry primed seeds must repeat phase I and often at least a short phase II before growth occurs.

Many studies have reported germination responses of primed seeds that were sown within a few hours or days of priming. Typically, seed germination or seedling emergence is delayed in proportion to the degree of seed drying after priming but before sowing. Slower and less synchronous emergence from primed and dried-back seeds (compared with primed seeds which had not been dried back) has been noted for carrot, celery, and onion (Brocklehurst and Dearman, 1983a, 1983b), leek (Brocklehurst, Dearman, and Drew, 1984), asparagus (Evans and Pill, 1989; Pill, Frett, and Morneau, 1991), parsley (Pill, 1986), and tomato (Pill, Frett, and Morneau, 1991). Delayed emergence of primed and dried-back seeds was, at least in part, a result of the increased time needed for seeds to re-imbibe. Finch-Savage, Gray, and Dickson (1991) noted that seed drying for two days (15°C, 45 percent relative humidity) after priming (compared to priming without drying) had little or no effect on mean germination time of *Impatiens*, *Verbena*, *Petunia*, and *Primula*. Even partial drying of *Salvia* seeds after priming damaged the seeds.

Successful long-term storage of primed seeds usually requires drying the primed seeds to some low moisture content, for example carrots, 7 percent, and leeks, 10 percent (Dearman, Brocklehurst, and Drew, 1987); onion, 9 percent (Dearman, Brocklehurst, and Drew, 1986); sweet pepper, 9.8 percent (Georghiou, Thanos, and Passam, 1987; Thanos, Georghiou, and Passam, 1989); or tomato, 6 percent (Alvarado and Bradford, 1988; Argerich, Bradford, and Tarquis, 1989).

Handling of primed seeds in the dry state has practical advantages, but reports on the effects of longer-term storage (at least two months) on seed germination have been conflicting. Primed onion seeds with 9 percent moisture stored at 10°C maintained a higher germination rate than untreated seeds and did not lose viability during 18 months of storage (Dearman, Brocklehurst, and Drew, 1986). These authors further found that seed priming before aging (elevated seed moisture content and temperature) delayed the loss of viability due to aging. Subsequent work (Dearman, Brocklehurst, and Drew, 1987) revealed that primed carrot (7 percent moisture) and leek seeds (10 percent moisture) stored at 10°C maintained high germination rates without loss of viability after 12 months of storage. This work revealed further that priming increased the sensitivity of carrot and leek seeds to accelerated aging. Primed *Impatiens* seeds dried to several moisture contents maintained a higher rate and synchrony of germination, and showed no loss of viability following storage at 25°C and 50 percent relative humidity for up to eight weeks (Frett and Pill, 1989).

Primed tomato seeds stored at 4°C retained viability and germination rate after one year at 4°C, but at 30°C viability and vigor were markedly reduced within six months (Argerich, Bradford, and Tarquis, 1989). These authors concluded that primed tomato seeds must be considered to be vigorous with a reduced shelf-life. In contrast, Georghiou et al. (1987) noted that sweet pepper seeds that were primed then dried to their original moisture contents had greater germination rate and viability than untreated seeds following six months of storage at 35°C. Thanos, Georghiou, and Passam (1989) found that while untreated sweet pepper seeds showed a marked loss in viability when stored at 25°C for up to three years, primed seeds showed little loss in viability under these conditions and maintained a higher germination rate. Storage of the primed seeds at 5°C versus 25°C gave a slight advantage in viability and germination rate, particularly as the storage period increased (Thanos, Georghiou, and Passam, 1989). These authors concluded that since viability and germination rate were enhanced by priming both before or after seed storage, priming was involved in both delaying the aging process and in repairing seed deterioration.

Pill and Frett (1989) prepared "seed sheets" by placing unprimed or primed (–1 MPa PEG, 15°C, one week) seeds of tomato or *Celosia* in drying solutions of 5 percent hydroxyethyl cellulose plasticized with 0.35 percent triacetin. Seedling emergence characteristics and shoot weights from unprimed and primed seeds stored in the sheets for four weeks were equal to those from seeds not stored in sheets. After four weeks of storage in the sheets, primed tomato seeds gave earlier seedling emergence and greater seedling shoot fresh weights than unprimed seeds stored for four weeks in the sheets.

Alvarado and Bradford (1988) observed a storage temperature effect on viability of osmotically primed seeds stored at constant moisture. Primed tomato seeds (dried to 6 percent moisture) stored at 10 or 20°C maintained high viability for at least 18 months. When stored at 30°C, however, the primed seeds lost vigor and viability within five months. These authors suggested that to maintain high quality in primed tomato seeds for extended storage periods, the seeds should be stored at low temperature and low moisture content. While storage of primed tomato seeds (6 percent moisture) at 4°C maintained high vigor and viability after 12 months (Argerich and Bradford, 1989), the low-temperature storage reduced root growth rate. Owen (1990) found that the combination of surface drying (high-moisture seeds) and 4°C storage resulted in greater germination percentage and rate of salt-primed asparagus seeds than dried-back (low moisture) primed seeds stored at 4 or 20°C for three months.

Matric Priming

The superior performance of coated lettuce seed (Royal Sluis Splitkote™) under dry conditions compared to untreated seeds probably was an effect of matric priming (Khan, 1992). Khan and Taylor (1986) had observed that a period of drought after planting untreated beet root seeds seemed to provide a "natural matric priming" that improved the emergence performance in a manner similar to that of primed seeds. Mimicking this natural effect by amending beet seed pellets with PEG (postplant matric priming) improved seedling rate and emergence (Khan and Taylor, 1986). Peterson (1976) primed onion seeds in a slurry of PEG 6000 (now 8000) and vermiculite. The mixture of three to four parts vermiculite moistened with PEG solution was combined with one part seed. Although this osmotic plus matric priming technique overcame the need for aeration of PEG, the seeds were difficult to separate from the vermiculite following priming.

The use of moist solid or semi-solid carriers (e.g., exfoliated vermiculite, expanded calcined clay, Agro-Lig [leonardite shale], bituminous soft coal, sodium polypropionate gel, or synthetic calcium silicates) to condition seeds for enhanced germination has been developed only recently (Bennett and Waters, 1984, 1987a, 1987b; Kubik et al., 1988; Taylor, Klein, and Whitlow, 1988; Zuo, Hang, and Zheng, 1988a, 1988b; Harman and Taylor, 1988; Berg et al.,

1989; Harman, Taylor, and Stasz, 1989; Khan, 1992; Khan et al., 1990, 1992; Khan, Abawi, and Maguire, 1992; Parera and Cantliffe, 1990, 1991, 1992). An extensive review of matric priming is available (Khan, 1992). Reduced free energy of water with matric priming (also termed matric conditioning, matriconditioning) is derived from the adsorptive, interfacial tension, attraction, and adhesion between matrix, matrix-air, and matrix-water interfaces (Hadas, 1981). The matric potential (Ψ_m) component of the carrier matrix depends upon the carrier texture, structure, and water content.

Seedling emergence from matrically primed seeds was superior to that from osmotically primed seeds (Kubik et al., 1988; Khan et al., 1990). Seeds of six vegetable species matrically primed in synthetic silicate had greater percentage and rate of emergence and greater shoot fresh weights than those primed in PEG, although drying the matrically primed seeds lessened the difference (Khan et al., 1990). Kubik et al. (1988) compared the emergence of pepper and tomato seeds primed in calcined clay to that of commercial osmotically primed seeds at 15/10°C and 25/20°C. Emergence synchrony was greater from the matrically primed seeds of both species at both temperatures. Matric priming gave greater emergence percentage of pepper and faster emergence of tomato than osmotic priming at 15/10°C.

Seedling emergence in early field plantings from matric-primed carrot seeds was earlier and more rapid than from untreated seeds (Khan et al., 1992). These researchers further noted that percentage seedling emergence as a result of matric priming was improved by matric priming, but to a greater extent in low-vigor (four-year-old) than in high-vigor (one-year-old) seedlots suggesting that matric priming might influence metabolic repair processes related to seed aging and deterioration. In this same study, seedling establishment and stand density from tomato and pepper seeds were improved by matric priming.

Priming Conditions

According to Khan (1992) and Khan et al. (1990; 1992), solids used for matric priming should have: (1) a proportionately high matric to osmotic component; (2) negligible water solubility; (3) low chemical reactivity; (4) high water-holding capacity; (5) variable particle size, structure, and porosity; (6) high surface area; (7) high bulk value and low bulk density; and (8) the ability to adhere to seed surfaces.

Based on moisture characteristic curves (matric potential versus water content) of a synthetic silicate (Micro-Cel E™) produced by hydrothermal reaction of diatomaceous silica, hydrated lime, and water, or of expanded fine-grade vermiculite (hydrated aluminum silicate), a small decrease in the water content (by evaporation or seed uptake) would not greatly influence the matric potential of the carrier or the moisture equilibrium between the seed and carrier during long periods of priming (Khan et al., 1992). For matrices with low water-holding capacity, a small decrease in water content would greatly alter its Ψ_m so that large volumes of matrix would be required to meet the seed water requirement. For matrices of low bulk density, the matrix to seed weight ratio can be small to facilitate treatment, handling, and transport of large amounts of seeds. Matrices with large surface area are efficient carriers of nutrients, pesticides, and growth regulators, and permit ready exchange and diffusion of toxic substances and inhibitors present in some seeds (Khan, 1992).

For each species, the weight ratio of seed to matrix to water must be determined empirically. After priming, the Ψ_m of the matrix can be determined from the matric water content, providing the moisture characteristic of the matrix has been determined. During the matric priming described by Khan (1992), Khan et al. (1990, 1992), and Khan, Abawi, and Maguire (1992), the containers with the matrix were loosely capped jars so that seed moisture content remained relatively constant throughout the priming period. During the matric priming (termed Solid Matrix Priming [SMP™], Kamterter Products, Inc., Lincoln, Neb), as described by Kubik et al. (1988) and Parera and Cantliffe (1991, 1992), the matrix (calcined clay) and seeds were placed in well-ventilated containers which were rotated continuously. Seeds reached maximum water content within one day and thereafter declined to their original contents after ten days (Kubik et al., 1988). Requirements of temperature, irradiance, and duration of matric priming were found to be similar to those for osmotic priming (Khan, 1980/81).

One way to avoid contamination of the seeds with matrix following matric priming was to enclose the seed in a semi-permeable membrane which then was submerged in the moist matrix (Khan and Maguire, 1990). Vermiculite used for matric priming of wildflower seeds also may serve as a bulking agent to facilitate broadcast sowing of the seeds, the similar bulk density of the seeds and matrix minimizing their separation during seed sowing.

Additives to the Matrix

Matric priming in the presence of ethephon has prevented thermoinhibition of lettuce and celery seeds at supraoptimal temperatures (Khan et al., 1990). At 37/30°C (day/night), percentage emergence values for lettuce were: matric primed + ethephon, 94; matric primed, 84; osmotic primed + ethephon, 71; osmotic priming, 24; and untreated, 0. The high percentage emergence values were retained for at least 120 days following matric priming with ethephon solution. This same study (Khan et al., 1990) revealed that the emergence rate and percentage of *Primula* and *Impatiens* seeds were increased by matric priming, the addition of 10 mM ethephon + 0.05 mM GA_3 further speeding emergence of *Primula* and increasing the shoot fresh weight of *Impatiens*.

A combined treatment of matric priming with fungicide mixtures (metalaxyl + tolclofos-methyl or metalaxyl + thiram) reduced the early emergence time and increased the marketable root yield of beet root (*Beta vulgaris* L.) over that with matric priming or fungicide treatments alone in a field plot artificially infested with *Rhizoctonia solani* Khun and *Pythium ultimum* Trow (Khan, Abawi, and Maguire, 1992). Root disease incidence was reduced maximally when priming was combined with metalaxyl + tolclofos-methyl, to a lesser extent by fungicides alone, and none by priming alone. Air-drying of the primed seeds was essential for obtaining maximum stand.

Matric priming of snap beans in synthetic calcium silicate or expanded vermiculite (two days at 15°C) increased the rate and percentage of seedling establishment in the greenhouse (Khan et al., 1990). However, the presence of the same volume of 0.001 mM GA_3 rather than water further increased the speed of seedling emergence and seedling shoot fresh weights. In later work (Khan et al.,

1992), earliness and speed of snap bean emergence in the field was enhanced by matric priming, but percentage emergence was depressed. Further increases in earliness and speed of emergence and restoration of percentage emergence to the control level were achieved when 0.001 mM GA_3 was added to the matrix. Adding fungicides (with or without matric priming) gave no additional emergence of improvement and adversely affected emergence in a few cases. Soybean seeds responded similarly, in that the combined effect of matric priming and GA_3 was greater than that of matric priming alone.

Solid matric priming thus may be more suitable than osmotic priming for preplant low water potential treatments of seeds susceptible to imbibitional and chilling injury. Further evidence of this was reported by Parera and Cantliffe (1991, 1992), who matrically primed sweet corn seeds carrying the mutant *shrunken-2* gene which typically have low seed vigor. The lower imbibitional rates and lower leachate electrical conductivity of seeds primed in calcined clay moistened with 0.5 percent sodium hypochlorite than of nonprimed seeds was associated with increased germination under stressful (cold soil test) conditions (Parera and Cantliffe, 1991). In a field study, Parera and Cantliffe (1992) noted that rate and percentage of emergence and seedling growth from matrically primed seeds treated with sodium hypochlorite were much greater than from untreated seeds but not different from seeds treated with a imazalil + captan + apron + thiram fungicide combination. Sodium hypochlorite treatment or matric priming alone was not effective in cold germination tests or in field experiments.

Solid matrices have characteristics that make them ideal carriers for liquid inoculants and for multiplying beneficial microorganisms (bacteria and fungi) around the seed and delivering them to target areas in the rhizosphere for the control of destructive soil-borne insects and pathogens during seedling establishment. Although an attempt to integrate soybean matric priming in calcined clay with *Bradyrhizobium japonicum* did not improve nodulation in a field trial (Berg et al., 1989), biopriming is only in its infancy. For instance, Callan, Mathre, and Miller (1990) were successful in protecting sweet corn seed from preemergence damping-off caused by *Pythium ultimum* by biopriming the seeds on germination blotters in the presence of a naturally occurring strain of *Pseudomonas fluorescens*.

Mechanisms of Priming

Presumably, ultrastructural, physiological, and biochemical events occurring during osmotic and matric priming are similar. These events include weakening of radicle restraining tissues, embryo development, and increases in macromolecule synthesis, activities of several enzymes, and in metabolic rate (Khan, 1992). Together, these events participate in the mobilization of seed reserves and improve the capacity for rapid cell differentiation and growth.

Heydecker, Higgins, and Turner (1975) suggested that osmotic priming synchronized the development of seeds, bringing then all to the same metabolic state. In carrot or celery seeds with rudimentary embryos, considerable embryo growth may occur during priming, even though radicle growth is prevented (Weibe and Tiessen, 1979; Van der Toorn, 1989). Welbaum and Bradford (1991) noted that less mature muskmelon seeds (harvested 40 days after anthesis) benefitted more by KNO_3-priming than more mature seeds (harvested 60 days after anthesis), consistent with the hypothesis that priming advances developmental processes. They concluded that the extended hydration phase of development of the seeds in fruit (which has a similar osmotic potential as the priming solution) may constitute *in situ* priming. Repair processes due to priming in these instances presumably were not involved since seed deterioration would occur only after seeds were mature. Since changes in gene expression are required in the early phases of imbibition to switch seeds from the developmental to the germinative mode (Kermode et al., 1986), the completion of this "switching" during priming could hasten growth initiation when the seeds are reimbibed.

In many species, the radicle is surrounded by an endosperm or testa which may act as a barrier to germination. A role of gibberellin in promoting germination of tomato and pepper seeds appeared to be the weakening of this restraining tissue (Watkins and Cantliffe, 1983; Groot and Karssen, 1987). During extended PEG-priming, however, no anatomical changes in lettuce endosperm surrounding the radicle tip which accompanies germination were observed (Georghiou, Psaras, and Mitrakos, 1983). Haigh (1988) observed that although the break-strength of isolated endosperm tissues decreased during priming of tomato seeds, the remaining resistance exceeded the force that could be generated by the embryo. He proposed that the endosperm must undergo a further weakening coincident with radicle growth for germination to occur.

There are very few reports on seed water content (Hegarty, 1978; Brocklehurst and Dearman, 1984), despite the importance of controlling seed water potential and seed water content during priming. Bradford (1986) reviewed the seed water relations of osmotic priming. During imbibition in water, seed water content reaches a plateau, and then changes very little until just before radicle emergence. When exposed to a reduced external water potential (Ψ_o), the water content plateau falls and germination is delayed. At this plateau, there is no net movement of water, so that Ψ_o is equal to the water potential of the cell (Ψ_c), which is the sum of cell solute (osmotic) potential (Ψ_s) and cell turgor potential (Ψ_p). During seed priming, Ψ_o is either set sufficiently low to prevent radicle expansion, or the priming duration at higher Ψ_o is shortened to fall within the plateau range. Several researchers (Hegarty, 1977; Bradford, 1986) showed that at high Ψ_o values which eventually led to germination, the primary effect of reduced water availability was increased duration of the lag phase between imbibition and growth. Regardless of Ψ_o, seed water content increased to a threshold level before germination occurred. At high Ψ_o values seed water content increased gradually, suggesting solute generation during the lag phase. Varying lag periods with different Ψ_o values may reflect the time required for solute accumulation to lower cell Ψ_s so that water influx leads to the threshold water content. Bradford (1986) employed a modification of the Lockhart (1965) model to relate germination to water potential:

$$dV/dt = \frac{mL\ \Delta\Psi + \Psi_p - \Psi_{p,th}}{m + L}$$

where dV/dt is the rate of volume increase, m is the plastic extensibility of the cell walls, L is the hydraulic conductance of the tissue, $\Delta\Psi$ is the water potential gradient between the external liquid and the tissue (cells that will undergo expansion, $\Psi_o - \Psi_c$), and $\Psi_{p,th}$ is a minimum threshold turgor that must be exceeded for growth to occur. At the water content plateau, Ψ_c is close to equilibrium with Ψ_o so that there is little net movement of water and L will be very high relative to m. Thus, the equation can be simplified to:

$$dV/dt = m(\Delta\Psi + \Psi_p - \Psi_{p,th}).$$

By substituting $\Psi_o + \Psi_c$ for $\Delta\Psi$, and in turn substituting $\Psi_p + \Psi_s$ for Ψ_c (assuming negligible cellular Ψ_m), the equation can be expressed as:

$$dV/dt = m(\Psi_o - \Psi_s - \Psi_{p,th})$$

Thus, Ψ_s establishes the gradient for water uptake into the cell, with the growth potential of the seed proportional to $\Psi_p - \Psi_{p,th}$. $\Psi_{p,th}$ is not only a characteristic of the radicle cell walls but also of enclosing tissue which must be weakened (lowering of $\Psi_{p,th}$) before radicle growth can occur. Based on this derivation, Bradford (1986) hypothesized that solutes accumulate during osmotic priming, but that Ψ_o is sufficiently low that Ψ_p never exceeds $\Psi_{p,th}$. If accumulated solutes remain during drying of the primed seeds, then a high Ψ_p would be quickly attained during re-imbibition.

Bradford (1986) reported that osmotic adjustment does occur, but that it required priming for a much longer time than was necessary to advance germination. Others have failed to detect any lowering of embryo osmotic potential prior to radicle emergence (Haigh and Barlow, 1987b; Haigh, 1988). Ions absorbed during priming contribute to osmotic accumulation and thus may be responsible for the faster germination noted with salt-primed compared to PEG-primed seeds (Alvarado, Bradford, and Hewitt, 1987; Frett, Pill, and Morneau, 1991). Bradford et al. (1988) concluded that osmotic adjustment (and increased turgor) was not the primary mechanism of the priming effect. This conclusion conforms to the response of seedling roots to osmotic stress in that as Ψ_o was lowered rapidly, root cell turgor also fell but osmotic adjustment to restore turgor was very slow compared to the rapid restoration of the growth rate (Hsaio and Jing, 1987; Itoh et al., 1987). The maintenance of cell growth despite decreases in turgor implicates changes in cell wall properties in regulating root growth. Lack of correlation between cell growth and turgor has been documented elsewhere (Nonami and Boyer, 1989).

Recent work by Finch-Savage and McQuistan (1991) has lent further credence to the concept that water relations are not of primary importance in seed priming. They found that tomato seeds soaked in 10^{-4}M ABA (15 days at 15°C) and then dried back to original moisture contents had increased rate, synchrony, and percentage of emergence. The enhancement of seedling emergence by ABA was similar to that achieved by PEG-priming for the same duration and at the same temperature, and yet seed relative water contents were very different (ABA 86 percent, PEG 60 percent). These workers suggested that it may be unnecessary for seeds to encounter an osmotic stress in order to be primed, rather the progress of germination needs only to be blocked during the lag phase. The nature of the block would appear to be unimportant provided it is truly reversible. ABA may block germination by preventing radicle cell wall "loosening" and thereby cell expansion (Schopfer and Plachy, 1984), or by antagonizing the weakening of the endosperm restraint (Karssen and Groot, 1987).

Respiratory activity is a triphasic process related to the degree of hydration of the seed (Bewley and Black, 1978). Oxygen consumption increases rapidly during the log phase and is associated with activation and hydration of mitochondrial enzymes. Phase II is characterized by a lag in O_2 uptake, while during phase III when the seed coat ruptures there is a secondary respiratory burst. This latter rise in O_2 consumption is associated with increased activity of enzymes synthesized in Phase II. A positive correlation between tomato seed O_2 consumption and primability (change in rate of germination due to priming, relative to the rate of germination of the control) was strongest during phase III (Suena, 1990). These results suggest that seeds with a greater capacity to prime have the potential to reestablish structural integrity and synthesize new compounds more rapidly during the early stages of imbibition.

Accelerated germination of maize seeds following salt-priming was associated with increased amounts of all embryo phospholipid fractions and sterols (Basra, Bedi, and Malik, 1988). The proportion of phospholipid which was diphosphotidylglycerol (DPG) also increased. These researchers suggested that the marked increase in DPG content of primed embryos may be due to enhanced internal organization of their mitochondrial membranes, and that the benefit of priming may be due at least partly to an increased potential for oxidative phosphorylation and ATP accumulation. This conclusion is supported by the report that increasing amounts of ATP accumulated during osmotic priming of seeds (Mazor, Perl, and Negbi, 1984).

Respiratory rates of water-imbibed or salt-primed pepper seeds were the same during the log phase of water uptake, but respiration during the lag phase was extended in the primed seeds, the extension being inversely proportional to the osmotic potential of the priming solution (Smith and Cobb, 1991b). Mazor, Perl, and Negbi (1984) likewise noted that ATP accumulation in PEG-primed seeds followed a pattern similar to that in seeds imbibed in water, however ATP accumulation reached a maximum after four to five hours in water and after about 24 hours in PEG. A decline of at least 80 percent of the accumulated ATP occurred during 18 days of post-priming drying. Zuo, Hang, and Zheng (1988b) noted increased respiratory activities in primed than in water-imbibed seeds of four vegetable species, however details regarding the period of measurement with respect to hydration were not mentioned. Sundstrom and Edwards (1989) noted that salt-primed seeds of jalapeno pepper had a slightly higher respiration rate than water-control seeds only during the first hour of hydration but were thereafter the same.

Bray et al. (1989) noted in low- and high-vigor seedlots of leek with similar viability that PEG-priming abolished differences between the lots in both germination performance and rates of protein biosynthesis. Lower levels of DNA synthesis were detected in leek embryos during the priming period in the absence of any cell division but were followed by a five-fold increase upon germination following priming. During germination, the rate of DNA synthesis in embryos of primed seeds was significantly greater than that found at any time over the first four days of germination in embryos of unprimed seeds. These results indicated that the benefits in germination performance of primed seeds was accompanied by marked increases in protein, DNA, and nucleotide biosynthesis after a lag period of six to 12 hours following priming. Khan (1980/81) similarly showed that little DNA synthesis occurred during priming but that large increases occurred after germination. Cordycepin, an inhibitor of RNA synthesis, strongly inhibited RNA synthesis but had no effect on germination when present either during osmotic priming, germination, or both (Khan and Karssen, 1981). Thus, it appears that RNA synthesized during priming or at the time of radicle protrusion does not participate in germination advancement.

Protein synthesis increased in primed seeds of lettuce (Khan et al., 1978), leek (Bray et al., 1989), and pea, tomato, and spinach (Zuo, Hang, and Zheng, 1988b). Nearly twice as much amino acid was incorporated into protein during the first 24 hours in PEG-primed versus water-imbibed pepper seeds (Mazor, Perl, and Negbi, 1984). While the uptake of labeled amino acids increased slowly in primed pepper seeds, it increased rapidly in water-imbibed seeds (Smith and Cobb, 1991b). This is in contrast to the similar rates of protein synthesis in primed and water-imbibed seeds observed by others (Mazor, Perl, and Negbi, 1984; Dell-Aquilla and Bewley, 1989). The results of Smith and Cobb (1991b) indicated that protein synthesis occurred at a reduced rate or over an expanded period in primed seeds compared to water-imbibed seeds.

Activities of alcohol dehydrogenase, glucose-6-phosphatase, aldolase, and isocitrate lyase decreased in water-imbibed seeds during radicle emergence but remained high or were increased during seed priming (Smith and Cobb, 1991b). Increased activity of several acid phosphatases and esterases were found in primed lettuce seeds as a result of activation or *de novo* synthesis, activities which were increased further by GA_3 addition to the priming solution (Khan et al., 1978). The presence of growth regulators in the priming solution has caused the appearance of new isoforms of acid phosphatase as well as other isozymes (Khan et al., 1978). Priming increased the amylase activity in pea, tomato, and spinach, and peroxidase activity in tomato and spinach seeds (Zuo, Hang, and Zheng, 1988b).

In summary, priming operates to enhance seed quality by a combination of processes that may include cellular repair and improved membrane integrity; decreased seed exudation and concomitant decreased growth of pathogenic organisms; enhanced mobilization of seed protein, lipid, and starch as a result of activation or synthesis of key enzymes; osmotic adjustment and increase in radicle turgor; advanced embryo development; weakened restraining tissues around the radicle; and increased potential for oxidative phosphorylation and ATP accumulation.

PREPLANT GERMINATION AND FLUID DRILLING

The rationale behind preplant germination (pregermination, chitting) is to lessen the effects of the seedbed environment on this phase of crop establishment so that seedling emergence can be more rapid, more synchronous, and reach higher levels than occur with conventional dry-seed sowing. Faster seedling emergence should lessen the likelihood of soil crust (cap) development or pathogen attack before the seedlings emerge. Fluid drilling (fluid sowing, gel seeding) is a technique in which germinated seeds are delivered to the seedbed in a protective carrier gel. Recently, however, it has been found that germinated seeds of certain species can withstand extensive desiccation with little loss in viability. Such seeds could be planted like natural, dry seeds in conventional machinery or they could be suspended in gel and fluid drilled.

Seed selection on the basis of newly emerged radicles can be used to improve both the viability and vigor of a seedlot since germination rate and subsequent performance are positively correlated (Finch-Savage, 1986; Finch-Savage and McQuistan, 1988a). The viability and vigor of newly germinated cabbage seed can be maintained down to about 14 percent seed moisture (Finch-Savage and McKee, 1988). The optimum drying regime for these low-moisture-content-germinated (LMCG) seeds was 20 to 30°C, 80 percent relative humidity flowing through the seeds at 0.25 meters per second. A preliminary surface drying was needed to avoid seed agglutination during the main drying cycle. Results of preliminary tests using a vacuum dry-seeding unit onto a moist substrate showed no appreciable damage to the LMCG seeds since percentage of abnormal seedlings was negligible. The LMCG seeds imbibed water rapidly to become fully hydrated in 40 to 50 minutes and resumed growth almost immediately. The moisture content of LMCG seeds prevented continued embryo axis growth and the seeds tolerated low temperatures for storage. For example, LMCG cabbage seeds were stored at –20°C for seven days without loss of viability, indicating their potential for prolonged storage at this temperature (Finch-Savage and McQuistan, 1988b).

Germinating rape seeds selected on the basis of newly emerged radicles were dried to an equilibrium moisture content (11 percent) in air at 20°C and 80 percent relative humidity without loss of viability (Finch-Savage and McKee, 1989). Storage life of these LMCG seeds at 15°C was limited to seven days before viability was reduced. However, viability of LMCG rape seeds was maintained for 84 days in storage at –20°C. Longer periods in storage reduced viability, but 96 percent of seeds still remained viable after 336 days at –20°C. Storage under reduced pressure or in nitrogen atmosphere had little effect on seed longevity. Reduction of seed moisture below 11 percent using vacuum drying reduced seed vigor.

In greenhouse and field trials, cabbage and cauliflower seedling emergence was greater, earlier, and more uniform from LMCG seeds than that from untreated seeds (Finch-Savage and McKee, 1990). Three seedlots of different quality of the same cultivar for each species were used. LMCG seed effectively eliminated the differences due to seed quality shown by untreated seeds at emergence suggesting an improvement in seed vigor as well as viability from the LMCG treatment.

Germinated spinach seeds with emerged radicles 3 to 12 mm long exposed to severe dehydration, with no attempt made to control humidity during drying, were able to remain viable and emerge upon rehydration when sown into moistened media (Dainello, 1989). This discovery extends the possibility of LMCG seed beyond the *Brassica* genus. Dainello found that dehydration of germinated seed for at least 15 days was required before seedling emergence was reduced by 50 percent of that produced by undehydrated germinated seed.

Fluid drilling (fluid sowing or gel seeding) is the sowing of seeds which have already been germinated, using a gel to suspend and transfer them to the seedbed. Comprehensive reviews of the historical background, technological development, and benefits of fluid drilling have been presented elsewhere (Salter, 1978; Gray, 1981, 1984; Pill, 1991).

Researchers at Horticulture Research International, Wellesbourne, U.K., have developed techniques for germinating vegetable seeds in bulk and then cold-storing them. They also developed prototype fluid drilling planters, and examined over 50 possible carrier gels. This work established fluid drilling as an integrated system involving (1) germination of seeds prior to sowing; (2) separation of the germinated from the nongerminated seeds; (3) storage of the germinated seeds; (4) preparation of the gel for suspending the seeds; and (5) drilling of the germinated seeds (Gray, 1981).

Fluid drilling can give (1) earlier, greater, and more uniform seedling emergence; (2) earlier and greater yields; and (3) more uniform maturity in some crops than conventional methods of sowing dry seeds, as summarized by Gray (1984). Some yield increases for fluid drilled compared to dry-seeded crops include carrot (22 percent), celery (36 percent), parsley (107 percent), and tomato (12 percent) (Gray, 1984).

Crop response to fluid drilling can be variable, in large part due to the low proportion of germinated seeds in a seed batch at time of sowing (Finch-Savage, 1987). In fact, seedling emergence from a fluid-drilled crop can extend over a longer period than from one that is dry-seeded when only a small proportion of the seeds are germinated at the time of sowing, and when fluid drilling increases the percentage emergence of a seedlot. The proportion of germinated seeds at the time of fluid drilling was increased by subjecting imbibed seeds to a cold treatment (1°C for six to 12 days) before transfer to 20°C for germination (Finch-Savage and Cox, 1982a). Separating germinated seeds from nongerminated seeds using a flotation technique based on density differences (Taylor and Kenny, 1985) likewise can increase the proportion of germinated seeds at the time of fluid drilling. Seed leachate removal and inclusion of phthalimide growth regulator in the priming osmoticum increased the percentage of germinated carrot seeds to more than 90 percent at the time of fluid drilling (Pill and Finch-Savage, 1988). Synchronization of carrot seed germination also can be achieved by soaking them in abscisic acid solution (Finch-Savage, 1989). Treating seeds with 10^{-4}M ABA solution at 15°C for 12 days gave 93 percent gemination of viable seeds on subsequent transfer to water before radicle lengths became too long for fluid drilling. This compared with only 31 percent without pretreatment (Finch-Savage and McQuistan, 1989).

For very small seeds such as petunia (Pill and Mucha, 1984) or tobacco (Csinos and Ghate, 1982), fluid drilling of pregerminated seeds increased the rate and percentage of seedling emergence. At least part of this beneficial response may be attributed to the gel keeping the seeds near the growth medium surface, preventing some of them from settling deeply as can occur with conventional dry-seed sowing.

It must be possible to store germinated seeds for extended periods with minimal radicle extension and without loss of viability if adverse weather or equipment failure prevents fluid drilling on a proposed date. Germinated seeds can be stored in cold aerated water or in cold, humid air (Finch-Savage, 1981), in plastic bags packaged either in vacuum or nitrogen at 7°C (Ghate and Chinnan, 1987), or in cold hydroxyethyl cellulose fluid drilling gels (Pill and Fieldhouse, 1982).

Germinated seeds react strongly with the seedbed environment (Finch-Savage, 1987; Finch-Savage and Pill, 1990). Seedbed conditions for germinated or germinating seeds that are fluid-drilled must be conducive to continued seed growth since most germinated seeds become desiccation intolerant (Finch-Savage, 1987). Germinated seeds are usually transferred immediately from the ideal environment of the seed germinator to the more adverse conditions of the seedbed. For example, fluid drilling germinated seeds gave inferior stands of fall-harvest broccoli and cauliflower crops (Kahn and Motes, 1988, 1989), compared to conventional dry-sown seeds when air temperature at and following planting exceeded 30°C. The pregerminated seeds likely succumbed to heat and desiccation. Techniques to improve the hardiness of pregerminated seeds would be useful, particularly under harsh seedbed conditions.

One successful approach to modification of the seedbed and fluid drilling technique is that of "Gel Mix" in planting bell peppers in Florida (Schultheis et al., 1988a, 1988b). Pregerminated seeds were mixed in peat-lite media containing hydrophilic polymer. This gel mix was then planted at 60 ml per hill. The advantages of fluid drilling were enhanced by the peat-lite which provided not only a high-moisture environment but anticrustant properties.

Fluid drilling of osmotically primed, nongerminated seeds has been effective, thereby avoiding the need to pregerminate seeds and assure a high proportion of germinated seeds at time of fluid drilling. Although asparagus seedling emergence rate and synchrony were significantly increased by fluid drilling germinated seeds rather than osmotically primed seeds, this benefit was insufficiently large to warrant seed pregermination (Evans and Pill, 1989). Seedling emergence from osmotically primed parsley seeds was more rapid when they were fluid drilled than when they were dry-sown (Pill, 1986). Thus, priming and fluid drilling may be considered as complementary processes.

Hydrating the primed seeds before fluid drilling (prehydration) may be accomplished by incorporating them in the gel for extended periods before sowing. However, the moisture content of the seed at drilling will be critical because when radicle extension begins at the end of the lag phase of water uptake, water is again rapidly taken up for growth. As the radicle extends, the seeds progressively lose desiccation tolerance (Finch-Savage, 1990). Thus, as the seed moisture content increases at the time of fluid drilling so does the importance of adequacy of soil moisture. Primed seeds of relatively low moisture content, when fluid drilled into a dry seedbed, may be able to survive until conditions become amenable for germination and seedling emergence, whereas pregerminated seeds might perish.

Gels generally fall into one of five chemical classes: synthetic mineral clays, starch-polyacrylonitrile polymers, cellulose polymers, polyacrylamide polymers, and copolymers of potassium acrylate and acrylamide (Orzolek, 1987), although natural gels such as potato or manioc starches can be used effectively (Pill and Rojas, 1987). Some of the essential characteristics of a desirable gel carrier were reported by Darby (1980): the gel should suspend seeds of various sizes for at least 24 hours, and yet be easily pumped through delivery tubing; it should be nonphytotoxic and should be easily mixed with water of a different pH, mineral content, and hardness; it should be relatively inexpensive; and it should not dry to form a skin, but should break down readily in the soil.

While the gels are usually 1 to 3 percent solids at the time of fluid drilling, the water supplied with the gel at a normal extrusion rate (20 to 30 ml per meter of row) is inadequate to sustain seedling growth in a dry seedbed (Pill and Rojas, 1987). The protective gel, in addition to carrying seeds to the seedbed, can carry additives such as fertilizer salts, plant growth regulators, pesticides, and microorganisms, thereby creating a packaged environment for the seed and seedling (Salter, 1978). However, addition of the beneficial additives must not adversely affect the gel rheology, and in turn the gel must not counter the efficacy of the additive. Since seedlings can interact only with the small volume of soil occupied by their roots, inclusion of additives in the gel represents an efficient, low-input option with obvious cost and environmental benefits.

Small quantities of phosphorus in gels have increased seedling weight of fluid-drilled carrot (Finch-Savage and Cox, 1982b), lettuce and onion (Finch-Savage and Cox, 1983), tomato (Espinosa and Pill, 1987) and collards (Pill, 1990), even in soils that have received traditional rates of N-P-K fertilizer. Incorporation of commercially available biostimulants Agro-Lig, Enersol (humic acids), and Ergostim (folic acid) in magnesium silicate gel increased carrot seedling vigor and percentage emergence compared to fluid drilled seeds without stimulants (Sanders, Ricotta, and Hodges, 1990). Plant growth regulators mixed in fluid drilling gels have been evaluated on tomato seedling emergence and growth (Pyzik and Orzolek, 1986). Fungicide incorporation into fluid drilling gels effectively controlled damping-off caused by *Pythium aphanidermatum* (Giammichele and Pill, 1984; Ohep, Bryan, and Cantliffe, 1984). Incorporation of the insecticide chlorfenvinphos in fluid drilling gels was effective in protecting carrots against the carrot fly (Thompson, Suett, and Percival, 1982).

Activated charcoal incorporated into gels protected fluid-drilled lettuce from herbicide damage (Taylor and Warholic, 1987). Taylor and Warholic noted that the least crop injury occurred when the charcoal was incorporated in hydroxyethyl cellulose (HEC) gel rather than in magnesium silicate or polyacrylamide gels since HEC had greater mobility in the seedbed. This greater mobility also may reduce the likelihood of water stress due to osmotically active additives (e.g., fertilizer salts) in the gel (Espinosa and Pill, 1987).

Various biological materials can be delivered to the seed bed in the fluid drilling gel. Jawson, Franzleubbers, and Berg (1989) noted that for *Bradyrhizobium japonicum* both adhesive and carrier properties were combined in the fluid drilling gel. Hayman, Morris, and Page (1981) reported that fluid drilling had the advantages over other methods of greatly reducing the amount of vesicular arbuscular mycorrhiza inoculum needed to cause a given level of host infection, and of readily combining seeds and inoculum in a single carrier. Conway (1986) noted that delivery of *Laetisaria arvalis* sclerotia in HEC gels increased plant stands and decreased the incidence of damping-off. He found that another biological control agent (*Trichoderma*) was effective in reducing damage to fluid drilled peas in *Pythium*-infested soil when incorporated in HEC gel.

As a polysaccharide, HEC seems especially suitable as a fluid drilling carrier for biological (microorganism) additives. The properties of N-Gel™, an HEC gel that has been developed specifically for fluid drilling, have been described in detail (Banyai, 1987). In addition to the desirable properties of a fluid drilling gel listed by Darby (1980), HEC seems to possess superior gas exchange since germinated seed viability subsequent to storage at low temperatures for up to 20 days was greater in HEC than in other gels (Pill and Fieldhouse, 1982). Other important characteristics of HEC include its transparency so that the suspended seeds can be seen easily, and its non-ionic nature so that additives are not complexed (deactivated).

REFERENCES

Akers, S. W., and K. E. Holley. SPS: A system for priming seeds using aerated polyethylene glycol or salt solution. *HortScience* 21 (1986): 529-531.

Akers, S. W., G. A. Berkowitz, and J. Rabin. Germination of parsley seed primed in aerated solutions of polyethylene glycol. *HortScience* 22 (1987): 250-252.

Alvarado, A. D., and K. J. Bradford. Priming and storage of tomato (*Lycopersicon lycopersum*) seeds. I. Effects of storage temperature on germination rate and viability. *Seed Sci. & Technol.* 16 (1988): 601-612.

Alvarado, A. D., K. J. Bradford, and J. D. Hewitt. Osmotic priming of tomato seeds. Effects on germination, field emergence, seedling growth, and fruit yield. *J. Amer. Soc. Hort. Sci.* 112 (1987): 427-432.

Argerich, C. A., and K. J. Bradford. The effects of priming and ageing on seed vigour in tomato. *J. Exp. Bot.* 40 (1989): 599-607.

Argerich, C. A., K. J. Bradford, and F. M. Ashton. Influence of seed vigor and preplant herbicides on emergence, growth and yield of tomato. *HortScience* 25 (1990): 288-291.

Argerich, C. A., K. J. Bradford, and A. M. Tarquis. The effects of priming and ageing on resistance to deterioration of tomato seeds. *J. Exp. Bot.* 40 (1989): 593-598.

Banyai, B. E. N-Gel™ polymers for agricultural fluid drilling. *Acta Hortic.* 198 (1987): 111-120.

Barlow, E. W. R., and A. M. Haigh. Effect of seed priming on the emergence, growth and yield of UC82B tomatoes in the field. *Acta Hortic.* 200 (1987): 153-164.

Basra, A. S., S. Bedi, and C. P. Malik. Accelerated germination of maize seeds under chilling stress by osmotic priming and associated changes in embryo phospholipids. *Ann. Bot.* 61 (1988): 635-639.

Benjamin, L. R. The relative importance of some sources or root-weight variation in a carrot crop. *J. Agric. Sci.* 102 (1984): 69-77.

Benjamin, L. R. Variation in time of seedling emergence within populations: A feature that determines individual growth and development. *Adv. Agron.* 44 (1990): 1-25.

Bennett, M. A., and L. Waters, Jr. Influence of seed moisture on lima bean stand establishment and growth. *J. Amer. Soc. Hort. Sci.* 109 (1984): 623-626.

Bennett, M. A., and L. Waters, Jr. Germination and emergence of high sugar sweet corn is improved by presowing hydration of seed. *HortScience* 22 (1987a): 236-238.

Bennett, M. A., and L. Waters, Jr. Seed hydration treatments for improved sweet corn germination and stand establishment. *J. Amer. Soc. Hort. Sci.* 112 (1987b): 45-49.

Berg P. K., Jr., M. D. Jawson, A. J. Franzleubbers, and K. K. Kubik. *Bradyrhizobium japonicum* inoculation and seed priming for fluid-drilled soybean. *Soil Sci. Soc. Amer. J.* 53 (1989): 1712-1717.

Bewley, J. D., and M. Black. *Physiology and Biochemistry of Seeds. I. Development, Germination and Growth.* (New York: Springer-Verlag, 1978).

Bradford, K. J. Seed priming improves germination and emergence of cantaloupe at low temperatures. *HortScience* 20 (1985): 598.

Bradford, K. J. Manipulation of seed water relations via osmotic priming to improve germination under stress conditions. *HortScience* 21 (1986): 1105-1112.

Bradford, K. J., J. J. Steiner, and S. E. Trawatha. Seed priming influence on germination and emergence of pepper seed lots. *Crop. Sci.* 30 (1990): 718-721.

Bradford, K. J., C. A. Argerich, D. Peetambar, O. Somasco, A. Tarquis, and G. E. Welbaum. Seed enhancement and seed vigor. *Proc. Intl. Conf. Stand Estab. Hort. Crops*. Lancaster, PA. 1988. p. 1-35.

Bray, C. M., P. A. Davison, M. Ashraf, and R. M. Taylor. Biochemical changes during osmopriming of leek seeds. *Ann. Bot.* 63 (1989): 185-193.

Brocklehurst, P. A. Factors affecting seed quality of vegetable crops. *Scientific Hort.* 36 (1985): 48-57.

Brocklehurst, P. A., and J. Dearman. Interactions between seed priming treatments and nine seed lots of carrot, celery and onion: I. Laboratory germination. *Ann. Appl. Biol.* 102 (1983a): 577-584.

Brocklehurst, P. A., and J. Dearman. Interactions between seed priming treatments and nine seed lots of carrot, celery and onion: II. Seedling emergence and plant growth. *Ann. Appl. Biol.* 102 (1983b): 585-593.

Brocklehurst, P. A., and J. Dearman. A comparison of different chemicals for osmotic treatment of vegetable seed. *Ann. Appl. Biol.* 105 (1984): 391-398.

Brocklehurst, P. A., J. Dearman, and R. L. K. Drew. Effects of osmotic priming on seed germination and seedling growth in leek. *Scient. Hortic.* 24 (1984): 201-210.

Bujalski, W., A. W. Nienow, and D. Gray. Establishing the large scale osmotic priming of onion seeds by using enriched air. *Ann. Appl. Biol.* 115 (1989): 171-176.

Callan, N. W., D. E. Mathre, and J. B. Miller. Bio-priming seed treatment for control of *Pythium ultimum* preemergence damping-off in *sh-2* sweet corn. *Plant Disease* 74 (1990): 368-372.

Callan, N. W., D. E. Mathre, and J. B. Miller. Field performance of sweet corn bio-primed and coated with *Pseudomonas fluorescens* AB254. *HortScience* 26 (1991): 1163-1165.

Cano, E. A., M. C. Bolarin, F. Perez-Alfocea, and M. Caro. Effects of NaCl priming on increased salt tolerance in tomato. *J. Hort. Sci.* 66 (1991): 621-628.

Cantliffe, D. J. Benzyladenine in the priming solution reduces thermodormancy of lettuce seeds. *HortTechnol.* 1 (1991): 95-97.

Cantliffe, D. J., J. M. Fischer, and T. A. Nell. Mechanisms of seed priming in circumventing thermodormancy in lettuce. *Plant Physiol.* 75 (1984): 290-294.

Cantliffe, D. J., K. D. Schuler, and A. C. Guedes. Overcoming seed dormancy in heat sensitive romaine lettuce by seed priming. *HortScience* 16 (1981): 196-198.

Carpenter, W. J. *Salvia splendens* seed germination and priming for rapid and uniform plant emergence. *J. Amer. Soc. Hort. Sci.* 114 (1989): 247-250.

Carpenter, W. J. Priming dusty miller seeds: Role of aeration, temperature and relative humidity. *HortScience* 25 (1990): 299-302.

Carpenter, W. J., and J. F. Boucher. Priming improves high-temperature germination of pansy seed. *HortScience* 26 (1991): 541-544.

Conway, K. E. Use of fluid-drilling gels to deliver biological control agents to soil. *Plant Disease* 70 (1986): 835-839.

Csinos, A. S., and S. R. Ghate. Fluid sowing of pregerminated tobacco seed. *Tob. Sci.* XXVI (1982): 32-34.

Dainello, F. J. Radicle dehydration of germinated seed on seedling emergence and vigor in spinach. *HortScience* 24 (1989): 935-937.

Danneberger, T. K., M. B. McDonald, Jr., C. A. Geron, and P. Kumari. Rate of germination and seedling growth of perennial ryegrass seed following osmotic conditioning. *HortScience* 27 (1992): 28-30.

Darby, R. J. Effects of seed carriers on seedling establishment after fluid drilling. *Exp. Agric.* 16 (1980): 153-160.

Dearman, J., P. A. Brocklehurst, and R. L. K. Drew. Effects of osmotic priming and ageing on onion seed germination. *Ann. Appl. Biol.* 108 (1986): 639-648.

Dearman, J., P. A. Brocklehurst, and R. L. K. Drew. Effects of osmotic priming and ageing on the germination and emergence of carrot and leek seed. *Ann. Appl. Biol.* 111 (1987): 717-722.

Dell-Aquilla, A., and J. D. Bewley. Protein synthesis in the axes of polyethylene glycol-treated pea seed and during subsequent germination. *J. Exp. Bot.* 40 (1989): 1001-1007.

Delouche, J. C., and C.C. Baskin. Accelerated ageing techniques for predicting the relative storability of seed lots. *Seed Sci. & Technol.* 1 (1973): 427-452.

Durrant, M. J., P. A. Payne, and J. M. Maclaren. The use of water and some inorganic salt solutions to advance sugar beet seed: II. Experiments under controlled and field conditions. *Ann. Appl. Biol.* 103 (1983): 517-526.

Espinosa, W. A., and W. G. Pill. Response of tomato seeds fluid-drilled in low-phosphorus growth media to phosphorus incorporation in the carrier gel. *Scient. Hortic.* 33 (1987): 37-47.

Evans, T. A., and W. G. Pill. Emergence and seedling growth from osmotically primed or pregerminated seeds of asparagus (*Asparagus officinalis* L.). *J. Hort. Sci.* 64 (1989): 275-282.

Finch-Savage, W. E. Effects of cold-storage of germinated vegetable seeds prior to fluid drilling on emergence and yield of field crops. *Ann. Appl. Biol.* 97 (1981): 345-352.

Finch-Savage, W. E. A study of the relationship between seedling characters and rate of germination within a seedlot. *Ann. Appl. Biol.* 108 (1986): 441-444.

Finch-Savage, W. E. Some effects of seed bed conditions on seedling establishment from fluid-drilled pre-germinated seeds. *Acta Hortic.* 198 (1987): 277-286.

Finch-Savage, W. E. The use of abscisic acid to synchronize carrot seed germination prior to fluid drilling. *Ann. Appl. Biol.* 102 (1989): 213-217.

Finch-Savage, W. E. The effects of osmotic seed priming and the timing of water availability in the seed bed on the predictability of carrot seedling establishment in the field. *Acta Hortic.* 267 (1990): 209-216.

Finch-Savage, W. E. Development of bulk priming/plant growth regulator seed treatments and their effects on the seedling establishment of four bedding plant species. *Seed Sci. & Technol.* 19 (1991): 477-485.

Finch-Savage, W. E., and C. R. Cox. A cold-treatment technique to improve the germination of vegetable seeds prior to fluid drilling. *Scient. Hortic.* 16 (1982a): 301-311.

Finch-Savage, W. E., and C. R. Cox. Effects of adding plant nutrients to the gel carrier for fluid-drilling early carrots. *J. Agric. Sci.* 99 (1982b): 295-303.

Finch-Savage, W. E., and C. R. Cox. Effects of adding plant nutrients to the gel carrier used for fluid drilling lettuce and onion seeds. *Ann. Appl. Biol.* 102 (1983): 213-217.

Finch-Savage, W. E., and J. M. T. McKee. A study of the optimum drying conditions for cabbage seed following selection on the basis of a newly-emerged radicle. *Ann. Appl. Biol.* 113 (1988): 415-424.

Finch-Savage, W. E., and J. M. T. McKee. Viability of rape (*Brassicus napus* L.) seeds following selection of newly-emerged radicles then subsequent drying and storage. *Ann. Appl. Biol.* 114 (1989): 587-595.

Finch-Savage, W. E., and J. M. T. McKee. The influence of seed quality and pregermination treatment on cauliflower and cabbage transplant production and field growth. *Ann. Appl. Biol.* 116 (1990): 365-369.

Finch-Savage, W. E., and C. I. McQuistan. Performance of carrot seeds possessing different germination rates within a seed lot. *J. Agric. Sci.* 110 (1988a): 93-99.

Finch-Savage, W. E., and C. I. McQuistan. The potential for newly-germinated cabbage seed survival and storage at sub-zero temperatures. *Ann. Bot.* 62 (1988b): 509-512.

Finch-Savage, W. E., and C. I. McQuistan. The use of abscisic acid to synchronize carrot seed germination prior to fluid drilling. *Ann. Bot.* 63 (1989): 195-199.

Finch-Savage, W. E., and C. I. McQuistan. Abscisic acid: An agent to advance and synchronise germination for tomato (*Lycopersicon esculentum* Mill.) seeds. *Seed Sci. & Technol.* 19 (1991): 537-544.

Finch-Savage, W. E., and W. G. Pill. Improvement of carrot crop establishment by combining seed treatments with increased seed-bed moisture availability. *J. Agric. Sci.* 115 (1990): 75-81.

Finch-Savage, W. E., D. Gray, and G. M. Dickson. The combined effects of osmotic priming with plant growth regulator and fungicide soaks on the seed quality of five bedding plant species. *Seed Sci. & Technol.* 19 (1991): 495-503.

Finnerty, T. L., J. M. Zajicek, and M. A. Hussey. Use of seed priming to bypass stratification requirements of three *Aquilegia* species. *HortScience* 27 (1992): 310-313.

Frett, J. J., and W. G. Pill. Germination characteristics of osmotically primed and stored *Impatiens* seeds. *Scient. Hortic.* 40 (1989): 171-179.

Frett, J. J., W. G. Pill, and D. C. Morneau. A comparison of priming agents for tomato and asparagus seeds. *HortScience* 26 (1991): 1158-1159.

Georghiou, K., G. Psaras, and K. Mitrakos. Lettuce endosperm structural changes during germination at high temperature. *Bot. Gaz.* 144 (1983): 207-211.

Georghiou, K., C. A. Thanos, and H. C. Passam. Osmoconditioning as a means of counteracting the ageing of pepper seeds during high-temperature storage. *Ann. Bot.* 60 (1987): 279-285.

Ghate, S. R., and S. C. Chinnan. Storage of germinated tomato and pepper seeds. *J. Amer. Soc. Hort. Sci.* 112 (1987): 645-651.

Giammichele, L. A., and W. G. Pill. Protection of fluid-drilled tomato seedlings against damping-off by fungicide incorporation in a gel carrier. *HortScience* 19 (1984): 877-879.

Globerson, D., and Z. Feder. The effect of seed priming and fluid drilling on germination, emergence and growth of vegetables at unfavorable temperatures. *Acta Hortic.* 198 (1987): 15-21.

Gray, D. Fluid drilling of vegetable seeds. *Hort. Rev.* 3 (1981): 1-27.

Gray, D. The role of fluid drilling in plant establishment. *Ann. Appl. Biol.* 7 (1984): 153-172.

Gray, D. Improving the quality of horticultural seeds. *Professional Hortic.* 3 (1989): 117-123.

Gray, D., P. A. Brocklehurst, J. R. A. Steckel, and J. Dearman. Priming and pregermination of parsnip (*Pastinaca sativa* L.) seed. *J. Hort. Sci.* 59 (1984): 101-108.

Groot, S. P. L., and C. M. Karssen. Gibberellins regulate seed germination in tomato endosperm weakening: A study with GA-deficient mutants. *Planta* 171 (1987): 525-531.

Guedes, A. C., and D. J. Cantliffe. Germination of lettuce seeds at high temperature after seed priming. *J. Amer. Soc. Hort. Sci.* 105 (1980): 777-781.

Hadas, A. Seed soil contact and germination. In *The Physiology and Biochemistry of Seed Development, Dormancy and Germination*, A. A. Khan, ed. (Amsterdam: Elsevier, 1981), pp. 507-527.

Haigh, A. M. Why do tomato seeds prime? Physiological investigations into the control of tomato seed germination and priming. PhD Dissertation, Macquarie University, North Ryde, Australia, 1988.

Haigh, A. M., and E. W. R. Barlow. Germination and priming of tomato, carrot, onion and sorghum seeds in a range of osmotica. *J. Amer. Soc. Hort. Sci.* 112 (1987a): 202-208.

Haigh, A. M., and E. W. R. Barlow. Water relation of tomato seed germination. *Aust. J. Plant Physiol.* 14 (1987b): 485-492.

Haigh, A. M., E. W. R. Barlow, F. L. Milthrope, and P. J. Sinclair. 1986. Field emergence of tomato (*Lycopersicon esculentum*), carrot (*Daucus carota*) and onion (*Allium cepa*) seeds primed in an aerated salt solution. *J. Amer. Soc. Hort. Sci.* 111 (1986): 660-665.

Harman, G. E., and A. G. Taylor. Improved seedling performance by integration of biological control agents at favorable pH levels with solid matrix priming. *Phytopathology* 78 (1988): 520-525.

Harman, G. E., A. G. Taylor, and T. E. Stasz. Combining effective strains of *Trichoderma harzianum* and solid matrix priming to improve biological seed treatments. *Plant Disease* 73 (1989): 631-637.

Hayman, D. S., E. J. Morris, and R. J. Page. Methods of inoculating field crops with mycorrhizal fungi. *Ann. Appl. Biol.* 99 (1981): 247-253.

Hegarty, T. W. Seed activation and seed germination under moisture stress. *New Phytol.* 78 (1977): 349-359.

Hegarty, T. W. The physiology of seed hydration and dehydration, and the relation between water stress and the control of germination: A review. *Plant Cell & Environ.* 1 (1978): 101-119.

Helsel, D. G., D. R. Helsel, and H. C. Minor. Field studies on osmoconditioning soybeans, *Glycine max*. *Field Crops Res.* 14 (1986): 291-298.

Heydecker, W. Germination of an idea: The priming of seeds. Univ. of Nottingham School of Agriculture Report. 1973/74. pp. 50-67.

Heydecker, W., and P. Coolbear. Seed treatments for improved performance–Survey and attempted prognosis. *Seed Sci. & Technol.* 5 (1977): 353-425.

Heydecker, W., J. Higgins, and Y. J. Turner. Invigoration of seeds. *Seed Sci. & Technol.* 3 (1975): 881-888.

Hsaio, T. C., and J. Jing. Leaf and root expansive growth in response to water deficits. In *Physiology of Cell Expansion during Plant Growth*, D. J. Cosgrove and P. G. Knievel, eds. (Rockville, MD: American Society of Plant Physiologists, 1987), pp. 180-192.

Itoh, K., Y. Nakamura, H. Kawata, T. Yamada, E. Ohta, and M. Sakata. Effect of osmotic stress on turgor pressure in mung bean root cells. *Plant & Cell Physiol.* 28 (1987): 982-994.

Jawson, M. D., A. J. Franzleubbers, and R. K. Berg. *Bradyrhizobium japonicum* survival and soybean inoculation with fluid gels. *Appl. Env. Microbiol.* 55 (1989): 617-622.

Kahn, B. A., and J. E. Motes. Comparison of fluid drilling with conventional planting methods for stand establishment and yield of spring and fall broccoli crops. *J. Amer. Soc. Hort. Sci.* 113 (1988): 670-674.

Kahn, B. A., and J. E. Motes. Comparison of fluid drilling with conventional planting methods for stand establishment and yield of spring and fall cauliflower crops. *J. Amer. Soc. Hort. Sci.* 114 (1989): 200-204.

Karssen, C. M., and S. P. C. Groot. The hormone balance theory of dormancy evaluated. In *Growth Regulators and Seeds, Monograph 15*, N. J. Pinfield and M. Black, eds. (Bristol: British Plant Growth Regulator Group, 1987), pp. 17-30.

Kermode, A. R., J. D. Bewley, J. Dasgupta, and S. Misra. The transition for seed development to germination: A key role for desiccation. *HortScience* 21 (1986): 1113-1118.

Khan, A. A. Cytokinins: Permissive role in seed germination. *Science* 171 (1971): 353-359.

Khan, A. A. Hormonal regulation of primary and secondary seed dormancy. *Israel J. Bot.* 29 (1980/81): 207-224.

Khan, A. A. Preplant physiological seed conditioning. *Hort. Rev.* 13 (1992): 131-181.

Khan, A. A., and X. L. Huang. Synergistic enhancement of ethylene production and germination with kinetin and 1-aminocyclopropane-1-carboxylic acid in lettuce seeds exposed to salinity stress. *Plant Physiol.* 87 (1988): 847-852.

Khan, A. A., and C. M. Karssen. Changes during light and dark osmotic treatment independently modulating germination and ribonucleic acid synthesis in *Chenopodium bonus-henricus* seeds. *Physiol. Plant.* 51 (1981): 269-276.

Khan, A. A., and J. D. Maguire. Isolation of vegetable seeds by semi-permeable membrane during matriconditioning. *HortScience* 25 (1990): 1156 (Abst.)

Khan, A. A., and J. Prusinski. Kinetin enhanced 1-aminocyclopropane-1-carboxylic acid utilization during alleviation of high temperature stress in lettuce seeds. *Plant Physiol.* 91 (1989): 733-737.

Khan, A. A., and C. Samimy. Hormones in relation to primary and secondary seed dormancy. In *The Physiology and Biochemistry of Seed Development, Dormancy and Germination*, A. A. Khan, ed. (Amsterdam: Elsevier, 1982), pp. 203-241.

Khan, A. A., and A. G. Taylor. Polyethylene glycol incorporation in table beet seed pellets to improve emergence and yield in wet soil. *HortScience* 21 (1986): 987-989.

Khan, A. A., G. S. Abawi, and J. D. Maguire. Integrating matriconditioning and fungicidal treatment of table beet seed to improve stand establishment and yield. *Crop Sci.* 32 (1992): 231-237.

Khan, A. A., J. D. Maguire, G. S. Abawi, and S. Ilyas. Matriconditioning of vegetable seeds to improve stand establishment in early field plantings. *J. Amer. Soc. Hort. Sci.* 117 (1992): 41-47.

Khan, A. A., H. Miura, J. Prusinski, and I. Ilyas. Matriconditioning of seeds to improve emergence. *Proc. Natl. Symp. Stand Establ. Hort. Crops*. Minneapolis, MN. 1990. pp. 19-40.

Khan, A. A., N. H. Peck, A. G. Taylor, and C. Samimy. Osmoconditioning of beet seeds to improve emergence and yield in cold soil. *Agron. J.* 75 (1983): 788-794.

Khan, A. A., K. Tao, J. S. Knypl, B. Borkowska, and L. E. Powell. Osmotic conditioning of seeds: Physiological and biochemical changes. *Acta Hortic.* 83 (1978): 267-278.

Knypl, J. S., and A. A. Khan. Osmoconditioning of soybean seeds to improve performance at suboptimal temperatures. *Agron. J.* 73 (1981): 112-116.

Kubik, K. K., J. A. Eastin, J. D. Eastin, and K. M. Eskridge. Solid matrix priming of tomato and pepper. *Proc. Int. Conf. Stand Est. Hortic. Crops. Lancaster, PA*. 1988. pp. 86-96.

Lockhart, J. A. An analysis of irreversible plant cell elongation. *J. Theor. Biol.* 8 (1965): 269-275.

Matthews, S., and W. T. Bradnock. Relationship between seed exudation and field emergence in peas and french beans. *Hort. Res.* 8 (1973): 89-93.

Matthews, S., and A. A. Powell. Environmental and physiological constraints on field performance of seeds. *HortScience* 21 (1986): 1125-1128.

Mazor, L., M. Perl, and M. Negbi. Changes in some ATP-dependent activities in seeds during treatment with polyethylene glycol and during the redrying process. *J. Exp. Bot.* 35 (1984): 1119-1127.

Mexal, J., J. T. Fisher, J. Osteryoung, and C. P. Reid. Oxygen availability in polyethylene glycol solutions and its implication in plant-water relations. *Plant Physiol.* 55 (1975): 20-24.

Michel, B. E. Evaluation of the water potentials of solutions of polyethylene glycol 8000 both in the presence and absence of other solutes. *Plant Physiol.* 72 (1983): 66-70.

Michel, B. E., and M. R. Kaufmann. The osmotic potential of polyethylene glycol 6000. *Plant Physiol.* 51 (1973): 914-916.

Money, N. P. Osmotic pressure of aqueous polyethylene glycols. Relationship between molecular weight and vapor pressure deficit. *Plant Physiol.* 91 (1989): 766-769.

Nonami, H., and J. S. Boyer. Turgor and growth at low water potentials. *Plant Physiol.* 89 (1989): 798-804.

Ohep, J., H. H. Bryan, and D. J. Cantliffe. Control of damping-off of tomatoes by incorporation of fungicides in direct-seeding gel. *Plant Disease* 68 (1984): 66-67.

Oliviera, M. A., S. Matthews, and A. A. Powell. The role of split seed coats in determining seed vigour in commercial seed lots of soybean as measured by the electrical conductivity test. *Seed Sci. & Technol.* 12 (1984): 421-427.

Orzolek, M. D. Gel seeding update. *Amer. Veg. Grower.* 35 (1987): 10-11.

Osburn, R. M., and M. N. Schroth. Effect of osmopriming sugar beet seed on germination rate and incidence of *Pythium ultimum* damping-off. *Plant Disease* 73 (1989): 21-24.

Owen, P. L. Germination responses of osmotically primed asparagus and tomato seeds. MS Thesis. University of Delaware, Newark, DE, 1990.

Parera, C. A., and D. J. Cantliffe. Improved stand establishment of *sh-2* sweet corn by solid matrix priming. *Proc. National Symp. Stand Estab. Hort. Crops.* Minneapolis, MN. 1990. p. 91-96.

Parera, C. A., and D. J. Cantliffe. Improved germination and modified imbibition of *shrunken-2* sweet corn by seed disinfection and solid matrix priming. *J. Amer. Soc. Hort. Sci.* 116 (1991): 942-945.

Parera, C. A., and D. J. Cantliffe. Enhanced emergence and seedling vigor in *shrunken-2* sweet corn via seed disinfection and solid matrix priming. *J. Amer. Soc. Hort. Sci.* 117 (1992): 400-403.

Passam, H. C., P. I. Karavites, A. A. Papandreou, C. A. Thanos, and K. Georghiou. Osmoconditioning of seeds in relation to growth and fruit yield of aubergine, pepper, cucumber and melon in unheated greenhouse cultivation. *Scient. Hortic.* 38 (1989): 207-216.

Perkins-Veazie, P., and D. J. Cantliffe. Need for high quality seed for effective priming to overcome thermodormancy. *J. Amer. Soc. Hort. Sci.* 109 (1984): 368-372.

Peterson, J. R. Osmotic priming of onion seeds–The possibility of a commercial scale treatment. *Scient. Hortic.* 5 (1976): 207-214.

Pill, W. G. Parsley emergence and seedling growth from raw, osmoconditioned and pregerminated seeds. *HortScience* 21 (1986): 1134-1136.

Pill, W. G. Seedling emergence and yield from hydrated collard seeds fluid-drilled in high-phosphorus gel. *HortScience* 25 (1990): 1589-1592.

Pill, W. G. Advances in fluid drilling. *HortTechnol.* 1 (1991): 59-65.

Pill, W. G., and D. J. Fieldhouse. Emergence of pregerminated tomato seed stored in gels up to twenty days at low temperatures. *J. Amer. Soc. Hort. Sci.* 107 (1982): 722-725.

Pill, W. G., and W. E. Finch-Savage. Effects of combining priming and plant growth regulator treatments on the synchronisation of carrot seed germination. *Ann. Appl. Biol.* 114 (1988): 383-389.

Pill, W. G., and J. J. Frett. Performance of seeds embedded in hydroxyethyl cellulose sheets. *Scient. Hortic.* 38 (1989): 193-200.

Pill, W. G., and C. F. Mucha. Performance of germinated, imbibed and dry petunia seed fluid drilled in two gels with nutrient additives. *Scient. Hortic.* 22 (1984): 181-188.

Pill, W. G., and J. E. Rojas. Response of fluid-drilled "Grand Rapids" lettuce seeds to high temperature and low moisture. *J. Agric. Sci.* 109 (1987): 411-414.

Pill, W. G., J. J. Frett, and D. C. Morneau. Germination and seedling emergence of primed tomato and asparagus seeds under adverse conditions. *HortScience* 26 (1991): 1160-1162.

Powell, A. A., and S. Matthews. The damaging effect of water on dry pea embryos during imbibition. *J. Exp. Bot.* 29 (1978): 1215-1229.

Powell, A. A., and S. Matthews. The influence of testa condition on the imbibition and vigour of pea seeds. *J. Exp. Bot.* 30 (1979): 193-197.

Pyzik, T. P., and M. D. Orzolek. The effect of plant growth regulators and other compounds in gel on the emergence and growth of germinated tomato seeds. *J. Hort. Sci.* 60 (1986): 353-357.

Rabin, J., G. A. Berkowitz, and S. W. Akers. Field performance of osmotically primed parsley seed. *HortScience* 23 (1988): 554-555.

Rivas, M., F. J. Sundstrom, and R. L. Edwards. Germination and crop development of hot pepper after seed priming. *HortScience* 19 (1984): 279-281.

Roberts, E. H. Loss of seed viability: Ultrastructural and physiological aspects. *Seed Sci. & Technol.* 1 (1973): 529-545.

Sachs, M. Priming of watermelon seeds for low-temperature germination. *J. Amer. Soc. Hort. Sci.* 102 (1977): 175-178.

Salisbury, F. B., and C. W. Ross. *Plant Physiology*, 3d ed. (Belmont, CA: Wadsworth, 1985), 540 pp.

Salter, P. J. Fluid drilling of pregerminated seeds; Progress and possibilities. *Acta Hortic.* 33 (1978): 245-249.

Salter, P. J. Crop establishment: Recent research and trends in commercial practice. *Scientific Hort.* 36 (1985): 32-47.

Sanders, D. C., J. A. Ricotta, and L. Hodges. Improvement of carrot stands with plant biostimulants and fluid drilling. *HortScience* 25 (1990): 181-183.

Schopfer, P., and C. Plachy. Control of seed germination by abscisic acid. II. Effect of embryo water uptake in *Brassica napus* L. *Plant Physiol.* 76 (1984): 155-160.

Schultheis, J. R., D. J. Cantliffe, H. Bryan, and P. Stoffella. Improvement of plant establishment in bell pepper with gel mix planting medium. *J. Amer. Soc. Hort. Sci.* 113 (1988a): 546-552.

Schultheis, J. R., D. J. Cantliffe, H. Bryan, and P. Stoffella. Planting methods to improve stand establishment, uniformity, and earliness to flower in bell pepper. *J. Amer. Soc. Hort. Sci.* 113 (1988b): 331-335.

Smith, P. T., and B. G. Cobb. Accelerated germination of pepper seed by priming with salt solutions and water. *HortScience* 26 (1991a): 417-419.

Smith, P. T., and B. G. Cobb. Physiological and enzymatic activity of pepper seeds (*Capsicum annuum*) during priming. *Physiol. Plant.* 82 (1991b): 433-439.

Sosa-Coronel, J., and J. E. Motes. Effect of gibberellic acid and seed rates on pepper seed germination in aerated water columns. *J. Amer. Soc. Hort. Sci.* 107 (1982): 290-295.

Suena, W. The role of vigour in the priming of tomato seeds. PhD Dissertation. Macquarie University, Sydney, NSW, Australia, 1990.

Sundstrom, F. J., and R. L. Edwards. Pepper seed respiration, germination and seedling development following seed priming. *HortScience* 24 (1989): 343-345.

Suzuki, H., S. Obayashi, J. Yamaguchi, and S. Inanaga. Effect of pH of tertiary phosphate solutions on radicle protrusion during priming of carrot seeds. *J. Jap. Soc. Hort. Sci.* 59 (1990): 589-595.

Szafirowska, A., A. A. Khan, and N. H. Peck. Osmoconditioning of carrot seeds to improve seedling establishment and yield in cold soil. *Agron. J.* 73 (1981): 845-848.

Taylor, A. G., and T. J. Kenny. Improvement of germinated seed quality by density separation. *J. Amer. Soc. Hort. Sci.* 110 (1985): 347-349.

Taylor, A. G., and D. T. Warholic. Protecting fluid drilled lettuce from herbicides by incorporating activated carbon into gels. *J. Hort. Sci.* 62 (1987): 31-37.

Taylor, A. G., D. E. Klein, and T. H. Whitlow. SMP: Solid matrix priming of seeds. *Scient. Hortic.* 37 (1988): 1-11.

TeKrony, D. M., and D. B. Egli. Relationship of seed vigor to crop yield: A review. *Crop Sci.* 31 (1991): 816-822.

Thanos, C. A., K. Georghiou, and H. C. Passam. Osmoconditioning and ageing of pepper seeds during storage. *Ann. Bot.* 63 (1989): 65-69.

Thompson, A. R., D. L. Suett, and A. L. Percival. The protection of carrots against carrot fly (*Psila rosae*) with granular and emulsifiable concentrate formulations of chlorfenvinphos incorporated in gels used for drilling pregerminated seed in a sandy loam. *Ann. Appl. Biol.* 101 (1982): 229-237.

Valdes, V. M., K. J. Bradford, and K. S. Mayberry. Alleviation of thermodormancy in coated lettuce (*Lactuca sativa*) cultivar "Empire" by seed priming. *HortScience* 20 (1985): 1112-1114.

Van der Toorn, P. Embryo growth in mature celery seeds. PhD Dissertation, Agricultural University, Waginengen, The Netherlands, 1989. 95 pp.

Watkins, J. T., and D. J. Cantliffe. Mechanical resistance of the seed coat and endosperm during germination of *Capsicum annuum* at low temperature. *Plant Physiol.* 72 (1983): 146-150.

Watkins, J. T., D. J. Cantliffe, D. J. Huber, and T. A. Nell. Gibberellic acid stimulated degradation of endosperm in pepper. *J. Amer. Soc. Hort. Sci.* 110 (1985): 61-65.

Weibe, H. J., and T. Muhyaddin. Improvement of emergence by osmotic seed treatments in soils of high salinity. *Acta Hortic.* 198 (1987): 91-100.

Weibe, H. J., and H. Tiessen. Effects of different seed treatments on embryo growth and emergence of carrot seeds. *Gartenbauwissenschaft* 44 (1979): 280-284.

Welbaum, G. E., and K. J. Bradford. Water relations of seed development and germination in muskmelon (*Cucumis melo* L.) VI. Influence of priming on germination responses to temperature and water potential during seed development. *J. Exp. Bot.* 42 (1991): 393-399.

Wolfe, D. W., and W. L. Sims. Effects of osmoconditioning and fluid drilling of tomato seed on emergence rate and final yield. *HortScience* 17 (1982): 936-937.

Wurr, D. C. E., and J. R. Fellows. The effect of grading and priming crisp lettuce cultivar "Saladin" on germination at high temperature, seed vigor and crop uniformity. *Ann. Appl. Biol.* 105 (1984): 345-352.

Yang, S. F., and N. E. Hoffman. Ethylene biosynthesis and its regulation in higher plants. *Annu. Rev. Plant Physiol.* 33 (1984): 155-189.

Zuo, W., C. H. Hang, and G. Zheng. Effects of osmotic priming with sodium polypropionate (SPP) on seed germination. *Proc. Int. Conf. Stand Estab. Hort. Crops*, Lancaster, PA. 1988a. pp. 114-123.

Zuo, W., C. H. Hang, and G. Zheng. Physiological effects of priming with SPP on seeds of pea, tomato and spinach. *Proc. Int. Conf. Stand Estab. Hort. Crops*, Lancaster, PA. 1988b. pp 124-133.

Chapter 11

Influence of Seed Quality on Crop Establishment, Growth, and Yield

W. E. Finch-Savage

The quality of seeds has a profound influence on the economic production of agricultural crops of all species. Consequently, there has been a wealth of papers reporting the result of differences in seed quality on seedling emergence and crop yield in a wide range of species. These have been reviewed and discussed by several authors (e.g., Perry, 1972, 1976, 1980a, 1982; Powell, Matthews, and Oliveira, 1984; Powell, 1988; TeKrony and Egli, 1991). It is not the aim of this chapter to provide a complete catalogue of seed quality effects, but to illustrate how seed quality influences crop establishment, growth, and yield with representative references from a wide range of crops.

The preceding chapters have introduced the concepts of viability and vigor (Chapters 1 and 2) to show the nature of seed quality and how differences in quality between seedlots arise during production (Chapter 4) and subsequent storage (Chapters 5, 6, 7, 8). The resulting quality of seedlots can be enhanced by physiological treatments such as hydration-dehydration and priming (Chapters 1 and 10 respectively) and by seed treatments to control microorganisms (Chapter 5). Performance may also be reduced by seed technology, for example, the delayed emergence that can result from pelleting; a necessary modification to seeds of many species to facilitate precision sowing. Heydecker and Coolbear (1977) have reviewed the many other types of seed treatment that can affect seed quality. The influence of these treatments on crop establishment and yield will be discussed alongside those of seed quality.

The period between sowing and crop harvest can be conveniently divided into the events leading up to and including seedling emergence from the soil (crop establishment), and postemergence growth. The crop establishment period can be further subdivided chronologically, into seed imbibition, the processes of germination resulting in radicle emergence from the seed, and post germination growth to emergence. The events in each of these periods can influence crop yield, and because each is uniquely affected by seed quality they are initially considered separately and then drawn together in the final section.

SEED QUALITY AND SEEDLING EMERGENCE

Seedling emergence is the result of a large number of preceding processes which occur against the often hostile background of the seed bed environment. Under these circumstances, the chances of successful seedling emergence are greatly influenced by seed quality. Laboratory germination tests reveal differences in seedlot viability which will inevitably result in differences in the levels of seedling emergence. However, even when seedlots of the same viability are sown at the same time and place, differences in seedling emergence occur (Heydecker 1972; Perry, 1982). Similarly, the proportion of viable seeds that emerge from seedlots of different viability sown on the same occasion varies due to differences in seed vigor, for example from 43 to 77 percent in 18 carrot (*Daucus carrota* L.) seedlots of the same cultivar (Gray and Steckel, 1983). Furthermore, if the same seedlot is sown on a number of occasions and locations, emergence will vary due to different environmental conditions (Hegarty, 1974, 1976). Seedling emergence therefore results from a complex interaction of seed quality and the seedbed environment (Perry, 1984).

In general, poorer-quality seeds will show symptoms typical of seed aging, such as low viability, reduced germination and emergence rates, poor tolerance to suboptimal conditions, and low seedling growth rates (Powell, Matthews, and Oliveira, 1984). Physiological seed treatments generally act to improve seed performance directly by reducing the time to germination and seedling emergence, or indirectly by improving the seeds' ability to cope with stresses such as limited water availability. For example, in 37 seedbed environments, primed carrot seeds gave earlier, more uniform, and higher-percentage seedling emergence than untreated seeds (Finch-Savage, 1990). Conversely, seed-borne pathogens can adversely affect germination and early seedling growth to reduce and delay seedling emergence. The level of seed-borne infection and the susceptibility to soil-borne infections can differ between seedlots and interact with physiological differences in seed quality. However, pathogen effects can be minimized with the application of the appropriate chemical seed dressings. Details of the effects and control of specific seed-borne diseases are beyond the scope of this chapter, but are discussed at length elsewhere (e.g., Neergaard, 1977; Agarwal and Sinclair, 1987; Powell, Matthews, and Oliveira, 1984; and Chapter 5).

If seedling emergence is inadequate, crop yield will be reduced, and in most situations no amount of effort and expense later in crop development can compensate for this effect. The remainder of this section will examine in detail the influence of seed quality and its interaction with environmental conditions, affecting seed imbibition, germination, and subsequent growth to seedling emergence. In the following section the consequences of the resulting patterns of seedling emergence for crop yield will be considered.

Imbibition

The first impact of seed quality on performance is seen in the initial phase of water uptake following sowing, which has been described as a period of peril. "The ability of the seed to traverse this period successfully and to emerge as an autotrophic self-sustaining plant depends on the inherent soundness and vigor of the seed" (Woodstock, 1988). Powell and Matthews (1978) showed that the rapid uptake of water by dry pea (*Pisum sativum* L.) embryos can result in cell death on the surface of the cotyledons. Subsequently they found that differences in the rate of water uptake between seedlots of peas was due to the integrity of the seed coat, and was related to the incidence of imbibition damage (Powell and Matthews, 1979). Thus, rapid imbibition in peas associated with seed coat damage led to poor field emergence (Powell and Matthews, 1980).

Imbibitional damage associated with low vigor has also been shown in soybean ([*Glycine max* (L.) Merr.], Woodstock and Taylorson, 1981; Oliveira, Matthews, and Powell, 1984), and is consistent with a reported negative correlation between the percentage of soybean seeds with broken seed coats and field emergence in this crop (Luedders and Burris, 1979). However, the imbibition damage reported for dwarf French bean (*Phaseolus vulgaris* L.) was not associated with seed coat damage, but with the degree of adherence of the testa to the cotyledons (Powell, Oliveira, and Matthews, 1986a, 1986b). In all these cases, microbial attack stimulated by enhanced leaching from the damaged cotyledons may act to further reduce field emergence (Woodstock, 1988).

Many warm-season crops suffer from reduced seedling emergence because they are sensitive to chilling injury during early imbibition, for example, cotton ([*Gossypium hirsutum* L.], Christiansen, 1963), lima beans ([*Phaseolus lunatus* L.], Pollock, 1969), soybeans (Obendorf and Hobbs, 1970), maize ([*Zea mays* L.], Cal and Obendorf, 1972) and possibly sorghum ([*Sorghum bicolor* (L.) Moench], Phillips and Youngman, 1971). Powell and Matthews (1978) proposed that this chilling injury resulted from cold-enhanced sensitivity to imbibition damage. Susceptibility to chilling injury was correlated positively with the rate of water uptake which increased with seed coat damage (e.g., pea, Tully, Musgrave, and Leopold, 1981, and snap bean seeds [*Phaseolus vulgaris* L.], Taylor and Dickson, 1987). Such seed coat related problems are likely to differ between seedlots and rate of imbibition can vary with seed age (Blacklow, 1972).

These imbibition-related injuries are not limited to large-seeded species and have now been found in cauliflower seeds ([*Brassica oleracea* L.], McCormac and Keefe, 1990). The damage was again associated with seed coat integrity, was seedlot related, and resulted in reduced seedling emergence. The seed coat therefore affords a protection against imbibitional injury, the efficacy of which varies with seedlot and could potentially be improved through seed technology and the application of polymeric coatings to seeds (Taylor and Dickson, 1987). Because the seed coat determines the rate of imbibition it also influences the timing of germination.

Germination

Although seed dormancy is common among species in a wide range of plant families, it has largely been overcome, with some notable exceptions, in most important commercial crops (Villiers, 1972; Maguire, 1984). In the absence of dormancy the basic germination requirements for crop species are simple: adequate temperature, water, and a favorable gaseous environment. When any of these basic requirements become limiting in the seedbed, seeds may fail to germinate. Seed quality determines the ability of seeds to cope with these suboptimal conditions and to compete with soil microorganisms for resources. A detailed account of the influence of the seedbed environment on seed germination is given by Hegarty (1984).

Under optimal conditions the cumulative germination curve of a seedlot is generally sigmoidal, illustrating the variation in germination times and therefore seed quality within the seed population. This variation in germination times increases with mean germination time as the seedlot deteriorates prior to viability loss (Ellis and Roberts, 1980). Seedlots with similarly high levels of viability can therefore have germination rates that differ greatly, for example, in cabbage (Perry, 1982) and wheat (Dell'Aquila, 1987). Both percentage and rate of germination can also be related to seed size in a wide range of crops (Kaufmann, 1984).

Under seedbed conditions both inter- and intraseedlot variation in germination interacts with a range of environmental stresses (Hegarty, 1984). Ellis and Roberts (1981) hypothesize that the distribution of potential life spans of individual seeds within a population is normal and that the response of individual seeds to environmental stresses is dependant on their proximity to death, so that when quantified correctly (e.g., probit scales), a given level of stress effects the performance of all the individual seeds to an equal extent. However, because of the nonlinear nature of seed deterioration, stress appears to affect seedlots differently. For example, at high viability, few seeds are close to death and thus few seeds will fail to germinate, whereas an increasing proportion of seeds are closer to death as viability declines toward 50 percent. and so a progressively larger proportion of seeds will fail to germinate when stressed. It is not possible to cover examples of all of the large number of different environmental stresses that can influence germination in the seedbed in this chapter. However, the most universally important environmental variables that limit germination, temperature, water availability, and aeration, are here considered in relation to seed quality.

Germination: Temperature Effects

Seeds germinate over a wide range of temperatures, but maximum percentage germination is typically reduced at the extremes of this range. The band of temperatures over which maximum percentage germination occurs varies with seed quality and is generally narrower as the seedlot deteriorates (Ellis and Roberts, 1981). As low temperatures can limit the germination of warm-season crops, the seedlot germination percentage in a cool test can be a good indicator of subsequent field performance, especially at early season sowings (e.g., cotton, Smith and Varvil, 1984: soybean, Kulik and Yaklich, 1982).

For a given percentile of the seed population, germination rate (e.g., the reciprocal of germination time) increases linearly from a base to an optimum temperature above which it decreases linearly to a ceiling temperature that indicates the limit of tolerance (Garcia-Huidobro, Monteith, and Squire, 1982). This linear relationship at suboptimal temperatures has been shown in many vegetable species (Wagenvoort and Bierhuizen, 1977), range grasses, and shrubs (Jordan and Haferkamp, 1989). Therefore at suboptimal temperatures, thermal time (day degrees above the base temperature) to germination of a given percentile is a constant. These variables of base, optimum, and ceiling temperature and thermal time can be used to illustrate the influence of seed quality on germination.

The base temperature varied little between the different percentiles of the seed population in pearl millet ([*Pennisetum typhoides* S. and H.], Garcia-Huidobro, Monteith, and Squire, 1982). Therefore differences in germination rate between seeds of different quality at suboptimal temperatures appeared largely due to differences in response to thermal time. This also held true for differences within and between seedlots having environmentally induced differences in quality, for different genotypes, and for osmotically primed sublots, for example in sugarbeet ([*Beta vulgaris* L.], Gummerson,

1986; Durrant and Gummerson, 1990), tomato ([*Lycopersicon eculentum* Mill], Dahal, Bradford, and Jones, 1990), four grain legumes (Covell et al., 1986; Ellis et al., 1986), and onion ([*Allium cepa* L.], Ellis and Butcher, 1988). However both optimum and ceiling temperature can vary between and within seedlots, for example in onions (Ellis and Butcher, 1988) and chickpea ([*Cicer arietinum* L.], Ellis et al., 1986).

Germination: Water Availability and Aeration Effects

Water stress can affect both the final level and rate of germination (Doneen and MacGillivray, 1943), and seedlots vary in their ability to overcome this stress (e.g., a range of vegetable and agricultural species; Guy, 1982). Gummerson (1986) developed the concept of hydrothermal time to quantify the response of seed germination rates of sugar beet to water potential (Ψ) and thermal time. Hydrothermal time is a combination of temperature above a base temperature, Ψ above a base Ψ (Ψ_b) and time. In general, hydrothermal time did not differ among percentiles of the seed population, but Ψ_b was negatively related to germination rate (e.g., later germinators have higher Ψ_b). This suggests that it is variation in Ψ_b which leads to differences in the time seeds take to germinate and therefore the shape of the cumulative germination curve (Gummerson, 1986). Similar relationships were also found for lettuce ([*Lactuca sativa* L.], Bradford, 1990) and tomato (Dahal and Bradford, 1990). However, delayed germination in an older seedlot and reduced germination time due to priming of tomato seeds was largely caused by changes to hydrothermal time rather than changes to Ψ_b (Dahal and Bradford, 1990).

Excessive water in the seedbed can restrict the supply of oxygen to seeds and thereby reduce and delay germination. The oxygen requirement for germination can vary among individuals in the population (Siegel and Rosen, 1962), and seedlots can vary in their ability to emerge from seedbeds which receive excessive water (e.g., barley [*Hordeum vulgare* L.], and sugarbeet; Hegarty and Perry, 1974; Perry, 1984).

Preemergence Growth

Following germination, seedling extension growth is exponential until seed reserves become limiting (Wanjura and Buxton, 1977), and proceeds at a rate which is linearly related to temperature in a wide range of crops (Wheeler and Ellis, 1991). In the absence of abnormal seedlings, differences in seed quality both within and between seedlots had no influence on the rate of preemergence cotyledon elongation in onion, even though there were considerable differences in the time taken to germinate (Wheeler and Ellis, 1991). However, differences in seedlot quality can influence the proportion of seedlings having abnormal growth (Ellis and Roberts, 1980). When the whole seedlot was considered, seedling growth rates immediately following germination were correlated with germination rates within seedlots of leek, cauliflower, onion (Finch-Savage, 1986), carrot (Finch-Savage and McQuistan, 1988), and Brussels sprouts (Finch-Savage, 1988).

Although seed deterioration in storage resulted in decreased early root growth rates in broad beans, peas, and barley (Abdalla and Roberts, 1969), much of the reported differences in reserve-dependant seedling growth both within and between seedlots may result from differences in seed size. Larger seeds have been shown to result in larger seedlings in many crops (e.g., lettuce, Smith, Welch, and Little, 1973; Wurr and Fellows, 1984; *Lolium perenne*, Arnott, 1975). Such differences in growth potential can greatly influence seedling emergence. For example, the different capacity for endosperm dependant growth from seeds of different sizes in *Lolium perenne* can result in progressively lower emergence from lighter seeds as sowing depth increases (Arnott, 1975), presumably due to exhaustion of seed reserves. Similarly, large seeds of birdsfoot trefoil (*Lotus corniculatus* L.) produced faster elongating seeds than did small seeds (Curtis and McKersie, 1984). This was in part due to the greater capacity for the axis from bigger seeds to accumulate dry weight, as well as the greater quantity of seed reserves. For further information on the effect of seed size on germination and emergence see Black (1959), Perry (1980a), Kaufmann (1984), and Powell (1988).

The effects of germination rate and subsequent growth rate are often confounded in the growth tests used to estimate seed vigor. For example, studies showing interlot differences in seedling size at a fixed time from imbibition in carrots (Gray and Steckel, 1983), lettuce (Wurr and Fellows, 1984), and cabbage ([*Brassica oleracea* L.], Perry, 1982) did not separate germination rate effects and subsequent growth effects. In most field experiments, it is also difficult to separate the effects of germination time and preemergence seedling growth. Finch-Savage and McQuistan (1988) compared the seedling emergence from pregerminated seeds, selected from the same carrot seedlot, but with different potential seedling growth rates to overcome this problem. At six sowing dates in the field, more seedlings emerged from seeds producing seedlings with greater potential growth rates in all four seedlots tested.

Under nonoptimal conditions in the field, reduced soil water potentials can influence early seedling growth rates directly (e.g., mungbean [*Vignia radiata* L.], Fyfield and Gregory, 1989), or indirectly by increasing soil impedance (Goyal, 1982). The differing ability of sugarbeet seedlots to overcome soil impedance is shown by the laboratory packed sand test, which has a greater correlation with field emergence than the standard germination test (Akeson and Widner, 1980). Seeds from larger seeded soybean varieties can generate a greater emergence force than small seeds to overcome soil impedance (Rathore, Ghildyal, and Sachan, 1981), however a greater emergence force is needed to overcome the greater soil resistance to the emergence of a larger seed with epigeal germination. There is no guaranteed benefit from sowing large seeds (Perry, 1980a), for example, small soybean seeds can germinate faster than large seeds and the resulting seedlings may therefore emerge before the soil crust fully forms (Rathore, Ghildyal, and Sachan, 1982). In this way rapid emergence can help to avoid problems due to seedbed deterioration and may account for why rate of germination was shown by Kulik and Yaklich (1982) to be a good indicator of field performance in soybean. Rapidly emerging seedlots are also recommended for cotton to minimize exposure to seedling diseases (Kerby, Keeley, and Johnson, 1987).

THE EFFECT OF SEEDLING EMERGENCE ON CROP YIELD

In practice, only part of the total biomass produced by a crop is harvested and this component is species specific. This economic yield may be determined by the whole plant population, as a weight of a particular plant part per unit area, for example in grain and sugarbeet crops. Alternatively, economic yield may be determined by individual plants within the population such as in many horticultural crops, for example, the number of individuals within closely defined "high value" size grades (e.g., carrots and onions) or the number of plants which are "mature" at a single harvest (e.g., lettuce or mechanically harvested green beans). There are many such examples of the influence of modern marketing and crop husbandry on the nature of economic yield and these influence the importance of seed quality.

From the preceding section it is clear that seed quality and seed treatments affect the ability of seeds to overcome the variable conditions experienced by the seed during crop establishment. The pattern of seedling emergence resulting from these interactions between the seedlot and the environment can be summarized by three parameters: the numbers of emerged seedlings (crop density), the mean time of seedling emergence, and the spread in times to emergence of individual seedlings within the population (uniformity of emergence). The potential effect of seed quality and seed treatments on economic yield through its influence on crop establishment will be illustrated using these parameters.

Crop Density

Clear relationships exist between crop density and yield (Bleasdale, 1967; Willey and Heath, 1969), with yield increasing asymptotically as density increases. In some crops, yield may decrease at higher densities to form a parabolic relationship (e.g., grains, Holliday, 1960). Size-graded yields of crops harvested in the vegetative phase will also have a parabolic relationship with density because individual plant size decreases as density increases (e.g., parsnips, Bleasdale and Thompson, 1966; onions, Bleasdale, 1966). The number of seedlings emerging therefore not only affects total crop yield, but the size of individual plants and therefore graded yields. In addition it can influence the time taken to reach maturity (e.g., onions; Mondal et al., 1986) and the uniformity of plants at maturity (e.g., cauliflower; Salter and James, 1975).

Target populations are therefore set to achieve maximum economic yields, for example, 90 plants m^{-2} in vining peas (Gane, 1985), and differing populations for onions depending on the size of bulbs required (Frappell, 1973). For the sugarbeet crop which is sown to a stand at low density, a minimum of 70 percent seedling emergence is required to maximize yield (Jaggard, 1979). Failure to meet these targets brings severe economic penalties. A greater ability to achieve the desired plant stands could also limit the need for hand thinning (e.g., sugarbeet, lettuce) or the labor cost of transplanting (e.g., cabbage and calabrese; Perry, 1980b), and reduce the problems of weed competition. During plant raising in cellular trays or other modular systems, high plant stands are essential to reduce the cost of hand labor in replacing empty cells, and to reduce the total space (heated glasshouse in many countries) required for production (Matthews and Powell, 1986; Finch-Savage and McKee, 1990).

Timing of Seedling Emergence

From the previous sections it is clear that high-quality seeds generally give more rapid seedling emergence than do lower-quality seeds. Because seedbeds tend to deteriorate with time, this more rapid emergence often results in improved seedling stands. However, there are also more direct influences of improved seedling emergence rates on yield, many of which have been identified through studies on seed-treatment techniques such as priming or the sowing of pregerminated seeds (e.g., fluid drilling, reviewed by Gray, 1981) which reduce time to seedling emergence.

Early seedling emergence is important in temperate climates where season length is limiting. Low rates of leaf growth in the spring caused by low temperatures are thought to limit the yields of many temperate crops (Monteith and Elston, 1971). Consequently, improved yields of onions should be possible if a full leaf canopy can be established earlier in the year so that bulb growth can coincide with days having the highest mean solar irradiance (Brewster and Barnes, 1981). For maximum yields, sugarbeet crops also need the longest growing season possible, and must achieve maximum leaf area before days that have the highest mean levels of solar irradiation (Scott et al., 1973). In the limited season of northern latitudes, earlier emergence can increase the yield of grain maize (Breeze and Milbourn, 1981) and sweet corn ([*Zea mays* L.], Cal and Obendorf, 1972), which can lead to higher prices per unit of product and more efficient use of processing factories (Bennett and Waters, 1987), whereas delayed emergence due to soil crusting can reduce the yield of cotton independently of its effect on plant stand (Wanjura, 1982).

Earlier emergence leading to the potential for earlier harvesting dates can enhance the value of a number of horticultural crops, such as bunching carrots in the UK (Finch-Savage, 1984). Rapid establishment of ground cover in grasses and forage crops is a major objective in their crop husbandry, although differences in yields diminish with time (Perry, 1980a). In arid conditions, plants are usually established from seeds that have the most rapid germination within the seedlot (Jordan, 1983), and rapid seedling emergence can result in better use of limited water resources by reducing the irrigation requirement for crop establishment. Early seedling emergence also increases the competitiveness of crops against weeds and may allow earlier application of herbicides when weeds are most susceptible; also, the reduced time taken for seedlings to reach marketable size can minimize the cost of plant raising under heated glass (Finch-Savage and McKee, 1990).

Uniformity of Seedling Emergence

The influence of the variation in individual seedling emergence times within a population on subsequent growth and yield has been discussed at some length by Benjamin and Hardwick (1986) and Benjamin (1990). They suggest that, in crop monocultures, the relative time of seedling emergence is an important criterion for the future development and growth of plants and that its importance to plants has not been fully recognized. Time of emergence accounts for much of the subsequent variation in plant size (Benjamin, 1990). A negative relationship between seedling emergence time and weight at harvest has been reported for a number of crop plants (e.g., carrot, Mann and MacGillivray, 1949; Benjamin, 1982, 1984a; Salter, Currah, and Fellows, 1981; subterraneum clover [*Trifolium subterraneum* L.], Black and Wilkinson, 1963; lettuce, Gray, 1976; leek [*Allium porrum* L.], Benjamin, 1984b; barley, Soetono and Donald, 1980; maize, Breeze and Milbourn, 1981). Therefore the ranking of seedling size at emergence changes little with time, but there are often changes in the relative sizes, for example, the smallest size grades tend to have below-average relative growth rates (RGR)

(Benjamin and Hardwick, 1986). This may be due to early emergers receiving a disproportionate amount of the available resources so that relative differences between plants tend to increase rather than decrease during growth.

For plants harvested as individuals, the variation in the size at harvest can have a large impact on the economic yield, for example, on the graded yield of root crops (Benjamin, 1984a). Variation in maturity times greatly affects the number of mature plants that can be taken at any one harvest and therefore the profitability of crops such as cauliflower and lettuce. For these types of crops which are normally grown at wide spacings, the variability in time to maturity is directly related to variability in seedling emergence times (Salter, 1985). For crops where it is the harvested weight per unit area at full reproductive maturity that is important, initial differences in mean plant size due to seed quality tend to diminish during crop development and have no impact on yield (TeKrony and Egli, 1991). However, a long time-spread of seedling emergence may increase the problems of timing herbicide applications as reported for sugar beet (Longden et al., 1979). Other agronomic problems associated with plant variability may also indirectly affect yield per unit of area, for example, the loss in yield due to variable root size during preharvest topping in sugar beet (Longden et al., 1979), and in the temperate maize grain crop, unripe plants can clog threshing machinery, and precise maturity dates to maximize yield are difficult to determine (Breeze and Milbourn, 1981).

SEED QUALITY AND POSTEMERGENCE GROWTH

There have been many studies made on the effects of seed quality and seed treatment on the size of seedlings at some point following seedling emergence, for example, the effect of seed size on early plant size has been reported for a wide range of crops (Kaufmann, 1984). However, in many studies it is not clear if the reported differences in plant size are due to differences in RGR, to different times of emergence and therefore different durations of growth, or to the different sizes of seedlings at emergence. Clearly, the greater proportion of abnormally growing seedlings that can persist after emergence from aged seedlots (Roos, 1980), from mechanically damaged seedlots (Powell, Matthews, and Oliveira, 1984), or from different levels of seed-borne infection (Neergaard, 1977), can result in reduced crop growth. Seedlots may also vary in their ability to withstand the adverse affects of preplant herbicides on plant growth (e.g., cotton; Bailey and Bourland, 1986). But, does the quality of seeds producing normal, uninfected seedlings influence the RGR of plants once seed-reserve dependant growth is complete?

Black (1956) found that in subterraneum clover, larger seeds produced a greater cotyledon surface area. He therefore hypothesized that, in seeds with epigeal germination and no endosperm, early differences in growth may result from larger seeds having a greater photosynthetic capacity. Coleoptile width within barley cultivars was also positively related to seed weight (Ceccarelli and Pegiati, 1980). However, Burris, Edje, and Wahab (1973) showed that in soybean, the smaller cotyledons produced from small seeds exhibited a higher rate of photosynthesis than cotyledons from larger seeds; this compensated for the difference in photosynthetic area. Moreover, Egli, TeKrony, and Wiralaga (1990) reported that, provided that seedlings were free from physical injury or necrotic lesions, soybean seed vigor had no effect on RGR.

Abdalla and Roberts (1969) showed that early differences in growth rate between differently aged seedlots of barley, broad beans (*Vicia faba* L.), and peas were not present at later stages of growth. There was also no influence of seed quality or priming treatments on RGR in onion (Ellis, 1989) or tomato (Argerich and Bradford, 1989). RGR was also unaffected by seed quality in carrot (Gray, 1984) and wheat ([*Triticum aestivum* L. em. Thell.], Khah, Roberts, and Ellis, 1989), so that any seed quality effects on yield, in these species at least, must result from differences in the timing of emergence and/or the size of seedlings at emergence. In carrot, onion, and celery, the mean plant weight from different seedlots and primed sublots during growth was inversely related to seedling emergence time (Brocklehurst and Dearman, 1983). Examples of similar relationships were shown within seedlots in the previous section for a wide range of other crops.

In subterraneum clover, Black and Wilkinson (1963) recorded the growth of seedlings from different sowings in the same sward that had considerable overlap in emergence times. They showed that time of sowing and preemergence growth rate were unimportant in determining final individual plant weight, as this was closely related to day of emergence. Plants with very different preemergence growth rates (e.g., from different sowing dates and most likely from seeds of different quality) that emerged on the same day essentially had the same subsequent growth. Therefore it was concluded that the reduction in final plant size due to delayed emergence resulted from the increased severity of competition from plants already emerged in this densely sown crop. Thus seed quality could indirectly affect postemergence RGR of individuals by affecting their emergence time and therefore their ability to compete. Growth rate may also be affected indirectly because seedlings emerging at different times will experience different environmental conditions. However, there are reports of direct effects of seed quality on plant growth, for example, the ability of cereal crops to provide compensatory growth through tillering to offset the effect of low plant stands on yield may differ between seedlots and the severity of seedbed stress (Perry, 1984).

Seed deterioration can lead to increased variability in initial growth rates between plants (Abdalla and Roberts, 1969). Such differences are likely to persist as smaller seedlings tend to have below-average RGR, probably as a result of interplant competition (Benjamin and Hardwick, 1986), as demonstrated with subterraneum clover (Black and Wilkinson, 1963). Such interplant competition may eliminate differences in mean plant size between lots to leave no effect on final yield of crops such as herbage legumes (Black, 1959) and seed crops like dry peas, broad beans, and barley (Abdalla and Roberts, 1969). Differences in individual plant sizes resulting from different growth rates can, however, have a large impact on the yield of more densely sown graded vegetable crops such as carrots, as shown in the previous section.

In summary, seed quality frequently affects seedling size soon after emergence, and these relative differences may remain until harvest to influence yield in crops with low plant density. However, there is little evidence that seed quality affects the autotrophic RGR of emerged seedlings directly when seedlings with abnormal growth are discounted. But, in competitive situations (e.g., high plant density), early emerged seedlings tend to suppress the RGR of smaller, later-emerging seedlings below their potential. Thus different RGRs can result from seeds of the same seedlot, which may result in yield differences between seedlots in some crops.

SEED QUALITY AND CROP YIELD

Clearly, when seed quality is low and the resulting plant density falls below a threshold level, yield will be reduced. In other circumstances, the influence of seed quality on yield is less obvious, however, general conclusions can be drawn from the preceding sections. In the literature there is little evidence for direct seed quality or seed treatment effects on RGR once seed reserves have been exhausted, except where significant numbers of abnormal seedlings persist beyond emergence or where there are high levels of seed-borne pathogens. In the absence of density differences, seed-quality effects on yield must therefore be due largely to effects on the timing of emergence and the size of individuals at emergence within the seedling populations, both of which are greatly influenced by seed vigor and seed treatments. When plant density is above that required to maximize yield, differences in yield per unit of area due to these emergence effects diminish with time because of plant competition. However, when the harvestable component relates to individual plants, yield is greatly influenced by the pattern of seedling emergence and the relative size of seedlings at emergence; competition in more densely sown crops of this type tends to increase these differences and to cause differences in RGR between individuals.

TeKrony and Egli (1991) provide evidence from a number of studies which indicates that in the absence of density effects there is a greater importance of seed vigor in crops harvested during vegetative growth (e.g., lettuce and carrot) or early reproductive growth (e.g., tomato and green peas) than in crops harvested at full reproductive maturity (e.g., soybean and wheat). They argue that at commercial densities seed vigor affects vegetative growth, but there is usually no affect of seed vigor at full reproductive maturity because dry seed yield is not closely associated with vegetative growth.

It has been common practice, when seedlots have viability above the statutory requirements that exist in many countries, to increase seeding rates to offset the lower percentage emergence expected from poorer-quality seedlots. This assumes that yield will not suffer as a result of this action and that quality has little influence beyond emergence. This would seem reasonable for grain crops harvested at full reproductive maturity, considering the evidence provided by TeKrony and Egli (1991). Indeed, Abdalla and Roberts (1969) suggest that for three such crops (barley, dry broad beans, and peas), viability can be as low as 50 percent before yield suffers. However, for this to be true in practice, normal commercial densities need to be achieved, seeds should be disease free or adequately treated to control seed-borne infections, and the season should not be limiting (Roberts, 1986).

For crops harvested before full reproductive maturity, yield differences will result from offsetting low viability by sowing more seed. For example, even though no differences in seedling RGR were found between different seedlots of carrots, a proportionate increase in the number of seeds sown to compensate for differences in initial viability did not completely eliminate differences in performance between lots (Gray, 1984). This was because differences in seed viability between lots were correlated with other germination and emergence characteristics which substantially influenced growth and yield, particularly those aspects associated with plant-to-plant variability. Similar results have been found with other crops harvested before full reproductive maturity. The importance of seed-vigor and seed-treatment-induced differences at emergence in these crops can vary with the market supplied, for example, it has less importance when supplying vegetables to traditional product outlets, compared to the supermarket trade which has strict requirements for uniform produce. Reduced variability is especially important when plants below a minimum size do not contribute to yield or when once-over machine harvesting, because late-producing plants may not contribute to yield (Pollock and Roos, 1972). There is also a greater importance when the season length is marginal for the crop, when space is at a premium such as in glasshouse plant raising, and when early harvests are made for specialist markets.

In many crops, seed vigor and seed treatments can have a major influence on economic yield despite little effect on total yield, and this is particularly true where strict requirements are imposed by modern production and marketing practice. However, seed vigor and seed treatments have a more general importance to all crops, because whatever their harvestable component, crops are usually established and grown under suboptimal conditions; seed treatments and high vigor are therefore needed to consistently achieve the required plant population to maximize yield.

REFERENCES

Abdalla, F. H., and E. H. Roberts. The effect of seed storage conditions on the growth and yield of barley, broad beans and peas. *Ann. Bot.* 33 (1969): 169-184.
Agarwal, V. K., and J. B. Sinclair. *Principles of Seed Pathology.* (Boca Raton, Florida: CRC Press, 1987).
Akeson, W. R., and J. N. Widner. Laboratory packed sand test for measuring vigor of sugar beet seeds. *Crop Sci.* 20 (1980): 641-644.
Argerich, C. A., and K. J. Bradford. The effects of priming and ageing on seed vigour in tomato. *J. Exp. Bot.* 40 (1989): 599-607.
Arnott, R. A. A quantitative analysis of the endosperm-dependant seedling growth in grasses. *Ann. Bot.* 39 (1975): 757-765.
Bailey, B. A., and F. M. Bourland. The influence of seed quality on response of cotton seedlings to the preplant herbicide Trifluralin. *Field Crops Res.* 13 (1986): 375-382.
Benjamin, L. R. Some effects of differing times of seedling emergence, population density and seed size on root-size variation in carrot populations. *J. Agric. Sci.* 98 (1982): 537-545.
Benjamin, L. R. The relative importance of some sources of root-weight variation in a carrot crop. *J. Agric. Sci.* 102 (1984a): 69-77.
Benjamin, L. R. The relative importance of some different sources of plant-weight variation in drilled and transplanted leeks. *J. Agric. Sci.* 103 (1984b): 527-537.
Benjamin, L. R. Variation in time of seedling emergence within populations: A feature that determines individual growth and development. *Adv. Agron.* 44 (1990): 1-25.
Benjamin, L. R., and R. C. Hardwick. Sources of variation and measures of variability in even-aged stands of plants. *Ann. Bot.* 58 (1986): 757-778.
Bennett, M., and L. Waters, Jr. Seed hydration and seed quality concerns for sweet corn stand establishment in cool soils. *Acta Hortic.* 198 (1987): 171-179.
Black, J. N. The influence of seed size and depth of sowing on pre-emergence and early vegetative growth of subterranean clover (*Trifolium subterraneum* L.). *Aust. J. Agric. Res.* 7 (1956): 98-109.
Black, J. N. Seed size in herbage legumes. *Herbage Abstr.* 29 (1959): 235-241.
Black, J. N., and G. N. Wilkinson. The role of time of emergence in determining the growth of individual plants in swards of subterranean clover (*Trifolium subterraneum* L.). *Aust. J. Agric. Res.* 14 (1963): 628-638.
Blacklow, W. M. Mathematical description of the influence of temperature and

seed quality on imbibition by seeds of corn (*Zea mays* L.). *Crop Sci.* 12 (1972): 643-646.

Bleasdale, J. K. A. The effects of plant spacing on the yield of bulb onions (*Allium cepa* L.) grown from seed. *J. Hort. Sci.* 41 (1966): 145-153.

Bleasdale, J. K. A. The relationship between the weight of a plant part and total weight as affected by plant density. *J. Hort. Sci.* 42 (1967): 51-58.

Bleasdale, J. K. A., and R. Thompson. The effects of plant density and the pattern of plant arrangement on the yield of parsnips. *J. Hort. Sci.* 41 (1966): 371-378.

Bradford, K. J. A water relations analysis of seed germination rates. *Plant Physiol.* 94 (1990): 840-849.

Breeze, V. G., and G. M. Milbourn. Inter-plant variation in temperate crops of maize. *Ann. Appl. Biol.* 99 (1981): 335-352.

Brewster, J. L., and A. Barnes. A comparison of relative growth rates of different individual plants and different cultivars of onion of diverse geographic origin at two temperatures and two light intensities. *J. Appl. Ecol.* 18 (1981): 589-604.

Brocklehurst, P. A., and J. Dearman. Interactions between seed priming treatments and nine seed lots of carrot, celery and onion. II. Seedling emergence and plant growth. *Ann. Appl. Biol.* 102 (1983): 585-593.

Burris, J. S., O. T. Edje, and A. H. Wahab. Effects of seed size on seedling performance in soybean. II. Seedling growth and photosynthesis and field performance. *Crop Sci.* 13 (1973): 207-210.

Cal, J. P., and R. L. Obendorf. Imbibitional chilling injury in *Zea mays* L. altered by initial kernel moisture and maternal parent. *Crop Sci.* 12 (1972): 369-373.

Ceccarelli, S., and M. T. Pegiati. Effect of seed weight on coleoptile dimensions in barley. *Can. J. Plant Sci.* 60 (1980): 221-225.

Christiansen, M. N. Influence of chilling upon seedling development of cotton. *Plant Physiol.* 38 (1963): 520-522.

Covell, S., R. H. Ellis, E. H. Roberts, and R. J. Summerfield. The influence of temperature on seed germination rate in grain legumes. I. A comparison of chickpea, lentil, soyabean and cowpea at constant temperatures. *J. Exp. Bot.* 37 (1986): 705-715.

Curtis, K., and B. D. McKersie. Growth potential of the axis as a determinant of seedling vigor in birdsfoot trefoil. *Crop Sci.* 24 (1984): 47-50.

Dahal, P., and K. J. Bradford. Effects of priming and endosperm integrity on seed germination rates of tomato genotypes. II. Germination at reduced water potential. *J. Exp. Bot.* 41 (1990): 1441-1453.

Dahal, P., K. J. Bradford, and R. A. Jones. Effects of priming and endosperm integrity on seed germination rates of tomato genotypes. I. Germination at sub-optimal temperature. *J. Exp. Bot.* 41 (1990): 1431-1439.

Dell'Aquila, A. Mean germination time as a monitor of seed ageing. *Plant Physiol. & Biochem.* 25 (1987): 761-768.

Doneen, L. D., and J. H. MacGillivray. Germination (emergence) of vegetable seed as affected by different soil moisture conditions. *Plant Physiol.* 18 (1943): 524-529.

Durrant, M. J., and R. J. Gummerson. Factors associated with germination of sugar-beet seed in the standard test and establishment in the field. *Seed Sci. & Technol.* 18 (1990): 1-10.

Egli, D. B., D. M. TeKrony, and R. A. Wiralaga. Effect of soybean seed vigour and size on seedling growth. *J. Seed Technol.* 14 (1990): 1-12.

Ellis, R. H. The effects of differences in seed quality resulting from priming or deterioration on the relative growth rate of onion seedlings. *Acta Hortic.* 253 (1989): 203-211.

Ellis, R. H., and P. D. Butcher. The effects of priming and "natural" differences in quality amongst onion seed lots on the response of the rate of germination to temperature and the identification of these characteristics under genotypic control. *J. Exp. Bot.* 39 (1988): 935-950.

Ellis, R. H., and E. H. Roberts. Towards a rational basis for testing seed quality. In *Seed Production*, ed. P. D. Hebblethwaite (London, UK: Butterworths, 1980), pp. 605-635.

Ellis, R. H., and E. H. Roberts. The quantification of ageing and survival in orthodox seeds. *Seed Sci. & Technol.* 9 (1981): 373-409.

Ellis, R. H., S. Covell, E. H. Roberts, and R. J. Summerfield. The influence of temperature on seed germination rate in grain legumes. II. Intraspecific variation in chickpea (*Cicer arietinum* L.) at constant temperatures. *J. Exp. Bot.* 37 (1986): 1503-1515.

Finch-Savage, W. E. The effects of fluid drilling germinating seeds on the emergence and subsequent growth of carrots in the field. *J. Hort. Sci.* 59 (1984): 411-417.

Finch-Savage, W. E. A study of the relationship between seedling characters and rate of germination within a seed lot. *Ann. Appl. Biol.* 108 (1986): 441-444.

Finch-Savage, W. E. A comparison of Brussels sprout seedling establishment from natural and low-moisture-content germinated seeds. *Ann. Appl. Biol.* 113 (1988): 425-429.

Finch-Savage, W. E. The effects of osmotic seed priming and the timing of water availability in the seedbed on the predictability of carrot seedling establishment in the field. *Acta Hort.* 267 (1990): 209-216.

Finch-Savage, W. E., and J. M. T. McKee. The influence of seed quality and pregermination treatment on cauliflower and cabbage transplant production and field growth. *Ann. Appl. Biol.* 116 (1990): 365-369.

Finch-Savage, W. E., and C. I. McQuistan. Performance of carrot seeds possessing different germination rates within a seed lot. *J. Agric. Sci.* 110 (1988): 93-99.

Frappell, B. D. Plant spacing of onions. *J. Hortic. Sci.* 48 (1973): 19-28.

Fyfield, T. P., and P. J. Gregory. Effects of temperature and water potential on germination, radicle elongation and emergence of mungbean. *J. Exp. Bot.* 40 (1989): 667-674.

Gane, A. J. The pea crop–Agricultural progress, past, present and future. In *The Pea Crop*, eds. P. D. Hebblethwaite, M. C. Heath, and T. C. K. Dawkins. (London, UK: Butterworths, 1985), pp. 3-15.

Garcia-Huidobro, J., J. L. Monteith, and G. R. Squire. Time, temperature and

germination of pearl millet (*Pennisetum typhoides* S. & H.). I. Constant temperature. *J. Exp. Bot.* 33 (1982): 288-296.
Goyal, M. R. Soil crusts vs. seedling emergence: A review. *Agricultural Mechanisation in Asia, Africa and Latin America* 13 (1982): 62-75.
Gray, D. The effect of time to emergence on head weight and variation in head weight at maturity in lettuce (*Lactuca sativa* L.). *Ann. Appl. Biol.* 82 (1976): 569-575.
Gray, D. Fluid drilling of vegetable seeds. *Hort. Rev.* 3 (1981): 1-27.
Gray, D. The performance of carrot seeds in relation to their viability. *Ann. Appl. Biol.* 104 (1984): 559-565.
Gray, D., and J. R. A. Steckel. A comparison of methods for evaluating seed quality in carrots (*Daucus carota* L.). *Ann. Appl. Biol.* 103 (1983): 327-334.
Gummerson, R. J. The effect of constant temperatures and osmotic potentials on the germination of sugar beet. *J. Exp. Bot.* 37 (1986): 729-741.
Guy, R. Effets du potentiel osmotique sur la germination des semences de treize especes agricoles et potageres. *Schweizerische Landwirtschaftliche Forschung* 21 (1982): 41-48.
Hegarty, T. W. Seed quality and field emergence in calabrese and leeks. *J. Hort. Sci.* 49 (1974): 189-196.
Hegarty, T. W. Field establishment of some vegetable crops: Response to a range of soil conditions. *J. Hort. Sci.* 51 (1976): 133-146.
Hegarty, T. W. The influence of environment on seed germination. *Aspects of Applied Biology* 7 (1984): 13-31.
Hegarty, T. W., and D. A. Perry. Predictable stands: Problems and prospects. *Arable Farming* (April, 1974): 14-17.
Heydecker, W. Vigour. In *Viability of Seeds*, ed. E. H. Roberts (London, UK: Chapman & Hall, 1972), pp. 209-252.
Heydecker, W., and P. Coolbear. Seed treatments for improved performance–Survey and attempted prognosis. *Seed Sci. & Technol.* 5 (1977): 353-425.
Holliday, R. Plant population and crop yield: Part 1. *Field Crops Abstr.* 13 (1960): 159-167.
Jaggard, K. W. The effect of plant distribution on yield of sugar beet. PhD Thesis, University of Nottingham, U.K. 1979.
Jordan, G. L. Planting limitations for arid, semi-arid and salt-desert shrublands. In *Managing Intermountain Rangelands, 15-17 September 1981, Twin Falls, Idaho and 22-24 June 1982, Elko, Nevada*, eds. S. Monsen and N. Shaw (USDA Forest Service, Intermountain Research Station General Technical Report INT-157, 1983), pp. 11-16.
Jordan, G. L., and M. R. Haferkamp. Temperature responses and calculated heat units for germination of several range grasses and shrubs. *J. Range Manag.* 42 (1989): 41-45.
Kaufmann, M. L. Optimum Seed Size. In *Crop Physiology: Advancing Frontiers*, ed. U. S. Gupta (Oxford, UK: IBH Publishing, 1984), pp. 1-22.
Kerby, T. A., M. Keeley, and S. Johnson. Predicting cotton seedling emergence. *Calif. Agric.* (March 1987): 24-26.

Khah, E. M., E. H. Roberts, and R. H. Ellis. Effects of seed ageing on growth and yield of spring wheat at different plant-population densities. *Field Crops Res.* 20 (1989): 175-190.

Kulik, M. M., and R. W. Yaklich. Evaluation of vigor tests in soybean seeds: Relationship of accelerated aging, cold, sand bench and speed of germination tests to field performance. *Crop Sci.* 22 (1982): 766-770.

Longden, P. C., M. G. Johnson, R. J. Darby, and P. J. Salter. Establishment and growth of sugar beet as affected by seed treatment and fluid drilling. *J. Agric. Sci.* 93 (1979): 541-552.

Luedders, V. D., and J. S. Burris. Effects of broken seed coats on field emergence of soybeans. *Agron. J.* 71 (1979): 877-879.

Maguire, J. D. Dormancy in seeds. *Advances in Research and Technology of Seeds* 9 (1984): 25-60.

Mann, L. K., and J. H. MacGillivray. Some factors affecting the size of carrot roots. *Proc. Amer. Soc. Hort. Sci.* 54 (1949): 311-318.

Matthews S., and A. A. Powell. Environmental and physiological constraints on field performance of seeds. *HortScience* 21 (1986): 1125-1128.

McCormac, A. C., and P. D. Keefe. Cauliflower (*Brassica oleracea* L.) seed vigour: Imbibition effects. *J. Exp. Bot.* 41 (1990): 893-899.

Mondal, M. F., J. L. Brewster, G. E. L. Morris, and H. A. Butler. Bulb development in onion (*Allium cepa* L.). I. Effects of plant density and sowing date in field conditions. *Ann. Bot.* 58 (1986): 187-195.

Monteith, J. L., and J. F. Elston. Microclimatology and crop production. In *Potential Crop Production*, eds. P. F. Wareing and J. P. Cooper (London, UK: Heinemann, 1971), pp. 23-42.

Neergaard, P. *Seed Pathology*. (London, UK: Macmillan, 1977).

Obendorf, R. L., and P. R. Hobbs. Effect of seed moisture on temperature sensitivity during imbibition of soybean. *Crop Sci.* 10 (1970): 563-566.

Oliveira, M. de A., S. Matthews, and A. A. Powell. The role of split seed coats in determining seed vigour in commercial seed lots of soyabean, as measured by the electrical conductivity test. *Seed Sci, & Technol.* 12 (1984): 659-668.

Perry, D. A. Seed vigour and field establishment. *Hortic. Abstr.* 42 (1972): 334-342.

Perry, D. A. Seed vigour and seedling establishment. *Advances in Research and Technology of Seeds* 2 (1976): 62-85.

Perry, D. A. Seed vigour and seedling establishment. *Advances in Research and Technology of Seeds* 5 (1980a): 25-40.

Perry, D. A. The concept of seed vigour and its relevance to seed production techniques. In *Seed Production*, ed. P. D. Hebblethwaite (London, UK: Butterworths, 1980b), pp. 585-591.

Perry, D. A. The influence of seed vigour on vegetable seedling establishment. *Scientific Hortic.* 33 (1982): 67-75.

Perry, D. A. Factors influencing the establishment of cereal crops. *Aspects of Applied Biology* 7 (1984): 65-83.

Phillips, J. C., and V. E. Youngman. Effect of initial seed moisture content on emergence and yield of grain sorghum. *Crop Sci.* 11 (1971): 354-357.

Pollock, B. M. Imbibition temperature sensitivity of lima bean seeds controlled by initial seed moisture. *Plant Physiol.* 44 (1969): 907-911.

Pollock, B. M., and E. E. Roos. Seed and Seedling Vigor. In *Seed Biology, Volume I*, ed. T. T. Kozlowski (New York, USA: Academic Press, 1972), pp. 313-387.

Powell, A. A. Seed vigour and field establishment. *Advances in Research and Technology of Seeds* 11 (1988): 29-61.

Powell, A. A., and S. Matthews. The damaging effect of water on dry pea embryos during imbibition. *J. Exp. Bot.* 29 (1978): 1215-1229.

Powell, A. A., and S. Matthews. The influence of testa condition on the imbibition and vigour of pea seeds. *J. Exp. Bot.* 30 (1979): 193-197.

Powell, A. A., and S. Matthews. The significance of damage during imbibition to the field emergence of pea (*Pisum sativum* L.) seeds. *J. Agric. Sci.* 95 (1980): 35-38.

Powell, A. A., S. Matthews, and M. de A. Oliveira. Seed quality in grain legumes. *Advances in Applied Biology* 10 (1984): 217-285.

Powell, A. A., M. de A. Oliveira, and S. Matthews. Seed vigour in cultivars of dwarf French bean (*Phaseolus vulgaris*) in relation to the colour of the testa. *J. Agric. Sci.* 106 (1986a): 419-425.

Powell, A. A., M. de A. Oliveira, and S. Matthews. The role of imbibition damage in determining the vigour of white and coloured seed lots of dwarf French bean (*Phaseolus vulgaris*). *J. Exp. Bot.* 37 (1986b): 716-722.

Rathore, T. R., B. P. Ghildyal, and R. S. Sachan. Germination and emergence of soybean under crusted soil conditions. I. Effect of crust impedance to seedling emergence. *Plant & Soil* 62 (1981): 97-105.

Rathore, T. R., B. P. Ghildyal, and R. S. Sachan. Germination and emergence of soybean under crusted soil conditions. II. Seed environment and varietal differences. *Plant & Soil* 65 (1982): 73-77.

Roberts, E. H. Quantifying seed deterioration. In *Physiology of Seed Deterioration*, eds. M. B. McDonald, Jr., and C. J. Nelson (Madison, USA: Crop Science Society of America, 1986), pp. 101-123.

Roos, E. E. Physiological, biochemical, and genetic changes in seed quality during storage. *HortScience* 15 (1980): 781-784.

Salter, P. J. Crop establishment: Recent research and trends in commercial practice. *Scientific Hortic.* 36 (1985): 32-47.

Salter, P. J., and J. M. James. The effect of plant density on the initiation, growth and maturity of curds of two cauliflower varieties. *J. Hort. Sci.* 50 (1975): 239-248.

Salter, P. J., I. E. Currah, and J. R. Fellows. Studies on some sources of variation in carrot root weight. *J. Agric. Sci.* 96 (1981): 549-556.

Scott, R. K., S. D. English, D. W. Wood, and M. H. Unsworth. The yield of sugar beet in relation to weather and length of growing season. *J. Agric. Sci.* 81 (1973): 339-347.

Siegel, S. M., and L. A. Rosen. Effects of reduced oxygen tension on germination and seedling growth. *Physiol. Plant.* 15 (1962): 437- 444.

Smith, C. W., and J. J. Varvil. Standard and cool germination tests compared with field emergence in upland cotton. *Agron. J.* 76 (1984): 587-589.

Smith, O. E., N. C. Welch, and T. M. Little. Studies on lettuce seed quality: I. Effect of seed size and weight on vigour. *J. Amer. Soc. Hort. Sci.* 98 (1973): 529-533.

Soetono, and C. M. Donald. Emergence, growth and dominance in drilled and square-planted barley crops. *Aust. J. Agric. Res.* 31 (1980): 455-470.

Taylor, A. G., and M. H. Dickson. Seed coat permeability in semi-hard snap bean seeds: Its influence on imbibitional chilling injury. *J. Hortic. Sci.* 62 (1987): 183-189.

TeKrony, D. M., and D. B. Egli. Relationship of seed vigor to crop yield: A review. *Crop Sci.* 31 (1991): 816-822.

Tully, R. E., M. E. Musgrave, and A. C. Leopold. The seed coat as a control of imbibitional chilling injury. *Crop Sci.* 21 (1981): 312-317.

Villiers, T. A. Seed dormancy. In *Seed Biology, Volume II*, ed. T. T. Kozlowski (New York, USA: Academic Press, 1972), pp. 220-281.

Wagenvoort, W. A., and J. F. Bierhuizen. Some aspects of seed germination in vegetables. II. The effect of temperature fluctuation, depth of sowing, seed size and cultivar, on heat sum and minimum temperature for germination. *Scient. Hort.* 6 (1977): 259-270.

Wanjura, D. F. Reduced cotton productivity from delayed emergence. *Transactions of the American Society of Agricultural Engineers* 25 (1982): 1536-1539.

Wanjura, D. F., and D. R. Buxton. A systematic method for studying seedling emergence. *J. Amer. Soc. Sugar Beet Technologists* 19 (1977): 207-218.

Wheeler, T. R., and R. H. Ellis. Seed quality, cotyledon elongation at suboptimal temperatures, and the yield of onion. *Seed Sci. Res.* 1 (1991): 57-67.

Willey, R. W., and S. B. Heath. The quantitative relationships between plant population and crop yield. *Adv. Agron.* 21 (1969): 281-321.

Woodstock, L. W. Seed imbibition: A critical period for successful germination. *J. Seed Technol.* 12 (1988): 1-15.

Woodstock, L. W., and R. B. Taylorson. Soaking injury and its reversal with polyethylene glycol in relation to respiratory metabolism in high and low vigor soybean seeds. *Physiol. Plant.* 53 (1981): 263-268.

Wurr, D. C. E., and J. R. Fellows. Further studies on the slant test as a measure of the vigour of crisp lettuce seed lots. *Ann. Appl. Biol.* 105 (1984): 575-580.

Index